Modeling Explosions and Blast Waves

K. Ramamurthi

Modeling Explosions and Blast Waves

Second Edition

Ane Books
Pvt. Ltd.

K. Ramamurthi
Department of Mechanical Engineering
Indian Institute of Technology Madras
Chennai, India

ISBN 978-3-030-74340-6   ISBN 978-3-030-74338-3   (eBook)
https://doi.org/10.1007/978-3-030-74338-3

Jointly published with ANE Books India
In addition to this printed edition, there is a local printed edition of this work available via Ane Books in
South Asia (India, Pakistan, Sri Lanka, Bangladesh, Nepal and Bhutan) and Africa (all countries in the
African subcontinent).
ISBN of the Co-Publisher's edition: 978-81-948918-6-4

1st edition: © TataMcGraw Hill Education Private Limited 2011
2nd edition: © The Editor(s) (if applicable) and The Author(s), under exclusive license to Springer
Nature Switzerland AG 2021

This Springer imprint is published by the registered company Springer Nature Switzerland AG
The registered company address is: Gewerbestrasse 11, 6330 Cham, Switzerland

Preface

Simple analytical models are developed for energy release in explosions and formation of blast waves from explosions. Explosions from chemical, mechanical, and physical processes are considered. The role of confinement is studied, and case studies of the different explosions are analyzed using the models.

The motivation of a book on simple analytical modeling of explosions and blast waves from basics in thermodynamics, mechanics, gas dynamics, and chemistry arises from the following. The nonlinear processes associated with the blast wave propagation from explosions and interaction with objects have been considered in several advanced textbooks using direct numerical integration of the partial differential equations of the unsteady flow in a blast wave, perturbation analysis, and through numerical simulations. The solution procedure is somewhat tedious with the result that the focus shifts to mathematical methods. Empirical equations, charts, and figures derived from the analysis and from practical observations continue to be used to analyze explosions, and the subject is treated more as an art than a science in which thermodynamics, fluid flow, and chemistry play a role.

Explosions are often encountered during the handling, transportation, and processing of materials and could be accidental. Explosions also occur in nature. In the recent past, a spate of intentional explosions using improvised explosive devices with readily available substances has taken place leading to loss of life and property. It appears desirable that an engineer is familiar with the causes of these inadvertent explosions so that steps could be taken to minimize their occurrence. Though specialized courses are conducted in fire and safety engineering and in hydrogen safety, they do not form part of an engineering curriculum in which the subject of explosions could be taught as arising from application of core courses in thermodynamics, fluid mechanics, and gas dynamics. The author introduced a three-credit course on 'Explosions and Safety' at the Indian Institute of Technology Madras in the Mechanical Engineering Department in 2006. The course was hugely popular with the students, and the lecture notes of the course were published as a book entitled 'Explosions and Explosion Safety' in 2011 and an online course under National Programme of Technology Enhanced Learning (NPTEL) in 2015. This was followed in the online course 'SWAYAM' in 2021.

The book 'Explosions and Explosion Safety,' however, had several shortcomings. While it considered the different types of explosions, it did not deal with the interaction of blast waves with bodies necessary to model the damages from reflected shocks and expansion waves. Case studies of the practically encountered explosions were also not adequately covered, and a number of errors were noticed in the book. The drawbacks are removed in the present book, and the subject is focused on modeling the explosions and blast waves. The quasi-donations, which are more likely to be encountered in accidental and low intensity explosions, rather than Chapman–Jouguet detonations, are modeled from first principles.

The book consists of 14 chapters. Chapter 1 deals with wave motion, the genesis of the blast wave, its departure from wave phenomenon, and the different types of explosions. Chapters 2 and 3 deal with modeling overpressure, wind and fragments in a blast, the reflection and transmission of blast waves from objects, spall failure and formation of craters, and the structure of an explosion. This is followed in Chap. 4 by energy release in an explosive that basically drives the blast wave. A simplified procedure of determining the energy is demonstrated rather than the involved thermodynamic equilibrium method. The thermodynamic equilibrium method is given in appendix. This is followed by rate of energy release, and a thermal lumped mass model for explosion is developed in Chap. 5. The preheat, non-dimensional representation of characteristic timescales and chain reactions and their inhibition are dealt with. Chapters 6 and 7 model detonations and deflagrations, the flammability limits, minimum oxygen concentrations, and the quenching and initiation of flames and detonations. The formation of a steady quasi-detonation is modeled for the lower Chapman–Jouguet point in the reaction Hugoniot as against the steady Chapman–Jouguet detonation at the upper Chapman–Jouguet point. The three-dimensional propagation of a detonation and turbulent flames and rates of energy release are determined.

The chemistry and energy release in condensed phase explosives and pyrotechnics along with their detonation in confinement or otherwise and the capacity to generate gaseous products are given in Chap. 8. Confined and unconfined explosions are modeled in Chap. 9, while explosions in dust–air mixtures are dealt with in Chap. 10. Physical explosions and burst of pressure vessels and cryogenic storage vessels follow in Chap. 11. Chapter 12 relates the different explosions through yield and TNT equivalence.

The dispersion of flammable gases and pollutants in atmosphere is dealt with in Chap. 13 using atmospheric stability and the standard deviations in the standard Gaussian distribution. The quantification of damages is briefly considered in Chap. 14.

Examples that deal with fundamental aspects along with the case studies are given in the different chapters. A set of exercises is also provided to help the reader in internalizing the concepts.

The level of this book is suited for senior undergraduate and graduate/research students in mechanical, chemical, aeronautical, and petroleum engineering. The book would also serve as a useful reference for professionals working in aeronautical, mechanical, and chemical industries, explosive manufacture and high-energy materials, armaments, defense, space, and industrial and fire safety.

Chennai, India   K. Ramamurthi

Contents

About the Author

K. Ramamurthi obtained Ph.D. in Mechanical Engineering in 1976 from McGill University, Montreal after a Master's degree in Mechanical Engineering from Indian Institute of Science (IISc), Bengaluru, in 1970. He worked on the development of solid, liquid, and cryogenic propellant rockets in the Indian Space Research Organization, and at Indian Institute of Technology (IIT), Madras, where he has been teaching courses on thermodynamics, combustion, propulsion, gas dynamics and explosion and safety, and guiding research work. He has been on several committees and panels on propulsion, combustion, and shock waves and chairman of the Combustion, Detonics, and Shock Wave Panel of the Defense Research and Development Organization. He is an Honorary Fellow of the High Energy Materials Society.

Chapter 1
Basic Concepts and Introduction to Blast Waves and Explosions

1.1 Noise and Disruption of Objects in an Explosion

The word 'explosion' is generally used to describe events associated with a loud noise and sudden disruption of the objects at the site of its occurrence and around it. The word is derived from the Latin expression *Explodere* meaning 'driving out by making loud noise such as clapping'. Explosion means violent bursting or driving out of the objects with a destructive force and a loud bang.

1.1.1 Sound Waves

The term 'loud bang' requires some clarification. Let us first consider a small level of sound such as noise or music, instead of a loud bang. The sound pertaining to the noise or to the music is communicated from a source through the medium of air by very small magnitude compression (condensation) and expansion (rarefaction) disturbances. The disturbances are localized at a given point and are communicated to the neighboring points at a velocity, which is equal to the speed of sound. It is this communication of the small compression and expansion disturbances from one point to the next in the medium that constitutes the propagation of the sound wave. The wave does not consist of any mass or particles as such but represents the progressive propagation of the small compression and rarefaction disturbances in the medium.

The small compression and expansion disturbances in the medium are the positive and negative pressure fluctuations about the mean pressure of the medium. The pressure fluctuations are associated with velocity, density, and temperature changes. The fluctuations of pressure in the medium, denoted by p', can be idealized as being sinusoidal and is represented by the equation:

$$p' = \hat{p} \sin\left(\frac{2\pi}{\lambda}x\right) \tag{1.1}$$

© The Author(s), under exclusive license to Springer Nature Switzerland AG 2021
K. Ramamurthi, *Modeling Explosions and Blast Waves*,
https://doi.org/10.1007/978-3-030-74338-3_1

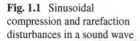

Fig. 1.1 Sinusoidal compression and rarefaction disturbances in a sound wave

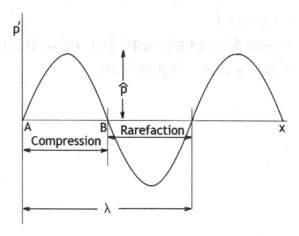

where x denotes the axis along the direction of propagation of the disturbance in the ambient. λ is the wavelength, and $\hat{p}$ is the maximum value of the pressure perturbation. Crests and trough of the fluctuation are sketched in Fig. 1.1. The portion denoting compression corresponds to the region where the pressure is locally higher than the steady ambient value, while the rarefaction is in the region where the pressure falls below the ambient value.

The disturbances are communicated in the medium at the speed of sound. Denoting the sound speed by 'a', the position of the wave, shown in Fig. 1.1, would have advanced by a distance 'at' over a time 't' in the direction x. This is shown in Fig. 1.2. The equation describing the pressure fluctuation for the sound wave propagating at speed 'a' in the direction x at time t is therefore given by

$$p' = \hat{p} \sin \frac{2\pi}{\lambda}(x - at) \qquad (1.2)$$

A medium of air or some other gas or solid or liquid substance is required for the sound wave to propagate unlike the electromagnetic waves, which can propagate in vacuum from the interacting electric and magnetic fields.

The magnitudes of the compression and rarefaction disturbances in the medium associated with a sound wave are so small that heating associated with compression and cooling associated with expansion are negligibly small. Further the medium of air is not a good conductor, and the negligibly small levels of heating and cooling do not transfer heat to the surrounding. The processes of compression and rarefaction can therefore be assumed to be adiabatic. The perturbations can also be considered to

Fig. 1.2 Traveling sound wave

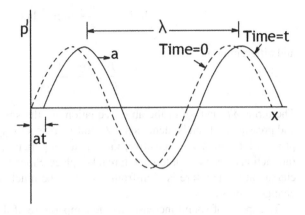

be reversible as they are small. The compression and rarefaction process is therefore said to be isentropic (reversible and adiabatic) and is given by the equation:

$$\frac{p}{\rho^{\gamma}} = \text{constant} \tag{1.3}$$

where ρ is the density of the medium, and γ is the ratio of the specific heats ($\gamma = C_P/C_V$, where C_P is the specific heat at constant pressure, and C_V is the specific heat at constant volume).

1.1.2 Finite Amplitude Waves

If the magnitude of the pressure perturbation is somewhat larger, then a finite though small increase in temperature during the compression or a decrease in temperature during rarefaction is possible. This can be seen as follows. For an ideal gas such as air at the ambient conditions, the equation of state relating pressure p, density ρ, and temperature T is written as

$$\frac{p}{\rho T} = \text{constant} \tag{1.4}$$

Solving Eq. 1.3 and the equation of state given by Eq. 1.4 for the dependence of temperature on pressure in an isentropic process, we get

$$\frac{T}{p^{(\gamma-1)/\gamma}} = C \tag{1.5}$$

where C is a constant. Taking logarithm of the above equation and differentiating, we have

$$\ln T - \frac{\gamma - 1}{\gamma} \ln p = \ln C$$

and differentiating

$$\frac{\delta T}{T} = \frac{\gamma - 1}{\gamma} \frac{\delta p}{p} \tag{1.6}$$

The terms δT and δp in the above equation are the small changes in temperature and pressure and can be denoted by T' and p', respectively. During the compression process for which p' is positive, there is a temperature increase while during the rarefaction a decrease of temperature takes place. Since γ is 1.4 for air, the fractional change in temperature is seen from Eq. 1.6 to be much smaller than the fractional change of pressure.

The speed of sound increases as the temperature of the medium increases. This can be seen from physical considerations that an increase of temperature enhances the energy of the molecules in the medium, which imparts higher velocity to the propagation of disturbances and hence increases the speed of the sound wave. The dependence of sound speed on temperature is derived in Annexure A.

1.1.3 Wave with a Steep Front

The variations of the sound velocity in a medium during the compression and rarefaction phases are negligibly small when the pressure amplitudes are small as in a sound wave. However, for sound generated by a high-intensity source, the resulting pressure amplitudes are much larger. The speed of sound in the compression region increases. The maximum sound speed is at $p' = \hat{p}$, which is the top or crest of the wave. Similarly, the sound speed decreases in the rarefaction region, and the minimum sound velocity is at bottom or trough of the wave ($p' = -\hat{p}$). We have assumed here that the isentropic process for compression and rarefaction continues to hold good though the pressure amplitudes are higher. The variation of the sound speed at the different points along the wave is shown in Fig. 1.3.

The point A at the crest has the highest value of pressure perturbation $\hat{p}$ and therefore has the maximum sound speed $a + \delta a$, while the point C in the trough has the minimum sound speed $a - \delta a$. Points on the wave with no perturbation (O, B, and D along the x-axis) continue with sound speed a. Since the sound speed increases from a to $a + \delta a$ between O and A and decreases back to a for B (Fig. 1.3), the sinusoidal waveform gets distorted with the point A at the crest of the wave traveling the farthest distance over a given value of time t to A', while B will travel a shorter distance to B'. The point C at the trough, having a minimum sound speed of $a - \delta a$, will travel the least distance shown by CC'. The sinusoidal waveform thus gets distorted with the portion $A'B'C'$ becoming steeper than the original sinusoidal shape ABC. This process of steepening due to the variation of the sound velocity in the compression and rarefaction regions continues until the waveform develops the

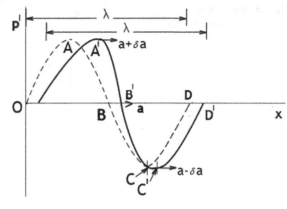

Fig. 1.3 Steepening of a finite amplitude pressure wave

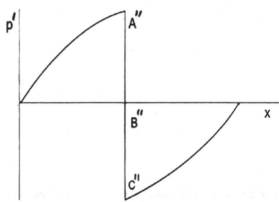

Fig. 1.4 Steep-fronted wave

shape shown by $A''B''C''$ in Fig. 1.4. Any further changes result in multiple values of pressure at a given location, which is not possible in practice.

Instead of the gradual compression and expansion process in a sound wave associated with very small pressure perturbations, a steep-front waveform is obtained when somewhat larger pressure amplitudes are generated from a high-intensity energy source. The front travels at a speed greater than the sound speed a. A sudden change of the pressure takes place across the front $A''B''C''$ compared to the gradual compression and the smooth variation in pressure, which was observed earlier. The sudden and large pressure rise is heard as a loud bang.

1.1.4 Shock Waves

Abrupt changes of pressure, temperature, density, and velocity are produced in a medium in which the energy is released impulsively. The word shock is used to denote the sudden or abrupt changes in the medium. The sudden changes are transmitted in

Fig. 1.5 Wave forms of
sound wave and shock wave

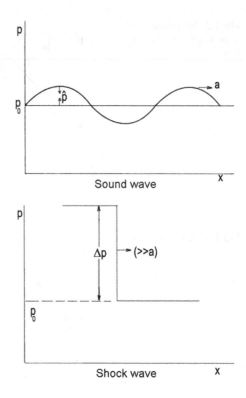

the medium from the zone or source of the impulsive energy release. The front of the sudden changes in the medium is referred to as a shock wave. The shock wave is similar to the steepened wave considered in the previous section; however, the abrupt changes in the properties across the wave front are much larger. The speed of the shock wave, so generated, is greater than the speed of the sound wave. Figure 1.5 gives a comparison of the waveform of a sound wave generated by a low intensity source, also referred to as an acoustic wave, with a shock wave formed by a high-intensity source.

1.1.5 Compression Disturbances Intensifying to a Shock Wave

The formation of a steep-fronted 'shock' wave from an impulsive release of energy in an explosion is similar to the steepening of finite amplitude sound waves into a shock wave. The release of energy in a medium causes compression wave packets or rather compression disturbances to be propagated in the medium. The medium could be a gas or a liquid or a solid. We can consider, for ease of understanding, the rapid release of energy from an impulsively accelerated piston instead of the

Fig. 1.6 Trajectory of an impulsively accelerated piston and compression disturbances propagating from it

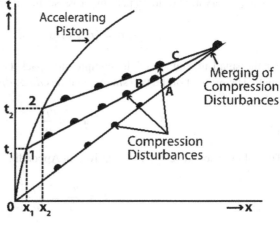

Fig. 1.7 Property changes across a wave propagating in a quiescent medium

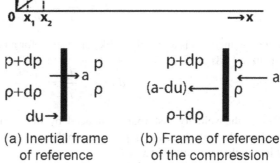

(a) Inertial frame of reference

(b) Frame of reference of the compression

energy release in the explosion. A piston is initially at rest at $x = 0$ and is started impulsively and accelerated along the X direction. The trajectory of the impulsively accelerated piston is shown in the time vs. distance diagram in Fig. 1.6. Compression disturbances are generated in the medium from the accelerating piston and propagate in the direction of the accelerating piston ahead of it.

The trajectory or path of the compression disturbance originating from the piston when it is impulsively started at $x = 0$ is shown by the line OA in Fig. 1.6. If the magnitude of the compression disturbance is small, the magnitude of the changes it causes in the medium can be assumed to be isentropic. The disturbance propagates at the sound velocity a in the undisturbed medium.

As the above compression packet moves along in the quiescent medium (O-A in Fig. 1.6), it drags the particles of the medium at a small velocity du as shown in Fig. 1.7. It also compresses the medium from its initial density ρ to $\rho + d\rho$ and increases the pressure of the medium from p to $p + dp$.

The velocity du induced in the medium can be determined by conserving the mass across the compression packet. This is done more easily in the frame of reference of the compression packet, i.e., the compression packet being held stationary. In this frame of reference, the quiescent medium at density ρ moves toward it at a velocity a, while the medium subjected to the compression at density $\rho + d\rho$ leaves the compression disturbance at velocity $a - du$ (Fig. 1.7b).

The conservation of mass per unit cross-sectional area gives

$$\rho a = (\rho + d\rho)(a - du) \tag{1.7}$$

For the small magnitude of the compression packets, the changes du and $d\rho$ are small. The product of the two small quantities $du \times d\rho$ can be neglected to give

$$du = a\frac{d\rho}{\rho} \tag{1.8}$$

The value of sound velocity a is derived in Appendix A to be given by

$$a^2 = \frac{dp}{d\rho} = \frac{\gamma p}{\rho} \tag{1.9}$$

when the medium is an ideal gas like air. Hence for a medium like air, we can write equation (1.8) as

$$du = \frac{a}{a^2}\frac{dp}{\rho} = \frac{a}{\gamma}\frac{dp}{p} \tag{1.10}$$

This is the velocity disturbance induced by the compression disturbance in the medium along its direction of propagation that originates at the piston surface when it is impulsively started at $x = 0$.

Since the piston is accelerating, a series of such compression packets get generated at the piston surface. Consider a subsequent compression disturbance 1-B from the accelerating piston when it is at a distance x_1 and at time t_1. This compression, shown in Fig. 1.6, now travels in a medium, which is no longer quiescent but is moving with a velocity du in the same direction as the compression disturbance. The compression packet will now travel with a velocity $a + du$. The pressure of the medium into which it moves is $p + dp$, and the velocity of sound a would therefore be greater than before. Hence the net velocity of travel of the compression packet shown by 1-B would be greater than the velocity of the initial compression shown by 0-A. The slope of 1-B would therefore be less than of 0-A on the $t - X$ diagram shown in Fig. 1.6.

A compression disturbance originating from the impulsive accelerating piston at a later instant of time would therefore travel faster than the earlier originating compression. For instance, the compression disturbance originating when the accelerating piston is at x_2 at time t_2 shown by line 2-C would travel even faster and therefore have a smaller slope than 1-B and 0-A. The compression disturbances from the piston at a later instant would therefore merge with the earlier compression disturbances to form a larger compression packet and are shown in Fig. 1.6.

The merging of the compression disturbances leads to the formation of a finite magnitude compression that travels faster than the sound speed in the quiescent medium. This merging of compression disturbances leads to the formation of a compression shock that is no longer isentropic.

1.1.6 Expansion Waves

Instead of a compression disturbance, we could have an expansion disturbance induced in the medium. Expansion disturbances could be induced in a stationary medium by a decelerating piston that moves away from the medium as shown in Fig. 1.8.

The expansion disturbance formed when the piston is at $x = 0$ and time $t = 0$ is shown by the line O-A. The particle velocity in the medium behind this expansion wave is $-du$ as it is pulled by the piston in the negative X direction. The pressure and density behind the wave decrease due to the expansion. Thus the wave that is subsequently formed when the piston is at $-x_1$ and time t_1 such as 1-B, in Fig. 1.8, would therefore travel in a medium of velocity $-du$. The velocity of the expansion disturbance would be $a - du$. The magnitude of pressure of the expansion disturbances can be determined from the mass continuity equation, viz.

$$(a - du)(\rho - d\rho) = \rho a \tag{1.11}$$

This is similar to Eq. 1.1 for the compression disturbance except that velocity is less and the density behind the wave has decreased. Simplifying the above expression, we get

$$ad\rho + \rho du = 0$$

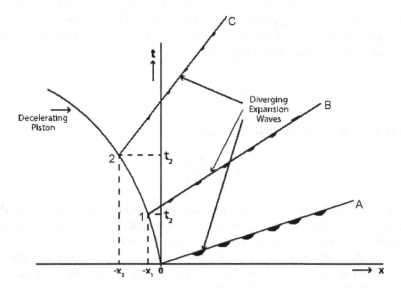

Fig. 1.8 Diverging expansion waves from a decelerating piston

With $a^2 = dp/d\rho$, we have

$$dp = -\frac{\rho a^2}{a}du = -\frac{\gamma p}{a}du \tag{1.12}$$

The magnitude of pressure of the expansion disturbance decreases. For a subsequent expansion wave originating from the piston at x_2 at time t_2, the magnitude of the expansion disturbance will further decrease as also its velocity and is shown by the line 2-C. The process of expansion waves subsequently formed by the decelerating piston therefore decreases both in the velocity of propagation and in the magnitude of the disturbance. The expansion waves in the $t - X$ diagram diverge out as a fan as shown in Fig. 1.8. They cannot merge or coalesce together as in the case of the compression disturbances. They continue to be of decreasing magnitude, and the propagation process is isentropic.

The motion of the compression and expansion disturbances causes the medium to move along or opposite to the direction of propagation. Such disturbances or waves are known as longitudinal waves. The particles in the medium do not physically carry the mass and energy of the disturbance from the initial position or source. The particles have limited motion and transmit the energy from one particle to the next.

1.1.7 Role of Compressibility of the Medium

The compressibility of a medium is defined as the decrease in volume dV of a volume V when a pressure dp is applied over the volume. The volume change can be done either isothermally, i.e., keeping the temperature constant or else adiabatically. We define the isothermal compressibility coefficient

$$\kappa_T = -\frac{1}{V}\frac{dV}{dp}\bigg|_T$$

and the isentropic compressibility coefficient

$$\kappa_S = -\frac{1}{V}\frac{dV}{dp}\bigg|_S$$

where dp is the change in pressure that causes the volume V to decrease by dV. The subscripts T and S denote that the temperature and entropy are maintained constant during the compression process. Since for a wave motion, the process is fast and near to adiabatic, we have isentropic compression for small magnitudes of the disturbances and use the isentropic compressibility κ_S. Further, with volume being inversely proportional to the density of the medium, we can write

$$\kappa_S = -\rho\frac{d(1/\rho)}{dp} = \frac{1}{\rho}\frac{d\rho}{dp}$$

But we have seen that $dp/d\rho = a^2$ where a is the sound speed in the undisturbed medium. We therefore get the isentropic compressibility as

$$\kappa_S = \frac{1}{\rho a^2} \tag{1.13}$$

When the medium is almost incompressible such as a hard metal, the compressibility tends to zero, and consequently the sound speed in it tends to be large. For a compressible medium like air, the sound speed is much smaller. It is also seen that for propagation of compression and expansion waves in air, much smaller values of force or pressure or energy would provide the volume change for the compression, shock, and expansion disturbances in it.

1.1.8 Wave Propagation and Not Matter Propagation

Though energy is transferred through the medium by the compression, shock, and expansion waves, matter itself containing the energy from the source of energy transfer is not communicated in the process. The motion of the medium, viz. the displacement of particles in the medium, is generally small, and the wave progresses due to the existence of the compressibility or elasticity of the medium, which provides a restoring force. The restoring force tends to bring the particles in the medium back to the original state. However, the inertia of the restoring force causes it to overshoot the original state, and the mean motion of the particles in the medium is small. The particle oscillates about the mean and is not transported with the wave. No mass from the source of the energy driving the wave is transferred nor does any matter or mass in the medium physically move during the wave motion.

1.1.9 Blast Wave and Disruption of Objects

The shock wave is associated with a sudden change in pressure, density, temperature, and velocity across it and therefore has extremely large value of gradients of pressure, temperature, and velocity. Values of thickness over which the changes occur are typically of the order of micrometers. Significant heat conduction corresponding to the temperature gradient and viscous shear from the velocity gradients is therefore present. These dissipative processes make the changes in the case of a shock wave to be irreversible unlike in the case of a sound wave. The magnitude of jump in the properties across the wave tends to get diminished as energy is dissipated due to the irreversible processes in the medium. The magnitude of jump in pressure and other parameters at the shock front including the speed of propagation of the front keeps diminishing as the shock wave propagates. The work done in the expansion process behind the shock compression also dissipates some energy. The shock cannot therefore travel with the same constant jump in pressure or equivalently as a constant velocity shock unless it is continually strengthened by further release of energy to compensate for the dissipation. The velocity of the shock decays as it progresses unless otherwise energized. The decaying shock wave is known as a 'blast wave'. The pressure changes across the shock front and the variations of the pressure as it propagates cause very strong winds, viz. a blast, and hence the name blast wave is given for the decaying shock wave. A constant velocity shock wave in a medium has a constant value of pressure behind it and therefore does not have the blast wind.

The trajectories of a sound wave traveling at sound speed a in a medium with very low-pressure amplitudes, a constant velocity shock wave with jump in pressure across it, and a blast wave, which is a decaying shock wave, are shown in Fig. 1.9.

The blast wave, which is a decaying shock wave, travels faster than the speed of sound just as a shock wave and therefore does not give any warning of its approach.

Fig. 1.9 Trajectories of acoustic wave, constant velocity shock, and blast wave

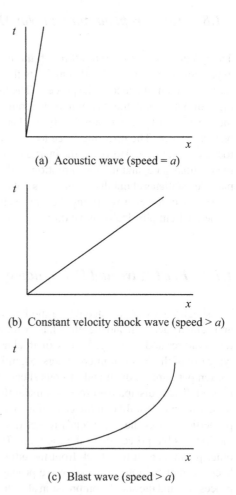

(a) Acoustic wave (speed = a)

(b) Constant velocity shock wave (speed > a)

(c) Blast wave (speed > a)

The initial pressure rise crushes the object, while the blast or strong winds behind it lead to the disruption of the crushed object from the place it has originally been.

A very rapid or impulsive energy release that results in a blast wave forms the basis of all explosions. The energy is dispersed from the zone of the release into the surrounding medium through the blast wave. We shall in the next section discuss different methods of sudden energy releases and accordingly classify explosions in different categories.

The word implosion is also associated with a shock front and strong pressure jump across it. The energy release in an implosion is, however, directed inwards rather than outward.

1.2 Types of Explosions

Explosions are associated with spontaneous release of energy, causing strong pressure waves and blast waves, and are categorized according to the mode of energy release. The energy release could take place naturally in the environment, and the energy release could be deliberately planned or take place accidentally. Accordingly explosions are categorized as follows:

1. Naturally occurring explosions
2. Intentional explosions and
3. Accidental explosions.

We consider a few examples of explosions in the above three categories with a view to understand the mechanisms leading to the explosions.

1.2.1 Naturally Occurring Explosions

The explosions occurring in nature are those related to lightning, volcanic eruptions and impact of meteors and asteroids with the atmosphere of the Earth. In the case of lightning, the charged mass of a water vapor cloud at a high voltage above the Earth's surface forms an arc discharge when the voltage difference between it and the Earth exceeds a threshold value. The discharge takes place with a very high current pulse and a spontaneous release of electrical energy. A blast wave is created from the localized region of the sudden and large energy release and is heard as a strong thunder. The rumble of thunder heard after the lightning is from the remaining decayed pressure of the powerful blast wave formed by the lightning.

Volcanoes spew hot gases and molten rocks from below the Earth's surface. The rapid release of hot gases and large volume of steam from the sudden evaporation of water, if present in the volcano, forms a blast wave. The greatest volcanic eruption recorded so far is the eruption at Krakatau, 40 km off the west coast of Java in Indonesia on August 26, 1883. A huge mass of seawater, which entered the volcano from a tidal wave, spontaneously got converted to superheated steam from the hot lava and resulted in an enormous volume and sudden release of very high-pressure superheated steam. The loud bang of the blast wave, thus created, was heard about 4500 km away. The blast waves damaged walls and smashed windows over a distance of 160 km. Charged particles of cloud released from a volcano also lead to arc discharge and an explosion.

The impact of a meteor or comet with atmospheric air causes an extreme and sudden compression and heating of air, the spontaneous energy of which forms a blast wave in the atmosphere. This is similar to the blast wave formed by spontaneous energy release from lightning or flash steam formed in the volcano. However, the intensity of the energy release is very much higher. On July 30, 1908, a meteor or perhaps a comet or an asteroid impacted the atmosphere at an altitude of 5 to 10 km above the surface of Earth over Tunguska in Siberia. The blast wave generated from the spontaneous energy was so strong that it knocked down people and broke windows of buildings hundreds of kilometers away. Trees were stripped off their branches and scorched by the heat over a distance of 50 km. Similarly, when a tiny meteorite impacted the atmosphere in the city of Yakaterinberg in Russia on February 13, 2013, the blast wave so formed caused damages. The blast was felt in the city of Chelyabinsk about 200 km away.

In addition to the explosions occurring in nature from lightning, volcanic eruptions and impact of meteors and asteroids with the Earth's atmosphere, stellar explosions such as novae and supernova take place in our Milky Way galaxy and other galaxies. A supernova explosion occurs when a star toward the end of its life, known as a white dwarf, has consumed all its hydrogen and helium in the thermonuclear reactions. It is left with carbon and oxygen in equal proportions. If the white dwarf accretes, i.e., acquires mass from a neighboring star or it collapses under its own gravity, its temperature increases to a very high value. The sudden burst of thermal energy is manifested in the form of brightness, and a blast wave is driven into the interstellar medium. The very high temperatures cause thermonuclear fusion and formation of elements with heavy nuclei. This has been the origin of gold, silver, platinum, uranium, etc., on Earth. The blast wave cannot travel through the vacuum in space, and it is the cosmic radiation from stellar explosions that is observed on Earth.

1.2.2 Intentional Explosions

Constructive Applications

Explosions are not always destructive. They have been constructively used for blasting rocks, making canals, tunnels, railroads and roads, demolishing old buildings, separating rocket stages, operating valves, providing safety features in automobiles, for manufacturing purposes, etc. Laser and electrical sparks in which energy is deposited rapidly in a very small volume in a medium are used in the laboratory experiments and in practice for ignition. The discharge of electrical energy stored in capacitors in a thin wire, known as exploding wire, creates a powerful blast wave and is useful for studying high-temperature phenomenon.

Technical advancement and the present standard of living would not have been achieved without the constructive use of explosions. Surgery is also carried out using very minute explosions. The safety air bag in automobiles makes use of explosive energy release for spontaneously inflating the air bag and protecting the driver and passengers in case of a collision.

Explosions in Engineering Applications

Explosions are sometimes formed in the combustion chambers of internal combustion engines when the engines are not properly tuned. They are known as 'engine knock'. The sonic boom, from the flight of an aircraft at supersonic speed, results in an explosion being heard as a sharp and loud bang. The energy release in rockets is explosive, and a loud bang is heard when a rocket takes off from the launch pad.

Destructive Applications

Explosions have unfortunately been used to inflict damages. An example is the concentrated energy release in the atom bomb over Hiroshima and Nagasaki on August 9, 1945. Terrorist strikes are worrisome as they cause damage to life and property. A typical case is the sabotage of the Murrah Federal Building in Oklahoma City, in the United States on April 15, 1995. The rapid energy release from chemicals stored in a truck parked adjacent to the building caused the destruction.

A number of such painful explosions have unfortunately occurred with the use of improvised explosive devices (IED) and other means of forming blast waves. Appropriate measures to avoid the damages from such explosions need to be evolved.

1.2.3 Accidental Explosions: Hazard

Accident denotes an unplanned event. Accidental explosions take place due to rapid and significant release of energy from certain class of substances during inadvertent leaks and spills, storage problems, or mishandling. The potential for an accident to take place is referred to as hazard, and substances that are likely to cause an explosion are termed as hazardous substances.

A few examples of accidental explosions are discussed in the following section.

1.3 Typical Examples of Accidental Explosions

1.3.1 Texas City Disaster (April 16, 1947)

This explosion involved 7700 tons of fertilizer ammonium nitrate (NH_4NO_3), stored in the hull of a ship in Texas City port and represents the worst industrial explosion ever recorded. NH_4NO_3 is used as a fertilizer for fixing nitrogen in the soil for improving agriculture yield. NH_4NO_3 involved in the explosion was coated with wax to reduce the possibility of it catching fire. The coated NH_4NO_3 is known as fertilizer grade ammonium nitrate (FGAN).

The storage of large quantities of FGAN in the ship's hull under the relatively warm conditions resulted in some heat generation from incipient chemical reactions. A fire therefore got initiated. Water and steam, available in the ship for fire fighting, were used to put out the fire. The excess heat from steam led to further decomposition of NH_4NO_3 to nitrous oxide and water vapor, and the reaction generated more heat. The increase in the rate of dissociation due to the enhancement of temperature caused a further increase of the reaction rate and promoted yet higher temperatures and larger rates of heat release. This bootstrap effect resulted in a runaway in temperatures and rate of chemical reactions and therefore in extremely rapid rates of energy release. An explosion driven by the escalating temperature, viz., a thermal explosion resulted.

The blast wave formed by the high rate of energy release blew off the heavy hull of the ship like a missile. The blast wave knocked down people 16 km away from the site of the explosion. Windows of buildings, 60 km from the site, were smashed off their positions. The blast wave was felt even at a distance of 400 km from the explosion, and 581 people were killed. The heat release caused FGAN stored in the hull of a second ship to explode the next day bringing about further damages.

The spontaneous release of energy from large quantities of ammonium nitrate contributed to the catastrophic explosion. Since the explosion involved solid NH_4NO_3, the explosion is termed as a condensed phase explosion. The explosion resulted from the bootstrapping effect of temperature and chemical reactions, leading to a runaway in the rate of chemical reactions and rate of energy release in the solid phase ammonium nitrate. It is therefore considered as *condensed phase thermal explosion*.

1.3.2 Beirut Explosion (August 4, 2020)

A massive explosion of about 2750 tons of ammonium nitrate stored over a prolonged period of about six years in a warehouse at Beirut port took place in August 2020. It was not as severe as the one at Texas City port though it wrecked part of the city over a distance of 3 km from the warehouse and killed over 200 people. A crater of diameter of approximately 120 m was formed at the site of the explosion.

Due to the warm weather and the large amounts of ammonium nitrate stored for long period under humid and almost confined conditions within the warehouse, some initial chemical decomposition occurred. It led to the formation of whitish vapor that was initially noticed. However, there are some reports that firecrackers stored in an adjacent warehouse initiated the decomposition of the ammonium nitrate. The heat generated in the decomposition led to a fire and a small explosion. This was followed after about 30 s by a second explosion that was very severe in which the ammonium nitrate detonated. The detonation of the ammonium nitrate under confined conditions led to the massive blast wave. As in the case of the Texas City disaster, the Beirut explosion is also a case of condensed phase thermal explosion.

Such accidental condensed phase explosions involving ammonium nitrate have been occurring at regular intervals. A notable one was at Tianjin harbor on August 12, 2015, where about 800 tons of the ammonium nitrate exploded after catching fire due to faulty storage conditions. Though the quantity that exploded was less than one-third of the quantity involved in the Beirut explosion, the destruction was more severe. It is not only the quantity that explodes which decides the destructive potential; rather the mode of explosion such as a fire, a detonation, or a quasi-detonation governs the destruction, and we shall deal with them. We shall study in the subsequent chapters about the formation of a fire, an explosion from a fire in confined and partially confined geometries, and the transition from a fire to a detonation and quasi-detonation under confined conditions.

1.3.3 Explosion of Fuel Tank of Aircraft During Flight

A Boeing 747 aircraft on flight TWA 800 took off from New York on July 17, 1996, and exploded eleven minutes after takeoff. The central fuel tank of the aircraft was almost empty at takeoff. When the aircraft reached a height between 4.5 and 5 km, a powerful explosion occurred in the central fuel tank, and a huge ball of fire engulfed the aircraft. All 230 persons aboard the flight lost their lives.

Volatile liquid fuel kerosene with small amount of additives and known as 'jet fuel' is used as fuel for aircrafts. The small amount of kerosene, present in the central fuel tank, formed an explosive mixture of kerosene vapor and air during the ascent of the aircraft. At ground level, the ambient pressure is higher than at the higher altitudes, and the fraction of kerosene fuel vapor in the vapor–air mixture was not very significant. At the higher altitude, the smaller mass of available air in the tank formed a fuel vapor–air mixture, which could easily catch fire and burn unlike on the ground where a combustible fuel vapor–air cloud could not form in the tank. An electrostatic spark perhaps ignited the flammable fuel vapor–air mixture in the tank. The release of energy during the combustion of the mixture in the near-confined volume of the tank ruptured the tank and formed a blast wave. The blast contributed to rupture the other fuel tanks and spilled the large amounts of kerosene contained therein and ignited them to form a huge fireball.

This example represents a *confined fuel vapor–air explosion*.

1.3.4   Largest Man-Made Explosion: Ural Mountains, June 4, 1989

A pipeline 0.7 m in diameter conveying natural gas (NG) in Siberia got ruptured over a length of about 1 and 2 km at a distance of 1400 km downstream of the supply station. The pipeline was designed for a maximum pressure of 5 MPa but was operated at a lower value of 2.5 MPa so that the possibility of a rupture in the pipeline was not foreseen. Large quantities of NG leaked from the ruptured pipeline, and huge quantities of vapor traversed the Siberian plains. Natural gas, it may be noted, contains mainly methane (CH_4).

The double railway lines of the Trans Siberian Railway between Vladivostok and Moscow passed through the region of this cloud. In the absence of mixing with air and formation of a flammable mixture and an ignition source to ignite the mixture, no fire or untoward incident occurred even though a large cloud of NG was present. Two electric trains happened to pass in opposite directions through the NG cloud. The turbulence generated between the trains enabled mixing of the methane gas and air, and the electric spark from the overhead electric line ignited this mixture. As a result of the explosion, 600 people were killed, and trees were flattened 4 km from the source of the blast. This represents an *unconfined fuel–air explosion.*

Many unconfined and partially confined fuel–air explosions are reported. An example, often cited, is the explosion of about 40 tons of cyclohexane in a chemical plant in Flixborough, England on June 1, 1974. The cyclohexane vapor mixed with air and was ignited from the heat in a furnace. There were 28 fatalities, and the blast wave could be heard 35 km away.

1.3.5   Fireball and Blast in the Explosion at Crescent City, Illinois: June 2, 1970

A train transporting volatile liquid fuel derailed at Crescent City, Illinois. It had ten tank cars, each of 125 m^3 capacity. The leak of the volatile fuel caused a fire. The heating of the tank cars from the fire increased the pressure in one of the tank cars as the vent provided in the tank was insufficient to allow the large quantities of vapor formed to escape. As a result, high pressure developed in the tank, and it ruptured. A huge fireball and blast developed from the burning of the accelerated release of fuel vapor. The business section of the Crescent City was destroyed in the explosion and fire.

In this explosion, the boiling of the volatile liquid fuel caused by heating due to fire increased the pressure in the tank and the bursting of the tank car. This was followed by the fireball and the blast. This type of explosion is therefore known as *boiling liquid expanding vapor explosion* (BLEVE). The accidental explosion of a petroleum storage plant in Jaipur, India, on October 29, 2009, was also a boiling liquid expanding vapor explosion and formed a huge fireball.

Such accidental explosions involving BLEVE are met with in the offshore plat-forms and petrochemical industries. Instead of the fire heating the fuel tank, a flame torch from one tank incident on the other can boil the liquid in the other tank and rupture it, thus leading to a BLEVE.

1.3.6   Explosion in a Bakery at Turin Involving Dust

This is the first recorded explosion involving dust. Late in the evening on December 14, 1785, a boy was sent by a baker in Turin, Italy, to fetch dry corn flour from a bakery store. The boy stirred the corn flour with one hand while holding onto a lamp in the other. The mixture of the stirred corn flour and air got ignited by the lamp and exploded.

The above example involves the *explosion of dust in a confined environment*. Grain dust explosions, such as the above, have been quite frequent. At Wichita, Kansas, in the United States, the explosion of dust–air mixture in a kilometer long grain elevator on June 8, 1998, resulted in the death of seven people in addition to injuring many. An entire port terminal was nearly destroyed in Parangua, Brazil, in November 2001 from the explosion of a grain elevator at the port.

The dust need not be of organic origin, and metal dust used in the paints and explosives industries is even more vulnerable to explosion. Dust explosions involving coal dust in the presence of flammable gases have occurred in mines. On July 1, 2020 a powerful explosion involving lignite coal dust occurred in a boiler unit of the Neyveli Lignite Corporation in South India killing 13 workers.

1.3.7   Explosion in a Copper Smelter at Flin Flon, Canada

On the evening of August 8, 2000, after the shutdown of a reverberatory furnace at a copper smelter in Flin Flon, Ottawa, Canada, water was poured into the furnace in order to cool it down. The large heat content in the furnace flashed the water to steam. The resultant sudden increase in pressure from the increased specific volume of steam resulted in an explosion.

No chemical energy release resulted from this process, but the explosion was a result of the physical process of change of state involving flash vaporization, causing a large increase of volume. The volcanic explosion at Krakatau was also seen to be the result of flash vaporization. Such *physical explosions* have also been encountered in the storage of cryogenic fuels such as liquid petroleum gas (LPG) and rupture of cryogenic vessels.

1.3.8   World's Worst Industrial Disaster: Bhopal Gas Tragedy (December 2/3, 1984)

The accidental ingress of water into the methyl isocyanide (MIC) tanks during the regular cleaning operations and the subsequent hydrolysis reactions between water and MIC led to buildup of temperature and pressure in the MIC tanks. The presence of iron and chloroform in the tank and the absence of refrigeration cooling of the tank further augmented the chemical reactions, leading to a runaway reaction and vigorous boiling of liquid MIC from the increase of temperature and pressure. The vent valve of the tank opened at the high pressure, and a two-phase flow of liquid and vapor MIC escaped through the vent valve. About 27 tons of MIC got released into the ambient from a stack.

The dense MIC vapor got transported by surface winds blowing at a low speed of 2.9 m/s. Being a cold winter night, the MIC vapor could not readily disperse into the atmosphere. As a result, it traversed along the ground and diffused into homes and poisoned a lot of people. A total of 20,000 people lost their lives. The runaway reactions in the tank and the *atmospheric dispersion* were the controlling elements of this tragedy.

A flammable gas leak could likewise disperse and result in a fire or an explosion. Pollutants, toxic gases, and reactive gases could under certain atmospheric conditions accumulate over the ground instead of dispersing away. In a major calamity in London called the Great Smog of London in 1952, 12,000 people died. The low temperature of the winter prompted the people to burn more coal, and the absence of sunlight resulted in lower temperatures at the ground level, viz. an inversion of temperature. The surface winds were negligibly small. The pollutant gases and smoke settled over the ground, entered the houses, and affected the people.

1.3.9   Nuclear Explosions: Chernobyl (April 26, 1986) and Fukushima (March 11, 2011)

The loss of coolant in the nuclear power plant at Chernobyl in Ukraine on April 26, 1986, led to formation of steam and an explosion and resulted in the spread of radioactive material. Hydrogen gas was generated from the metal-steam reaction in this loss of coolant accident (LOCA) and formed an explosive gas mixture with ambient air. Such loss of coolant accidents (LOCA) also took place in the nuclear plant at the Three Mile Island in United States on March 28, 1979.

A major LOCA occurred at the Fukushima Daiichi power plant in Japan on March 11, 2011. The Fukushima Daiichi nuclear power plant was designed to take care of LOCA for inadvertent earthquakes and tsunamis. Backups with redundancy to ensure coolant supply to the reactor with pumps, power generators, and batteries were provided should there be a shutdown of the reactors following an earthquake or tsunami. However, following a massive earthquake for which the reactors did automatically shut down, the coolant supply could not be maintained due to the flooding of seawater from the tsunami and disruption of power following the earthquake. The heat in the reactor core melted the zirconium alloy used for cladding the reactor. The reaction between the very hot zirconium and water generated hydrogen gas. This happened a few hours to a day after the disruption. The hydrogen–air mixture in the confined space exploded releasing radioactive material and contaminated the township and the water in the Pacific ocean.

In addition to LOCA, nuclear explosions could also originate from *nuclear fission and fusion*. These are associated with very rapid release of energy.

1.4 Classification of Explosions

Based on the examples considered in Sects. 1.3.1 to 1.3.8, explosions could be classified into the following eight groups:

1. Condensed phase thermal explosions
2. Gas phase confined explosions
3. Unconfined explosions
4. Boiling liquid expanding vapor explosions
5. Dust explosions
6. Physical explosions
7. Atmospheric dispersion of flammable gases, vapors, and pollutants
8. Nuclear explosions.

We shall deal with them, except for nuclear fission and fusion, in the subsequent chapters. However, before doing so, we consider the characteristic features of blast waves formed in explosions, their decay, and their damage potential in the next two chapters. We follow this up with general discussions on the magnitude and rate of energy release from chemical reactions involving explosives in Chap. 4. An explosive could be a gas, liquid, or solid and is defined as a substance capable of producing spontaneous energy release for an explosion to occur.

Examples

1.1 *Merging of Waves*: If a compression wave of 10 kPa merges with a compression front of 5 kPa, what would be the net pressure in the merged compression wave? If an expansion front of 2 kPa merges with the compression front, what would be the pressure of the merged wave?

Solution
When compression packets of 10 kPa and 5 kPa merge together the pressure becomes $5 + 10 = 15$ kPa.

The expansion wave has a negative pressure. The pressure in the expansion wave is -2 kPa. The pressure in the merged wave of the expansion wave and compression wave would therefore be $+5 - 2 = 3$ kPa.

1.2 *Sound Speed*: Determine the sound speed at a height of 10 km above the surface of the Earth wherein the temperature is −50°C. The sound speed in air at the surface of Earth for which the temperature is 27°C is 330 m/s.

Solution

Assuming air to be an ideal gas, the sound speed a in it is given by $a = \sqrt{\gamma RT}$, where γ is the specific heat ratio, R is the specific gas constant in J/(Kg K), and T is its temperature in K. The sound speed a_1 at the height of 10 km is therefore given by

$$\frac{a_1}{a} = \sqrt{\frac{T_1}{T}}$$

where the temperature $T_1 = 273 - 50 = 223$ K. The temperature of air at the surface is $273 + 27 = 300$ K. The sound velocity $a = 330$ m/s at $T = 300$ K and hence

$$a_1 = 330\sqrt{\frac{223}{300}} = 284.5 \text{ m/s}$$

The sound speed at a height of 10 km over the surface of Earth is 284.5 m/s.

1.3 *Wave Velocity in Finite Amplitude Waves*: A finite amplitude sinusoidal wave with maximum pressure amplitude of 2 kPa traverses in the atmosphere having a pressure of 100 kPa and a temperature of 27 °C. The sound velocity in the atmosphere is 330 m/s. Determine the sound velocity at the crest and trough of the finite amplitude wave assuming its propagation to be isentropic. The specific heat ratio of the atmospheric air can be assumed as 1.4.

Solution

Taking air to be an ideal gas, we have its equation of state as $pV/T = $ constant. For an isentropic process $pV^\gamma = $ constant where γ is the specific heat ratio. Solving the equation of state with the isentropic process equation, we get $p^{[(\gamma-1)/\gamma]}/T = $ constant.

Denoting the atmospheric pressure and temperature by p_0 and T_0 and the pressure and temperature at the crest as p_1 and T_1, respectively, we have

$$\frac{T_1}{T_0} = \left(\frac{p_1}{p_0}\right)^{(\gamma-1)/\gamma}$$

The value of $p_1 = 100 + 2 = 102$ kPa while $p_0 = 100$ kPa. The temperature T_0 is $273 + 27 = 300$K. Hence the value of temperature at the crest T_1 is

$$T_1 = 300 \times \left(\frac{102}{100}\right)^{0.4/1.4} = 301.7 \text{ K}$$

Similarly the temperature at the trough is

$$T_2 = 300 \times \left(\frac{98}{100}\right)^{0.4/1.4} = 298 \text{ K}$$

If the sound velocity of the atmosphere, which is the mean velocity corresponding to no fluctuations in the temperature and pressure, is denoted by a_0, we have the value of sound velocity at the crest as

$$\frac{a_1}{a_0} = \sqrt{\frac{T_1}{T_0}}$$

The sound velocity a_0 is given as 330 m/s. We therefore get the sound velocity at the crest as

$$a_1 = a_0\sqrt{\frac{T_1}{T_0}} = 330 \times \sqrt{\frac{301.7}{300}} = 330.93 \text{ m/s}$$

Similarly, the sound velocity at the trough is

$$a_2 = a_0\sqrt{\frac{T_2}{T_0}} = 330 \times \sqrt{\frac{298}{300}} = 328.9 \text{ m/s}$$

The sound velocity at the crest increases by 0.93 m/s, while at the trough the sound velocity decreases by 1.1 m/s.

1.4 *Sound Speed in Different Media*: The following table gives the values of bulk modulus and density of air, water, and steel. Compare the sound velocity in air, water, and steel.

S. No.	Material	Bulk Modulus	Density
1	Air	140 kPa	$1.2\,\text{kg/m}^3$
2	Water	2.1 GPa	$1000\,\text{kg/m}^3$
3	Steel	135 GPa	$7800\,\text{kg/m}^3$

Solution

Bulk modulus for any material denotes its ability to withstand changes in its volume. Denoting a volume of material as V and the change in volume as ΔV for an applied pressure Δp, the bulk modulus is $\beta = -((\Delta p/\Delta V)/V)$, i.e., applied pressure by the relative deformation. It is therefore the inverse of compressibility, which was defined as $\kappa = -(\Delta V/V)/\Delta p$.

The sound speed was given by $a^2 = 1/\rho\kappa = \beta/\rho$. Substituting the values from the table, we get the sound speed in air, water, and steel to 341, 1500, and 4160 m/s, respectively.

Nomenclature

a	Speed of sound (m/s)
C	Constant
p	Pressure (Pa)
p'	Pressure fluctuation (Pa)
$\hat{p}$	Maximum amplitude of pressure fluctuation (Pa)
T	Temperature (K)
U	Velocity (m/s)
V	Volume
d, δ	Small increment
Δ	Finite increment
γ	Specific heat ratio
λ	Wavelength (m)
ρ	Density (kg/m^3)
κ_S	Isentropic compressibility coefficient
κ_T	Isothermal compressibility coefficient

Further Reading

1. Baker, W. E., *Explosions in Air*, University of Texas Press, Austin, 1973.
2. Courant, R. and Freidricks, K. O., *Supersonic Flow and Shock Waves*, Interscience, New York, 1948.
3. Croft, W. M., Fires Involving Explosions - A Literature Review, *Fire Safety Journal*, 3, 3–24, 1980.
4. Crowl, D. A. and Louvar, J.F., *Chemical Safety: Fundamentals with Applications*, Prentice Hall, Englewood Cliffs, NJ, 2002.
5. Kinney, G.F. and Graham, K. J., *Explosive Shocks in Air*, Springer, Berlin, 1985.
6. Kinsler, L. E., Frey, A.R., Coppens, D.B. and Sanders, J.V., *Fundamentals of Acoustics*, 4th ed., John Wiley and Sons, New York, 2000.
7. Lee, J.H., Quivavo, C.M. and Grierson, D. E., (eds.), Fuel-Air Explosions, Proc. International Conference on Fuel–Air Exolsions held at McGill University, Montreal, University of Waterloo Press, Waterloo, Ontario, 1982.
8. Strehlow, R.A., Unconfined Vapor–Cloud Explosions—An Overview, Proc.14th Int. Symposium on Combustion, The Combustion Institute, Pittsburg, PA, 1973, pp. 1189–1200.
9. Strehlow, R. A. and Baker, W.E., The Characterization and Evaluation of Accidental Explosions, NASA CR-134779, National Aeronautics and Space Administration, Cleveland, OH, June 1965.
10. Stull, D. R., *Fundamentals of Fire and Explosion*, AIChE Monograph Series, Vol. 73, No. 10, 1977.

Chapter 2
Blast Waves in Air

The blast wave, generated by a rapid release of energy, disperses and dissipates the energy into the surrounding medium. It was seen in the last chapter that the blast wave consists of a shock front, which progressively decays in its strength as it travels away from the source of the energy release. The characteristic features of the blast wave, the rate of its decay, and the parameters that make it destructive are addressed in this chapter.

2.1 Ideal Blast Wave

We first introduce certain simplifying assumptions to help in developing a rational understanding. Energy E_0, in Joules, is assumed to be impulsively released in a very small spherical volume of radius R_{S0} in the ambient atmosphere having density ρ_0. A spherical blast wave, symmetric about the energy source of radius R_{S0}, propagates into the surrounding atmosphere. The radius of the spherical blast wave is taken to be R_S at time t after release of the energy E_0. Figure 2.1 illustrates the spherical blast wave formed at radius R_S.

The trajectory of the blast wave is shown in Fig. 2.2. The wave progresses a smaller distance for the same time interval as time increases. This is due to the decrease of its speed of propagation. A schematic of the variation of the velocity of the blast wave, denoted by $\dot{R}_S (= dR_S/dt)$ is shown in Fig. 2.3.

A typical profile of the rate of energy released $\dot{E}_0$ in the source of radius R_{S0} with time is shown in Fig. 2.4. The profile for the energy release is discussed in Chap. 4. The major part of the energy is shown in Fig. 2.4 to be released over a small duration τ. By sudden or impulsive energy release, we imply that the period τ is very much

Fig. 2.1 Blast wave formed at radius R_S from energy source of radius R_{S0}

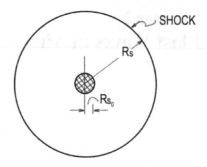

Fig. 2.2 Trajectory of blast wave

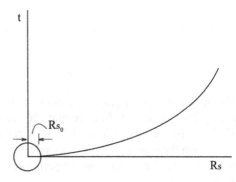

Fig. 2.3 Velocity of blast wave

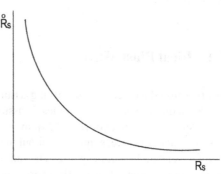

smaller than the time t of travel of the blast wave. In the limit of the duration τ tending to zero ($\tau \rightarrow 0$), the blast wave is on its own after it is initiated. If τ is comparable to t, the energy release continues to push the wave.

The spherical blast wave is said to be ideal for $\tau = 0$ and $R_{S0} = 0$, implying that an ideal blast wave is formed for instantaneous energy release at a point. We would like to determine the trajectory of such an ideal spherical blast wave, obtained by an energy release E_0 in a medium of density ρ_0. A functional relationship between R_S, E_0, ρ_0, and t is addressed in the form

$$R_S = f(E_0, \rho_0, t) \tag{2.1}$$

Fig. 2.4 Energy release rate

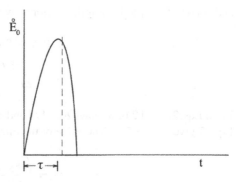

2.1.1 Ideal Blast Trajectory from Dimensional Considerations

A dimensional analysis provides the likely combination of the four variables in Eq. 2.1. G. I. Taylor carried out the dimensional analysis in 1950 and showed the results to agree well with the blast from the nuclear explosion over Hiroshima and Nagasaki during World War 2. The dimensions of the four parameters in Eq. 2.1 are

$$
\begin{aligned}
R_S &: \ L \\
E_0 &: \ ML^2/T^2 \\
\rho_0 &: \ M/L^3 \\
t &: \ T
\end{aligned}
\tag{2.2}
$$

L, M and T in the above equation, denote dimensions of length, mass, and time respectively. Three of the four parameters are seen to have independent dimensions. The number of non-dimensional groups that can be formed from the four parameters as per Buckingham π theorem (parameters-independent dimensions) is $4 - 3 = 1$. The non-dimensional parameter can be denoted by

$$
\pi_1 = R_S E_0^a \rho_0^b t^c
\tag{2.3}
$$

where a, b, and c are exponents such that π_1 is non-dimensional.

Substituting the dimensions from Eq. 2.2 in Eq. 2.3, we get

$$
\pi_1 = L \left(\frac{ML^2}{T^2} \right)^a \left(\frac{M}{L^3} \right)^b T^c
\tag{2.4}
$$

and solving for π_1 to be non-dimensional, we obtain

$$1 + 2a - 3b = 0 \tag{2.5}$$
$$a + b = 0 \tag{2.6}$$
$$-2a + c = 0 \tag{2.7}$$

From Eqs. 2.5 and 2.6, we have $b = 1/5$ and $a = -1/5$. Substituting these values in Eq. 2.7 gives $c = -2/5$. The value of the dimensionless group π_1 becomes

$$\pi_1 = \left(\frac{R_S \rho_0^{1/5}}{E_0^{1/5} t^{2/5}} \right) \tag{2.8}$$

The dimensionless parameter can also be written as

$$\left(\frac{R_S^5 \rho_0}{E_0 t^2} \right)$$

and is referred to as a similarity parameter.

For a given value of the dimensionless parameter π_1 equal to C, the value of R_S from Eq. 2.8 is written as

$$R_S = C \left(\frac{E_0}{\rho_0} \right)^{1/5} t^{2/5} \tag{2.9}$$

A given source energy E_0 released in an ambient having density ρ_0 gives a constant value of the ratio $(E_0/\rho_0)^{1/5}$. We therefore write Eq. 2.9 as

$$R_S = A t^{2/5} \tag{2.10}$$

The value A is a constant for a given value of E_0 and ρ_0. The velocity of the blast wave $\dot{R}_S$ is determined from Eq. 2.10 as

$$\dot{R}_S = \frac{dR_S}{dt} = \frac{2}{5} A t^{-3/5} \tag{2.11}$$

Simplifying the above expression and denoting:

$$t^{-3/5} = (t^{2/5})^{-3/2} = \left(\frac{R_S}{A} \right)^{-3/2}$$

we get

$$\dot{R}_S = \frac{2}{5} A \frac{1}{A^{-3/2}} R_S^{-3/2} \tag{2.12}$$

Fig. 2.5 Decay of blast wave velocity with distance

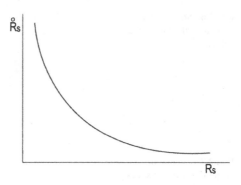

Representing a constant B as $(2/5)A^{5/2}$, we get

$$\dot{R}_S = BR_S^{-3/2} \tag{2.13}$$

The velocity $\dot{R}_S$ of the blast wave therefore decays with distance as $R_S^{-3/2}$. The trend of the variation is shown in Fig. 2.5.

The change of the pressure, velocity, and density of the medium processed by the blast wave requires that the changes across the shock front be known. We therefore first address the jump in the pressure, density, and velocity across a shock wave traveling at a constant velocity $\dot{R}_S$ in the next section and thereafter determine these parameters for a blast wave.

2.2 Modeling of Parameters Across A Constant Velocity Shock

2.2.1 Rankine–Hugoniot Equations

Consider a planar shock wave traveling at constant velocity $\dot{R}_S$ in a stationary gaseous medium having density ρ_0, pressure p_0 and temperature T_0. The medium is assumed as a perfect gas. This implies that the ideal gas equation $p = \rho RT$, where R is the specific gas constant is valid and the specific heats at constant pressure and constant volume in the gaseous medium are constants.

Let the values of pressure, density, and temperature of the medium behind the shock wave be denoted by p, ρ, and T respectively. The velocity of the medium behind the shock wave is denoted by u. Figure 2.6 shows the propagation of the shock at velocity $\dot{R}_S$ in the stationary medium and the properties of the medium ahead of the shock wave and behind it.

We write the equations for conservation of mass, momentum, and energy across the shock wave by fixing the frame of reference to be the shock wave. In this frame

Fig. 2.6 Shock wave
moving in the medium at
velocity $\dot{R}_S$

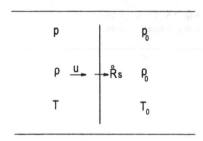

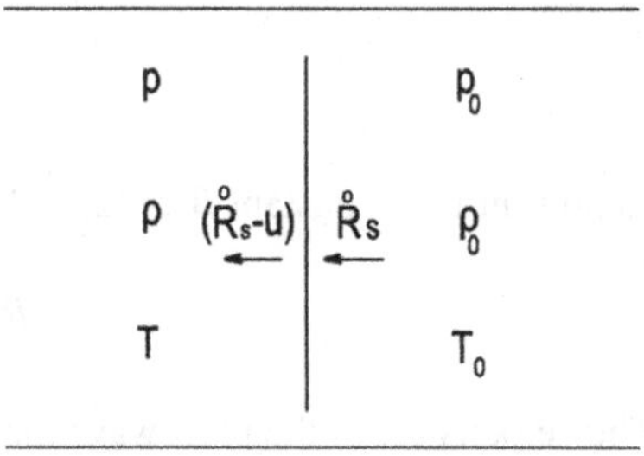

Fig. 2.7 Change of
properties across a shock
wave in frame of reference of
the shock

of reference, the ambient gases at p_0, ρ_0, and T_0 move toward the shock at velocity
$\dot{R}_S$. The conditions behind the shock are p, ρ, and T, respectively, while the velocity
of the gases is $\dot{R}_S - u$ (Fig. 2.7).

The mass, momentum, and energy conservation equations are

$$\text{Mass:} \quad \rho_0 \dot{R}_S = \rho (\dot{R}_S - u) \tag{2.14}$$

$$\text{Momentum:} \quad p - p_0 = \rho_0 \dot{R}_S [\dot{R}_S - (\dot{R}_S - u)] \tag{2.15}$$

$$\text{Energy:} \quad h_0 + \frac{\dot{R}_S^2}{2} = h + \frac{(\dot{R}_S - u)^2}{2} \tag{2.16}$$

Here h_0 and h denote the specific enthalpies of the gas ahead and behind the shock,
respectively. From Eqs. 2.14 and 2.15, we get

$$p - p_0 = \rho_0 \dot{R}_S u \tag{2.17}$$

$$= \rho_0 \dot{R}_S^2 \left(\frac{u}{\dot{R}_S} \right)$$

$$= \rho_0^2 \dot{R}_S^2 \left(\frac{1}{\rho_0} - \frac{1}{\rho} \right)$$

We rewrite the above as

$$\rho_0^2 \dot{R}_S^2 = \frac{p - p_0}{1/\rho_0 - 1/\rho} \tag{2.18}$$

From energy conservation Eq. 2.16, we have

$$h - h_0 = \frac{1}{2}[\dot{R}_S^2 - (\dot{R}_S - u)^2]$$
(2.19)

Denoting the term $(\dot{R}_S - u) = u_1$ and the enthalpy $h_1 - h_0 = C_P(T - T_0)$ and noting that

$$C_P = \frac{\gamma R}{\gamma - 1}$$

where γ is the specific heat ratio and R is the specific gas constant, the energy equation given by Eq. 2.19 simplifies to

$$\frac{\gamma}{\gamma - 1}(RT - RT_0) = \frac{1}{2}(\dot{R}_S^2 - u_1^2)$$
(2.20)

Using the ideal gas equations $p/\rho = RT$ and $p_0/\rho_0 = RT_0$ and substituting the values of $\dot{R}_S$ and u_1 from Eqs. 2.18 and 2.14 in Eq. 2.20, we obtain

$$\frac{\gamma}{\gamma - 1}\left(\frac{p}{\rho} - \frac{p_0}{\rho_0}\right) = \frac{1}{2}\left(\frac{p - p_0}{\rho - \rho_0}\right)\left(\frac{\rho}{\rho_0} - \frac{\rho_0}{\rho}\right)$$
(2.21)

The above equation relates pressure and density behind the shockwave with the pressure and density ahead of it. It simplifies to give

$$\frac{\gamma}{\gamma - 1}\left(\frac{p}{\rho} - \frac{p_0}{\rho_0}\right) - \frac{1}{2}(p - p_0)\left(\frac{1}{\rho_0} + \frac{1}{\rho}\right) = 0$$
(2.22)

Equation 2.22 is known as the shock Hugoniot. All possible states behind a shock given by p and ρ are determined by this equation for the given values of pressure and density of the medium p_0 and ρ_0. Figure 2.8 shows the values of p and ρ obtainable in a plot of p versus $1/\rho$ for a given ambient value of p_0 and ρ_0 shown by point O.

Fig. 2.8 Conditions behind shock as given by Eq. 2.22

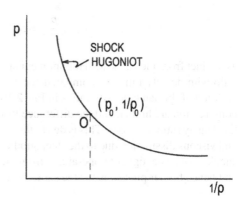

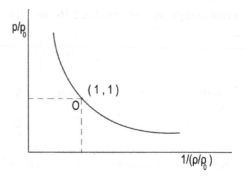

Fig. 2.9 Plot of
Rankine–Hugoniot equation

Equation 2.22 is further simplified to give values of pressure ratio p/p_0 as function of ρ/ρ_0 and also ρ/ρ_0 in terms of p/p_0. We get from Eq. 2.22

$$\frac{2\gamma}{\gamma - 1}\left(\frac{p}{p_0} - \frac{\rho}{\rho_0}\right) = \left(\frac{p}{p_0} - 1\right)\left(\frac{\rho}{\rho_0} + 1\right) \tag{2.23}$$

which on simplifying gives

$$\frac{p}{p_0} = \frac{\frac{\gamma+1}{\gamma-1}\frac{\rho}{\rho_0} - 1}{\frac{\gamma+1}{\gamma-1} - \frac{\rho}{\rho_0}} \tag{2.24}$$

Similarly, we can obtain

$$\frac{\rho}{\rho_0} = \frac{\frac{\gamma+1}{\gamma-1}\frac{p}{p_0} + 1}{\frac{\gamma+1}{\gamma-1} + \frac{p}{p_0}} \tag{2.25}$$

Eqs. 2.24 and 2.25 are known as Rankine–Hugoniot equations. They represent property changes across a shock. A typical plot of the pressure ratio p/p_0 as a function of $1/(\rho/\rho_0)$ is shown in Fig. 2.9.

When the changes are isentropic such as associated with a sound wave (see Chap. 1), the pressure and density variations are given by

$$\frac{p}{\rho^\gamma} = \text{constant} \tag{2.26}$$

as distinct from the Rankine–Hugoniot equation (Eq. 2.24). The changes of pressure ratio with density ratio are compared for a shock with that for an isentropic compression of the same density ratios in Fig. 2.10. The pressure ratios across the shock compression are higher, and the difference from isentropic compression increases as the density ratio increases. This is due to the irreversible processes of heat conduction and viscous dissipation due to the steep gradients in the temperature and velocity in a shock wave. The higher temperatures from the heating result in higher temperatures and hence higher pressures for the same value of density ratio.

Fig. 2.10 Shock compression and isentropic compression

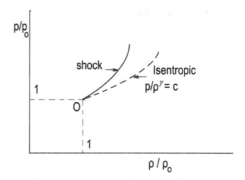

2.2.2 Rayleigh Line and Properties Across a Shock Wave of Given Velocity $\dot{R}_S$

The Rankine–Hugoniot equations relate pressure and density changes across a shock wave. It represents the changes between pressure and density for all possible shock velocities. The specific changes in pressure or density for a given shock velocity $\dot{R}_S$ are to be obtained from it. In the following, we relate the pressure and density changes with the shock velocity.

The shock velocity $\dot{R}_S$ from the solution of the mass and momentum conservation equations given by Eq. 2.18 is

$$\dot{R}_S^2 = \frac{1}{\rho_0^2} \frac{p - p_0}{1/\rho_0 - 1/\rho}$$

The above equation can be expressed as

$$\dot{R}_S^2 = \frac{p_0}{\rho_0} \frac{p/p_0 - 1}{1 - \rho_0/\rho} \tag{2.27}$$

It gives an indication of the slope of the line joining the initial state with the final state behind the shock in the plot of p/p_0 as a function of $1/(\rho/\rho_0)$. The line is illustrated in Fig. 2.11 and is known as the Rayleigh line.

If we consider a shock traveling at velocity $\dot{R}_S$, the values of pressure and density behind it should obviously be on the Rayleigh line corresponding to velocity $\dot{R}_S$ in addition to being on the shock Hugoniot (Fig. 2.9) which represents all possible states behind a shock. The point of intersection of the Rayleigh line and the shock Hugoniot would therefore give the state behind the shock of velocity $\dot{R}_S$. This is shown by point S in Fig. 2.12.

The pressure and density of the medium behind a shock traveling at a given velocity is thus obtained. Knowing the pressure and density, the temperature is determined from the equation of state. Expressions for the change in the properties of the medium across a shock traveling with a velocity $\dot{R}_S$ are derived in the next section.

Fig. 2.11 Rayleigh line

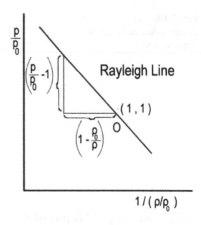

Fig. 2.12 Pressure and density behind shock traveling at velocity $\dot{R}_S$

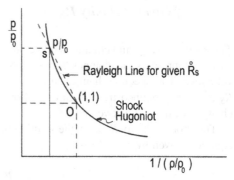

2.2.3   Properties Behind a Shock of Mach Number M_S

The Rankine–Hugoniot and the Rayleigh line equations are solved to determine the properties behind a shock wave. The properties comprise the pressure p, the density ρ, and the velocity u_1. It must be noted that u_1 is the velocity in the medium behind the shock in the frame of reference of the shock with the medium ahead of the shock moving toward the stationary shock with a velocity $\dot{R}_S$. The shock velocity $\dot{R}_S$ is expressed in a non-dimensional form using the speed of sound a_0 in the ambient medium to give its Mach number $M_S (M_S = \dot{R}_S / a_0)$. Similarly, the velocity of the medium behind the shock u_1 is divided by the sound speed corresponding to the shocked gases a to give the Mach number of the shocked medium ($M = u_1 / a$). The momentum Eq. 2.15 is written noting that $u_1 = \dot{R}_S - u$ as:

$$p + \rho u_1^2 = p_0 + \rho_0 \dot{R}_S^2 \tag{2.28}$$

The above equation is expressed in terms of Mach number by using the expression for sound speed a_0 in the undisturbed medium and in the medium behind the shock a to be (See Appendix A)

$$a_0^2 = \frac{\gamma p_0}{\rho_0} \tag{2.29}$$

$$a^2 = \frac{\gamma p}{\rho} \tag{2.30}$$

Substituting in Eq. 2.28 gives

$$p(1 + \gamma M^2) = p_0(1 + \gamma M_S^2)$$

and

$$\frac{p}{p_0} = \frac{1 + \gamma M_S^2}{1 + \gamma M^2} \tag{2.31}$$

$$\frac{\rho}{\rho_0} = \frac{1 + \gamma M_S^2}{1 + \gamma M^2} \frac{a_0^2}{a^2} \tag{2.32}$$

The ratio of the sound speeds a_0/a in the above equation is obtained from the energy Eq. 2.16 to give

$$\frac{\gamma}{\gamma - 1} RT_0 + \frac{\dot{R}_S^2}{2} = \frac{\gamma}{\gamma - 1} RT + \frac{u_1^2}{2}$$

or

$$\frac{a_0^2}{2} \left(\frac{2}{\gamma - 1} + M_S^2 \right) = \frac{a^2}{2} \left(\frac{2}{\gamma - 1} + M^2 \right) \tag{2.33}$$

Here use has been made of the dependence of sound speed on temperature, viz. $a_0^2 = \gamma RT_0$ and and $a^2 = \gamma RT$ (See Appendix A). From Eq. 2.33, we get

$$\frac{a_0^2}{a^2} = \frac{1 + \frac{\gamma - 1}{2} M^2}{1 + \frac{\gamma - 1}{2} M_S^2} \tag{2.34}$$

We need to obtain Mach number of the shocked medium M for a given value of M_S. We use the mass conservation Eq. 2.14 to give

$$\frac{\rho}{\rho_0} = \frac{\dot{R}_S}{u_1} = \frac{M_S}{M} \frac{a_0}{a} \tag{2.35}$$

The above value of ρ/ρ_0, substituted in the momentum equation (2.32), yields

$$\frac{M_S}{M} \frac{a_0}{a} = \frac{1 + \gamma M_S^2}{1 + \gamma M^2} \left(\frac{a_0}{a} \right)^2$$

Simplifying and using the ratio of a_0/a from Eq. 2.34, we get

$$\frac{2 + (\gamma - 1)M^2}{2 + (\gamma - 1)M_S^2} = \frac{M_S^2}{M^2} \left(\frac{1 + \gamma M^2}{1 + \gamma M_S^2}\right)^2 \tag{2.36}$$

For a given value of shock Mach number M_S, the Mach number behind the shock M can therefore be determined. The equation is quadratic in M^2, giving two roots. One solution is imaginary and is neglected. The real root is

$$M^2 = \frac{M_S^2 + \frac{2}{\gamma - 1}}{\frac{2\gamma}{\gamma - 1} M_S^2 - 1} \tag{2.37}$$

The above equation can also be readily derived directly from the mass, momentum and energy conservation relations across a shock and is given in Appendix B.

Equation 2.37 can be rearranged as

$$M^2 = 1 - \frac{\gamma + 1}{2\gamma} \frac{M_S^2 - 1}{(M_S^2 - 1) + (\gamma + 1)/2\gamma} \tag{2.38}$$

We observe from the above equation that for $M_S > 1$ (the Mach number of the shock M_S is always greater than 1), the value of $M < 1$; i.e., the gas flow behind the shock wave is subsonic in the frame of reference of the shock. The gas behind the shock therefore gets compressed. We must not forget that M represents the Mach number of the medium behind the shock in the frame of reference of the shock being stationary and the flow upstream of the shock moves toward the stationary shock at a Mach number M_S.

When the shock is very weak, such as obtained when it has decayed down to a sound wave ($M_S \rightarrow 1$), we find from Eq. 2.38 that $M \sim 1$.

In order to find the Mach number behind the shock, in the case of a strong shock, i.e., for large M_S, we rearrange Eq. 2.38 to give

$$M^2 = 1 - \frac{\gamma + 1}{2\gamma} \left(\frac{1}{1 + \frac{\gamma + 1}{2\gamma(M_S^2 - 1)}}\right) \tag{2.39}$$

We note from the above equation that the Mach number behind a shock M in the frame of reference of the shock decreases as its Mach number M_S increases. Figure 2.13 shows the variations of Mach number M behind the shock wave as its Mach number M_S increases from 1 to 10. At $M_S = 10$, $M = 0.38$ for $\gamma = 1.4$. In the limit of $M_S \rightarrow \infty$

$$M^2 = \sqrt{\frac{\gamma - 1}{2\gamma}} \tag{2.40}$$

which for $\gamma = 1.4$ gives a value of $M \sim 0.38$.

Fig. 2.13 Variation of Mach number behind a shock

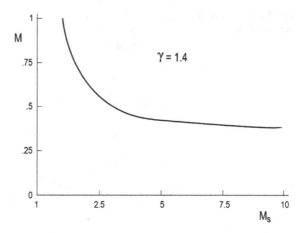

It may be noted that Eq. 2.39 predicts the value of M to be greater than 1 when the value of M_S is less than 1. However, it is not possible to have $M_S < 1$. This expansion solution is also physically not realizable as it leads to a decrease of entropy.

The pressure ratio across a shock can be determined by substituting the value of M determined in Eq. 2.37 in the Rayleigh line Eq. 2.31 to give

$$\frac{p}{p_0} = \frac{1 + \gamma M_S^2}{1 + \gamma M^2} = \frac{1 + \gamma M_S^2}{1 + \gamma \left(\frac{M_S^2 + 2/(\gamma+1)}{2\gamma M_S^2/(\gamma-1) - 1} \right)} \tag{2.41}$$

The above equation simplifies as

$$\frac{p}{p_0} = \frac{1 + \gamma M_S^2}{\left(\frac{(\gamma+1)(1+\gamma M_S^2)}{2\gamma M_S^2 - (\gamma-1)} \right)}$$

to give

$$\frac{p}{p_0} = \frac{2\gamma}{\gamma+1} M_S^2 - \frac{\gamma-1}{\gamma+1} \tag{2.42}$$

The pressure ratio is seen to be greater than one, and hence, a compression process takes place as observed earlier. The pressure ratio increases rapidly as the shock Mach number M_S increases. For $M_S = 1$, the pressure ratio is unity as in an acoustic or sound wave. Figure 2.14 shows the increasing trend of pressure ratio as the shock Mach number increases. For $M_S = 10$, the pressure ratio for $\gamma = 1.4$ is about 120.

The density ratio ρ/ρ_0 is determined from the pressure ratio using the Rankine–Hugoniot equation (Eq. 2.25). Substituting the value of p/p_0 from Eq. 2.42 in Eq. 2.25, we get

Fig. 2.14 Pressure ratio across a shock

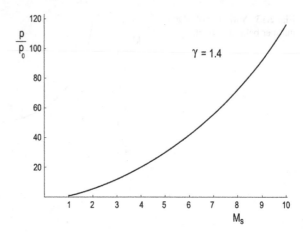

Fig. 2.15 Density ratio across shock

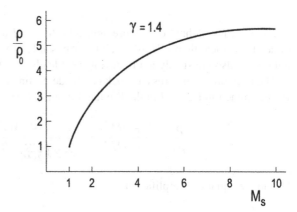

$$\frac{\rho}{\rho_0} = \frac{\frac{\gamma+1}{\gamma-1}\left(\frac{2\gamma M_S^2}{\gamma+1} - \frac{\gamma-1}{\gamma+1}\right) + 1}{\frac{\gamma+1}{\gamma-1} + \frac{2\gamma M_S^2}{\gamma+1} - \frac{\gamma-1}{\gamma+1}}$$

$$= \frac{(\gamma+1)M_S^2}{(\gamma-1)M_S^2 + 2} = \frac{\gamma+1}{\gamma-1+2/M_S^2} \qquad (2.43)$$

For $M_S = 1$, the density ratio is unity. When $M_S \rightarrow \infty$, the density ratio is $(\gamma + 1)/(\gamma - 1)$, which for $\gamma = 1.4$ is 6. A plot of the density ratio as a function of the shock Mach number is shown in Fig. 2.15.

The temperature ratio across a shock is deduced from the pressure and density ratios given in Eqs. 2.42 and 2.43 along with the equation of state ($p = \rho RT$). The temperature ratio is given by

$$\frac{T}{T_0} = \frac{p}{p_0}\frac{\rho_0}{\rho} = \left(\frac{2\gamma}{\gamma+1}M_S^2 - \frac{\gamma-1}{\gamma+1}\right)\left(\frac{\gamma-1+2/M_S^2}{\gamma+1}\right) \qquad (2.44)$$

The velocity ratio across a shock is determined directly from the density ratio using the mass conservation equation as

$$\frac{u_1}{\dot{R}_S} = \frac{\rho_0}{\rho} = \frac{\gamma - 1 + /2M_S^2}{\gamma + 1} \tag{2.45}$$

The pressure p behind the shock can also be expressed in terms of the shock speed $\dot{R}_S$ and initial density ρ_0 as

$$\frac{p}{\rho_0 \dot{R}_S^2} = \frac{p}{p_0} \frac{p_0}{\rho_0 \dot{R}_S^2}$$

Since $\gamma p_0 / \rho_0 = a_0^2$ and $M_S^2 = \dot{R}_S^2 / a_0^2$, and on using the value of p/p_0 from Eq. 2.42, we have

$$\frac{p}{\rho_0 \dot{R}_S^2} = \frac{2}{\gamma + 1} - \frac{\gamma - 1}{\gamma(\gamma + 1)} \frac{1}{M_S^2} \tag{2.46}$$

For strong shocks, i.e., for M_S being large, we get from Eqs. 2.44, 2.45 and 2.46, the following:

$$\frac{\rho}{\rho_0} = \frac{\gamma + 1}{\gamma - 1} \tag{2.47}$$

$$\frac{u_1}{\dot{R}_S} = \frac{\gamma - 1}{\gamma + 1} \tag{2.48}$$

$$\frac{p}{\rho_0 \dot{R}_S^2} = \frac{2}{\gamma + 1} \tag{2.49}$$

The above expressions, which provide the values of density, velocity, and pressure for constant velocity shocks, will be used for studying the decay of blast waves and the properties behind a strong blast wave.

2.3   Change of Properties in a Blast Wave

The blast wave was seen to consist of a shock front whose velocity continuously decreases unlike the constant velocity shock considered in the last section. In the region near to the source of the sudden energy release, wherein the Mach number is significantly large, the increase of the pressure, temperature, and density in the medium across the blast wave would be much higher than in the points further away from the source. The different points along the propagation of the blast would therefore experience different values of properties. Depending on the value of the velocity of the blast wave $\dot{R}_S$ at a specified location of R_S, the magnitudes of the density, velocity, and pressure of the medium behind the blast wave are given by Eqs. 2.43, 2.45, and 2.46, respectively.

Fig. 2.16 Medium
processed by a blast wave

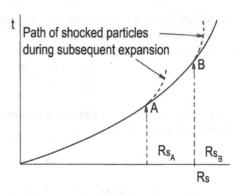

The compressed medium behind the shock front is free to expand and therefore
undergoes an expansion. The expansion process, we have seen in Chap. 1, to be
isentropic. This is illustrated in Fig. 2.16. Here, a parcel of gas that enters the blast
wave at A (radius R_{SA}) is raised to higher values of pressure and density than a parcel
entering at B (radius R_{SB}). The compressed parcel after entering at A expands as
shown. The expansion processes is seen to be also governed by the pressure behind
the blast wave as the front moves beyond A, i.e., R_{SA} to R_{SB}. The parcel of gas entering
at A is therefore influenced by the subsequent propagation of the shock wave beyond
A. The initial change, associated with the compression, however, depends on the
distance from the source, R_{SA}, since the Mach number of the blast wave decays with
the distance. It therefore becomes necessary to consider the initial position R_S of each
parcel of gas in the medium and the position of the blast wave while determining the
flow properties behind a blast wave.

2.3.1   Concentration of Mass at the Wave Front

Consider a spherical blast wave generated by the rapid deposition of energy E_0
instantaneously at a point. Let the blast wave so formed be at a distance $R_S(t)$ at time
t after the release of energy E_0. The mass of the medium within the spherical blast
wave between radius r and $r + dr$ is $4\pi r^2 dr \times \rho(r)$ where $\rho(r)$ is the density of the
medium at radius r when the blast wave is at radius R_S and $R_S > r$. This is illustrated
in Fig. 2.17.

Since the medium entering the blast wave expands after the initial compression
(Fig. 2.17), the density of the medium $\rho(r)$ at a distance r when the blast wave it at
R_S can be assumed to be of the form:

$$\rho(r) = \rho_S \left(\frac{r}{R_S}\right)^q \tag{2.50}$$

Fig. 2.17 Blast wave at
distance R_s at time t

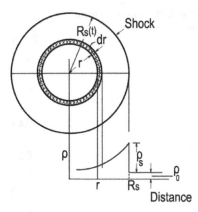

Here ρ_S is the density behind the blast wave when it is at R_S. The exponent q would
be greater than zero ($q > 0$) since the density at $r(r < R_S)$ would be lower than that
behind the shock at radius R_S.

   The total mass enclosed by the spherical blast wave is obtained by integrating the
mass of the annular element at radius r and thickness dr between $r = 0$ and $r = R_S$
(see Fig. 2.17) to give

$$\int_0^{R_S} 4\pi r^2 \rho(r) dr \tag{2.51}$$

This must equal the total mass enclosed by the blast wave, viz.

$$\frac{4}{3}\pi R_S^3 \rho_0$$

where ρ_0 is the initial density of the medium. This condition gives

$$\int_0^{R_S} 4\pi r^2 \rho_S (r/R_S)^q dr = \frac{4}{3}\pi R_S^3 \rho_0 \tag{2.52}$$

The value of ρ_S depends on $\dot{R}_S$ and is given by Eq. 2.43 as a function of M_S and γ.
For a strong shock, the value is fairly independent of M_S and was seen to be given
in Eq. 2.47 as

$$\frac{\rho_S}{\rho_0} = \frac{\gamma + 1}{\gamma - 1}$$

Restricting our attention to strong blast waves, we get on substituting the above
expression for ρ_S/ρ_0 in Eq. 2.52 and simplifying

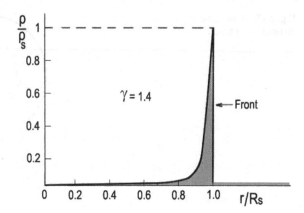

Fig. 2.18 Fraction of mass at the wave front

$$4\pi \rho_0 \frac{\gamma+1}{\gamma-1} \frac{1}{R_S^q} \int\limits_0^{R_S} r^{q+2} dr = \frac{4}{3}\pi R_S^3 \rho_0$$

$$\frac{\gamma+1}{\gamma-1} \frac{R_S^3}{q+3} = \frac{R_S^3}{3}$$

giving

$$q = 3\left(\frac{\gamma+1}{\gamma-1} - 1\right) \tag{2.53}$$

The value of q works out to be 15, when the specific heat ratio γ is 1.4. For such large values of q, the density profile behind the strong blast wave is very steep. The variation of density is plotted as a function of the distance from the shock front (shock front is at $r/R_S = 1$) in Fig. 2.18.

The steep gradient suggests that most of the mass of the medium enclosed by the wave is located just behind the wave front. The blast wave picks up the mass of the medium that it processes and accumulates it at the front just as a snow-plow removes the snow and accumulates it at the plow. With the mass being concentrated at the wave front and the front moving at high velocities, significant impulse from the large momentum change will result in considerable impact load in a strong blast wave.

When the blast wave weakens to give smaller values of M_S, the mass enclosed by the blast wave is no longer concentrated at the wave front.

2.3.2   Deviation from Wave Phenomenon

The piling up of mass at the front of a strong blast wave implies that the medium processed by it is transported and concentrated at its front. The strong blast wave can therefore no longer be considered as a wave in which the medium, viz. the particles in the medium transfers the energy without significantly moving away from their initial position. But then it is also seen that a strong blast wave induces significant velocity to the medium. As an example, if the blast wave is traveling at a Mach number of 5 in a medium of air having a sound velocity of 340 m/s, the velocity of the medium in the frame of reference of the blast front is given by

$$\frac{u_1}{\dot{R}_S} = \frac{\gamma - 1 + 2/M_S^2}{\gamma + 1}$$

For air the value of γ is 1.4 and $u_1/\dot{R}_S$ works out to be 0.2. The value of $\dot{R}_S$ is $340 \times 5 = 1700$ m/s and $u_1 = 1700 \times 0.2 = 340$ m/s. The particles in the medium behind the blast front therefore follow the blast wave at a velocity of $1700 - 340 = 1450$ m/s. This velocity is in the inertial frame of reference and is seen to be extremely large and cannot be assumed to be negligibly small as in a sound wave.

The high velocities of the medium following the blast wave result in a significant departure from the insignificant velocities of the medium associated with isentropic sound waves and also finite amplitude compression waves. The assumption of small movements of particles behind the wave is no longer true, and there is physical movement of the medium for a strong blast wave.

The mass thrown out from the high-energy source in the region near to the source of the explosion is also carried by the blast wave. In this near field region of the source, the strong blast wave sweeps the mass processed by it and keeps it at the front.

As the blast wave propagates further into the medium, the mass swept by it and accumulated at the front decreases. The energy of the source is radiated out in accord with the wave process.

The region of a strong blast wave, wherein we can assume the medium processed by it to be fairly concentrated near the front, is valid only in the near field of the energy release. The Mach number of the blast wave is significantly high in this region.

2.3.3   Decay of Blast Waves

The decrease in the propagating speed of a blast wave was addressed using dimensional considerations in Sect. 2.1.1. The decay can also be determined by conserving the energy used to drive the blast wave with the increase of the kinetic and potential energy of the medium enclosed by the blast wave. Consider an ideal case of instantaneous energy release E_0 at a point for which the spherical blast wave so formed was referred to as an ideal blast wave. Let the radius of the spherical blast wave formed be $R_S(t)$ and its speed be $\dot{R}_S(t)$ at time t after the energy release. Figure 2.19 shows a schematic diagram of the blast wave at radius $R_S(t)$ traveling with velocity $\dot{R}_S(t)$.

Let the medium be initially quiescent with specific internal energy e_0. If the velocity and specific internal energy of the shocked parcel of gas at any arbitrary

Fig. 2.19 Blast wave
formed by instantaneous
energy release E_0 at a point

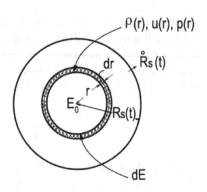

radius r within the shock, i.e., $r < R_S$ (when the shock front is at $R_S(t)$ at time t) are
u and e, respectively, the increase of energy dE in the spherical shell between radius
r and $r + dr$ is:

$$dE = \left(e - e_0 + \frac{u^2}{2}\right) 4\pi r^2 dr \times \rho \tag{2.54}$$

Here ρ denotes the density at the radius r.

When the blast wave is particularly strong and causes a significant change in the
internal energy (i.e., $e \gg e_0$), the initial internal energy e_0 of the medium can be
neglected. Equation 2.54 simplifies to give

$$dE = \left(C_V T + \frac{u^2}{2}\right) 4\pi r^2 dr \times \rho \tag{2.55}$$

The change in the specific internal energy $e - e_0$ is written as $C_V(T - T_0) \approx C_V T$.
Further, expressing $C_V = R/(\gamma - 1)$, where R is the specific gas constant, and noting
that the equation of state is $p = \rho R T$, the energy increase in the elemental volume
of width dr is

$$dE = 4\pi r^2 \left(\frac{p}{\gamma - 1} + \frac{\rho u^2}{2}\right) dr \tag{2.56}$$

The energy from the source which drives the blast wave increases the kinetic and
potential energy of the medium enclosed by the blast wave. The total increase of
energy ΔE enclosed by the blast wave, when it is at radius R_S, is obtained by integrating Eq. 2.56 between the point of energy release and R_S to obtain

$$\Delta E = 4\pi \int_0^{R_S} \left(\frac{p}{\gamma - 1} + \frac{\rho u^2}{2}\right) r^2 dr \tag{2.57}$$

Fig. 2.20 Non-dimensional
distance

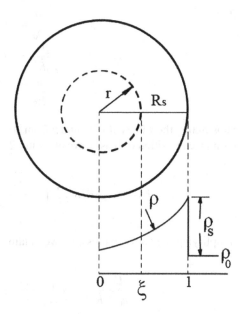

The above increase must come from the energy supplied E_0. We therefore have

$$E_0 = 4\pi \int_0^{R_S} \left(\frac{p}{\gamma - 1} + \frac{\rho u^2}{2} \right) r^2 dr \tag{2.58}$$

The values of p, ρ, and u at the front, viz. at R_S, are available only when the velocity $\dot{R}_S$ of the blast wave is specified. However, their values at the radius r when the blast wave front or shock front is at R_S are not known. Denoting a radial location r when the shock front is at R_S by a non-dimensional parameter $\xi = r/R_S$, we have the non-dimensional radial location ξ varying between 0 and 1 for the different particles in the medium from 0 to R_S when the blast wave is at R_S (Fig. 2.20).

If we denote the density ρ, pressure p, and velocity u at a given radial location r in the region between 0 and R_S in the following non-dimensional form:

$$\psi = \frac{\rho}{\rho_0}, \quad f = \frac{p}{\rho_0 \dot{R}_S^2} \quad \text{and} \quad \varphi = \frac{u}{\dot{R}_S} \tag{2.59}$$

and represent the values of these non-dimensional parameters as a function of $\xi (= r/R_S)$, the values of ρ, p, and u at any ξ can be expressed as

$$\frac{\rho}{\rho_0} = \psi(\xi)$$

$$\frac{p}{\rho_0 \dot{R}_S^2} = f(\xi) \tag{2.60}$$

$$\frac{u}{\dot{R}_S} = \varphi(\xi)$$

Substituting the values of ρ, p, and u from above equation in Eq. 2.58 and noting that for a given position of blast wave at radius R_S, $dr = R_S d\xi$ (since $\xi = r/R_S$), we get

$$E_0 = 4\pi\rho_0 R_S^3 \dot{R}_S^2 \int\limits_0^1 \left(\frac{f(\xi)}{\gamma - 1} + \frac{\psi(\xi)\varphi(\xi)^2}{2} \right) \xi^2 d\xi \tag{2.61}$$

Simplifying the above expression, we obtain

$$\frac{E_0}{4\pi\rho_0 R_S^3 \dot{R}_S^2} = \int\limits_0^1 \left(\frac{f(\xi)}{\gamma - 1} + \frac{\psi(\xi)\varphi(\xi)^2}{2} \right) \xi^2 d\xi \tag{2.62}$$

The parameters $f(\xi)$, $\psi(\xi)$ and $\varphi(\xi)$ denote the non-dimensional profiles of pressure, density and velocity behind the blast wave for $0 \leq \xi \leq 1$ (Fig. 2.21).

We have seen in the last section that the profile for density for a strong spherically symmetric blast wave is such that most of the mass within it is concentrated in the near vicinity of the wave front and the profile $\psi(\xi) \sim \xi^q$. For the particular case of a strong blast wave, the profiles of the non-dimensional parameters, $f(\xi)$, $\psi(\xi)$ and $\varphi(\xi)$, may be similarly represented as a function of ξ irrespective of the velocity $\dot{R}_S$. The right side of Eq. 2.62 under the above assumption works out to be near to a constant value. It is known as energy integral and is denoted by I. Equation 2.62 reduces to

$$\frac{E_0}{4\pi\rho_0 R_S^3 \dot{R}_S^2} = I \tag{2.63}$$

The energy integral I is representative of the fractional increase of energy in the medium processed by the blast wave to its kinetic energy if the entire mass within the blast wave were to travel at velocity $\dot{R}_S$. Taking the value I as a constant for a strong blast wave, we obtain from Eq. 2.63:

Fig. 2.21 Profiles of non-dimensional density, velocity, and pressure behind the front for ξ between 0 and 1

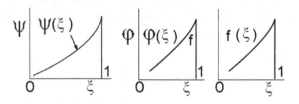

$$R_S^3 \dot{R}_S^2 = \text{constant} \tag{2.64}$$

for a given value of energy release E_0 in a medium of density ρ_0. Differentiating Eq. 2.64 gives

$$3R_S^2 \dot{R}_S^3 + 2\ddot{R}_S \dot{R}_S R^3 = 0$$

or

$$\frac{R_S \ddot{R}_S}{\dot{R}_S^2} = -\frac{3}{2} \tag{2.65}$$

$\ddot{R}_S$ is the acceleration of the blast wave. The negative sign signifies that the speed of the wave decays as it progresses.

If we are to denote the constant in Eq. 2.64 as C, we have:

$$\frac{dR_S}{dt} = CR_S^{-3/2}$$

and

$$R_S^{3/2} dR_S = Cdt$$

Integrating over a distance R_S and noting that at time $t = 0$, $R_S = 0$, we get:

$$R_S = At^{2/5}$$

where A is a constant. The result is the same as obtained from dimensional analysis given in Sect. 2.1.1. However, it is to be kept in mind that the profiles in density and pressure do not conform to a compression wave or for any wave process. It applies only for a strong blast wave near to the zone of the explosion.

2.3.4 Characteristic Length of Energy Release; Explosion Length

Equation 2.63 can be written in terms of the Mach number M_S of the propagating blast wave instead of its velocity $\dot{R}_S$ by multiplying the numerator and denominator by γp_0 to give

$$\frac{E_0}{4\pi \gamma p_0 (\rho_0/\gamma p_0) R_S^3 \dot{R}_S^2} = I$$

Since $\gamma p_0/\rho_0 = a_0^2$ and $M_S^2 = \dot{R}_S^2/a_0^2$, the above expression reduces to:

$$\frac{E_0/p_0}{4\pi \gamma R_S^3 M_S^2} = I \tag{2.66}$$

The dimensions of E_0/p_0 in the above equation is $(J/(N/m^2))$, which reduces to $(Nm/(N/m^2))$ and is m^3. The source energy E_0 can therefore be represented as a length scale by the term $(E_0/p_0)^{1/3}$. The characteristic length scale of the source energy is called as explosion length and is denoted by R_0

$$R_0 = \left(\frac{E_0}{p_0}\right)^{1/3} \tag{2.67}$$

Substituting the above expression for explosion length R_0 in Eq. 2.66 gives

$$\left(\frac{R_S}{R_0}\right)^3 = \frac{1}{4\pi I \gamma} \frac{1}{M_S^2} \tag{2.68}$$

With the energy integral I being assumed as constant, the value of R_S/R_0 is

$$\left(\frac{R_S}{R_0}\right)^3 \propto \frac{1}{M_S^2} \tag{2.69}$$

An unique value of M_S would therefore be obtained for a particular value of R_S/R_0. The distance travelled by the blast wave R_S, if made non- dimensional by dividing it by the explosion length R_0, would provide the Mach number of the blast wave at the given distance R_S. The explosion length is therefore a very useful parameter for representing the energy release. Based on the explosion length, the Mach number of the blast wave at any distance from the source of energy release can be determined. The trend of variation of M_S^2 with R_S/R_0 is shown in Fig. 2.22. It is seen that that when $R_S \rightarrow 0$, M_S is high and corresponds to a strong blast wave.

The pressure ratio p/p_0 across a shock was derived to be directly proportional to M_S^2 in Eq. 2.42. Since the value M_S^2 is inversely proportional to $(R_S/R_0)^3$, the increase of pressure over and above the ambient pressure across a blast wave, referred

Fig. 2.22 Variation of Mach number with scaled distance

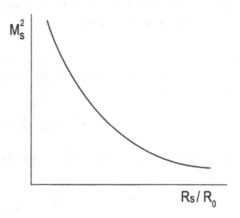

to as overpressure, would also depend on the ratio R_S/R_0. Different energy releases would give the same value of overpressure for the same value of the scaled distance R_S/R_0. The use of explosion length for scaling is very useful for estimating the blast properties. The scaling is called as Sachs' scaling and R_S/R_0 is also known as Sachs' scaled distance.

2.4 Predictions for Overpressures

2.4.1 Cranz–Hopkinson Scaling Law for Overpressure

The overpressure $(p - p_0)$ depends on the scaled distance R_S/R_0 and is expressed by the Cranz–Hopkinson scaling law. It states that an observer at a distance λR_S from the source of an explosion of explosion length λR_0 would experience the same peak overpressure as an observer at a distance R_S from an explosion having explosion length R_0. In terms of energy release, the overpressure from an explosion having energy release λE_0 would be the same at a distance $\lambda^{1/3} R_S$ as the overpressure at a distance R_S from an energy release E_0. The scaling law holds good for strong blast waves.

2.4.2 Overpressure from a Strong Blast Wave

The Mach number for the particular case of a strong blast wave was determined in Eq. 2.68 as

$$M_S^2 = \frac{1}{4\pi I \gamma} \frac{1}{(R_S/R_0)^3} \tag{2.70}$$

while the pressure ratio across a strong shock from Eq. 2.42 is

$$\frac{p}{p_0} = \frac{2\gamma}{\gamma + 1} M_S^2 \tag{2.71}$$

Here the term $(\gamma - 1)/(\gamma + 1)$ in Eq. 2.42 is neglected since it is much smaller than the term $2\gamma M_S^2/(\gamma + 1)$ for larger values of M_S. Combining Eqs. 2.70 and 2.71, the pressure ratio can be written as

$$\frac{\gamma + 1}{2\gamma} \frac{p}{p_0} = \frac{1}{4\pi I \gamma} \frac{1}{(R_S/R_0)^3} \tag{2.72}$$

The overpressure $(p - p_0)$ expressed in a non-dimensional form by dividing by the ambient pressure p_0 therefore becomes

Fig. 2.23 Overpressure for strong blast assumptions

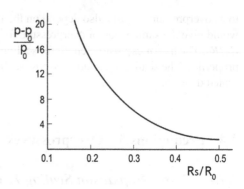

$$\frac{p - p_0}{p_0} = \frac{1}{2\pi I(\gamma + 1)} \frac{1}{(R_S/R_0)^3} - 1 \qquad (2.73)$$

The value of energy integral I is required to determine the values of the dimensionless overpressure with the scaled distance R_S/R_0. Based on the solution of the mass, momentum, and energy conservation equations for a strong blast wave, the value of I is obtained to be about 0.423 for ambient atmospheric medium for which γ is 1.4. The profiles of the dimensionless pressure f, density ψ, and velocity φ, schematically shown in Fig. 2.21, are taken to be independent of the Mach number at the higher values of Mach number of the blast wave. Substituting the value of $I = 0.423$ in Eq. 2.73 gives the overpressure in the ambient medium of air as

$$\frac{p - p_0}{p_0} = \frac{0.1568}{(R_S/R_0)^3} - 1 \qquad (2.74)$$

The variation of the dimensionless overpressure with the scaled distance R_S/R_0 is shown in Fig. 2.23. It decreases from a large value of about 19 at R_S/R_0 of 0.2 to about 1.25 at R_S/R_0 of 0.5. The value of the Mach number M_S from Eq. 2.70 is 4.1 at R_S/R_0 of 0.2 and 1.037 at R_S/R_0 of 0.5. M_S of 1.037 does not correspond to a strong blast wave and the strong blast wave assumption based on which the overpressure is calculated would therefore not be valid. The strong blast wave assumption would not be valid beyond a scaled distance R_S/R_0 of about 0.3 for which the Mach number M_S is 3.86.

2.4.3 Smaller Values of Overpressures

Values of overpressures much smaller than those determined from the strong blast wave assumption are required in practice. As an example, an overpressure of about 0.7 kPa over the ambient pressure of 100 kPa is sufficient to break a glass windowpane or cause injury to humans. This corresponds to a dimensionless overpressure of 0.007.

We need to be able to predict the lower overpressures which occur at much larger values of R_S/R_0 for which the strong blast wave assumption does not hold.

The strong blast wave formulation assumed that the ambient internal energy could be neglected compared to the internal energy behind the blast wave. The jump conditions across the shock for density and velocity, given by Eqs. 2.47 and 2.48, considered their dependence only on the specific heat ratio γ and not on the shock Mach number M_S^2. If the initial internal energy of the medium is of the same order as the energy released by the source, we have

$$\frac{4}{3}\pi R_S^3 \rho_0 e_0 = E_0 \tag{2.75}$$

Substituting the value of

$$e_0 = C_V T_0 = \frac{p_0}{(\gamma - 1)\rho_0}$$

where T_0 is the initial temperature of the medium, we get

$$\frac{E_0}{p_0} = \frac{4}{3}\frac{\pi R_S^3}{(\gamma - 1)} \tag{2.76}$$

and

$$\frac{R_S}{R_0} = \left(\frac{3(\gamma - 1)}{4\pi}\right)^{1/3} \tag{2.77}$$

When R_S/R_0 exceeds the above value, the initial internal energy becomes significant and cannot be neglected. This threshold value of R_S/R_0, from Eq. 2.77 for $\gamma = 1.4$, is 0.46. Even at $R_S/R_0 < 0.46$, the initial internal energy becomes comparable to the source energy and the strong blast wave assumption is strictly not valid.

The incorporation of the Mach number in the jump conditions across the shock and the initial internal energy of the medium in the predictions of the Mach number of the blast wave and overpressure are rather involved, and we shall not deal with it. Several methods such as:

(i) perturbation expansion of the scaled distance in terms of $1/M_S^2$,
(ii) closed form solutions of the differential equations for mass, momentum and energy along the blast trajectory,
(iii) asymptotic expansion analysis, and
(iv) numerical computations
  v have been investigated to determine the overpressure as a function of R_S/R_0. A few results are cited in the following.

The asymptotic expansion analysis gave reasonably good results for moderately strong blast waves up to $M_S \sim 1.2$. The value of R_S/R_0 is correlated with $1/M_S^2$ by the expression:

$$\left(\frac{R_S}{R_0}\right)^3 = 2.36246\eta + 4.53155\eta^2 + 6.44813\eta^3 \tag{2.78}$$

where $\eta = 1/M_S^2$. Once the value of M_S is known, the overpressure is determined from equation of the pressure ratio in terms of M_S^2 given in Eq. 2.46.

For very weak blast waves with Mach numbers near to unity, the overpressure is estimated from acoustic theory as

$$\frac{p - p_0}{p_0} = \frac{\gamma A}{(R_S/R_0)[B + \ln(R_S/R_0)]^{1/2}} \tag{2.79}$$

Here A and B are constants. Values of A between 0.23 and 0.246 and B between 0.36 and 0.45 give reasonable values of overpressures for $R_S/R_0 > 10$ when the value of $\gamma = 1.4$.

In general, the overpressure $(p - p_0)/p_0$ expressed as a function of R_S/R_0 based on numerical solution for a spherical blast wave propagating in air agrees reasonably well with the strong blast solution for $(p - p_0)/p_0 > 9$. For $0.09 \le (p - p_0)/p_0 \le 9$, the dependence of $(p - p_0)/p_0$ does not vary directly with $1/(R_S/R_0)^3$. The value of $(p - p_0)/p_0$ is:

$$\frac{p - p_0}{p_0} = \frac{0.145}{(R_S/R_0)^3} \quad \text{for} \quad \frac{p - p_0}{p_0} \ge 9 \tag{2.80}$$

and

$$\frac{p - p_0}{p_0} = \frac{0.27}{(R_S/R_0)} + \frac{0.113}{(R_S/R_0)^2} + \frac{0.126}{(R_S/R_0)^3} - 1.2 \quad \text{for } 0.09 \le \frac{p - p_0}{p_0} \le 9 \tag{2.81}$$

It is observed that significant blast overpressure does not exist beyond $R_S/R_0 \ge 0.55$.

2.5 Non-idealities of Source Influencing Overpressure

A spherical blast wave was considered in the earlier sections to be generated by the impulsive energy E_0 instantaneously released at a point. The volume and shape of the energy source and the rate of energy release are likely to influence the blast wave. If the energy is released in a spherical volume of radius R_E and at a pressure p_E, we have following Eq. 2.76:

$$E_0 = \frac{4}{3} \frac{\pi R_E^3 p_E}{(\gamma - 1)}$$

Here we have assumed the internal energy change due to the energy release E_0 is very much greater than the initial value of the internal energy of the medium. The explosion length R_0 being $(E_0/p_0)^{1/3}$, the above equation can be written as

$$\left(\frac{R_E}{R_0}\right)^3 = \frac{3(\gamma - 1)}{4\pi}\frac{p_0}{p_E} \qquad (2.82)$$

If the pressure generated in the volume of energy release is large ($p_E \gg p_0$), the value of $R_E \ll R_0$ and we can assume a point source of energy release. We shall see in later chapters that the explosion of typical solid explosives such as trinitrotoluene (TNT) generates pressures of about 19,000 MPa. The value of the specific heat ratio γ of the combustion products is about 1.2. The value R_E/R_0 from Eq. 2.81 works out to be about 0.03.

The Mach number of a blast wave, based on the strong blast wave assumption in Sect. 2.3.5, was seen to be about 12 for a value of $R_S/R_0 = 0.1$. The dimensionless overpressure was 156. The value of $R_E/R_0 (= 0.03)$ for the explosive TNT is therefore very much smaller than the value of R_S/R_0 at which a very strong blast wave is formed. The strong blast assumption (Mach number greater than about 4) was seen to be valid for R_S/R_0 less than about 0.3. We can therefore surmise that the ideal solution considered for a decaying spherical blast wave would be valid for:

$$R_E < R_0 < R_S$$

TNT is used as a reference for blast damages, and we shall deal with this in Chap. 12.

When the geometry of the energy release is not spherical, as is usually the case, the blast wave near the energy release zone cannot be spherical. The large values of sound speed behind the blast wave resulting from the high temperatures, however, smear out the asymmetry quite fast and a spherically symmetric blast wave is soon formed. The non-ideality with respect to the finite volume of the energy source and the shape of the energy source is very much localized to the region very near to the source. Far from the source, the behavior of the source does not influence its propagation.

Figure 2.24, taken from Strehlow and Baker's report on 'The Characterization and Evaluation of Accidental Explosions' compares the dimensionless overpressures obtained for an ideal point source as a function of the scaled distance R_S/R_0 with the measurements for solid explosives such as TNT and nuclear explosions. The overpressures compare well with those derived for an ideal point source.

2.6   Pressure Variations in a Blast Wave: Impulse

The jump in the pressure across a blast wave is seen to decrease as the scaled distance R_S/R_0 is increased. The pressure behind the blast at different values of R_S/R_0 is sketched in Fig. 2.25. At each value of R_S/R_0, there is a jump in pressure as shown by the dotted vertical lines and the pressure thereafter falls as shown by the dotted curves. A point in a medium, processed by a blast wave, would first experience the overpressure. The pressure at the point falls subsequently from the expansion. It is

Fig. 2.24 Dependence of blast wave overpressures on scaled distance for ideal blast wave and TNT and nuclear explosions

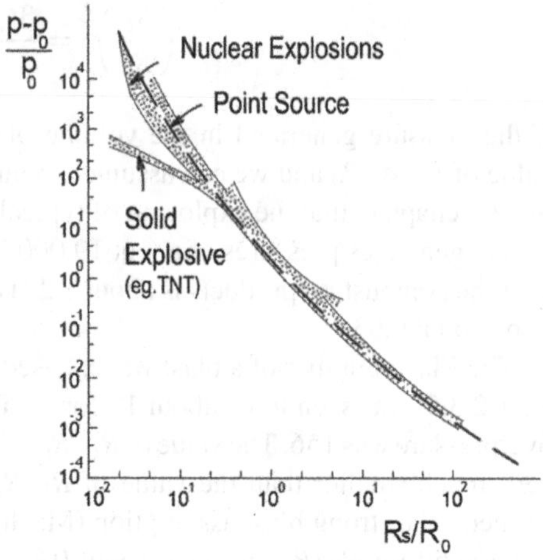

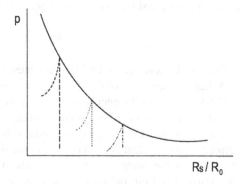

Fig. 2.25 Overpressure and expansion of medium processed by blast wave

the decrease of pressure with increase of R_S/R_0 that drives the wind in the direction of the blast wave.

Figure 2.26 illustrates the passage of a spherical blast wave originating from a source and an observer standing at some lateral distance watching the evolution of pressure at a point A situated at a distance L from the source. The observer sees the side view of the pressure jump at point A and the subsequent decay of the pressure at this point with time (Fig. 2.27).

The pressure at A is the ambient pressure p_0 before the arrival of the blast wave at the point. If the blast wave reaches the point A at time t_a after energy is released at the source, then at time t_a, the pressure at A jumps to p_S, the value of p_S depending on the Mach number of the blast wave when it reaches A. Subsequently, the pressure starts to decrease from the expansion process until it attains the ambient pressure p_0 at time $t_a + t^+$; t^+ is the time duration over which the pressure is greater than the ambient. This is seen by the observer in the side view of the blast wave and is illustrated in Fig. 2.27.

Fig. 2.26 Schematic
diagram showing an observer
watching the side view of
changes at a point due to
passage of blast wave past
point A

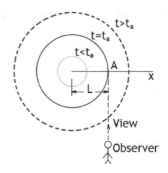

Fig. 2.27 Side view of
pressure time history

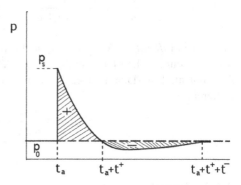

The pressure further drops below the ambient pressure due to the inertia of the expanding flow to sub-atmospheric pressures and gets back to the value of p_0 at time $t_a + t^+ + t^-$. The pressure at any point in the medium processed by the blast wave changes with time, viz. $p(t)$ between time t_a and $t_a + t^+ + t^-$. This is shown in the side view of the pressure profile in Fig. 2.27. Over a time period t^+ between t_a and $t_a + t^+$ the pressure is greater than the ambient, while over period t^- the pressure is less than p_0.

2.6.1 Arrival Time and Mach Number of The Blast Wave at a Distance L from the Source

The arrival time t_a of the blast wave at a point A in the medium is proportional to its distance L from the source (Fig. 2.26). Denoting the ratio of the distance to the arrival time as β, we have

$$\beta = \frac{L}{t_a} \tag{2.83}$$

or

$$L = \beta t_a$$

Differentiating with respect to L gives:

$$\beta \frac{dt_a}{dL} + t_a \frac{d\beta}{dL} = 1$$

and

$$\beta \frac{dt_a}{dL} = 1 - t_a \frac{d\beta}{dL} \tag{2.84}$$

Rearranging the terms, we get

$$\frac{dL}{dt_a} = \frac{\beta}{1 - t_a(d\beta/dL)} = \frac{\beta}{1 - (L/\beta)(d\beta/dL)} \tag{2.85}$$

The value of $dL/dt_a = \dot{R}_L$, where $\dot{R}_L$ is the velocity of the blast wave at distance L from the source. The Mach number $M_{S,L} = \dot{R}_{SL}/a_0$, where a_0 is the sound speed in the ambient. The Mach number of the blast wave at the distance L from the source is therefore

$$M_{S,L} = \frac{\beta}{a_0} \left(\frac{1}{1 - \frac{d \ln \beta}{d \ln L}} \right) \tag{2.86}$$

2.6.2 Impulse

The difference in the pressure of the medium subject to the blast wave $p(t)$ between time t_a and $t_a + t^+ + t^-$ and the ambient pressure p_0 will provide a force on the objects stationed in the medium. These forces act over the duration of the positive and negative pressures given by t^+ and t^-, respectively (Fig. 2.27).

Impulse is defined as change of momentum, viz. $I = \Delta(Mu)$ where I is the impulse and $M \times u$ is the momentum. However, rate of change of momentum is force $[F = \frac{d}{dt}(Mu)]$. Therefore, the impulse $I = \int F dt$ integrated over the duration of the force. For unit surface area over which the pressure acts, we define impulse per unit area as $I_S = \int [p(t) - p_0] dt$. Hence, the positive and negative impulses acting over unit surface area over the duration t^+ and t^-, respectively, are

$$I_S^+ = \int_{t_a}^{t_a + t^+} (p(t) - p_0) dt \tag{2.87}$$

$$I_S^- = \int_{t_a + t^+}^{t_a + t^+ + t^-} (p_0 - p(t)) dt \tag{2.88}$$

The negative impulse I_S^- given by Eq. 2.88 is generally small and is not considered for evaluating damages from blast waves. However, it does play a role in the spread of damages caused by the blast wave. The glass splinters from windowpanes, which get ruptured by the overpressure of a blast wave hitting it from outside the building often falls outwards when the negative impulse is appreciable.

The values of the positive impulse per unit area I_S^+ are determined from the peak overpressure $p_S - p_0$ and the decay of the pressure with time behind the blast wave. The impulse therefore decreases with increase of the scaling distance R_S/R_0. The conservation equations describing the blast wave are solved for this purpose. We shall not carry out the predictions but will reproduce the values of I_S^+ determined for a strong blast wave from a point explosion and for a very weak blast wave. The impulse per unit area (N-s/m^2) is expressed in a non-dimensional form $\bar{I}_S^+$ as

$$\bar{I}_S^+ = \frac{I_S^+ a_0}{p_0 R_0} \tag{2.89}$$

where a_0 is the speed of sound in the ambient atmosphere.

For a blast wave in atmospheric air from a point explosion

$$\bar{I}_S^+ = \frac{0.322}{R_S/R_0} \tag{2.90}$$

for overpressure pressure $p_S - p_0$ of 0.01 MPa $< p_S - p_0 <$ 150 MPa.

For a very weak blast wave

$$\bar{I}_S^+ = \frac{\gamma(\gamma + 1)A^3}{4R_S/R_0} \tag{2.91}$$

where A is typically between 0.23 and 0.246.

A plot of the dimensionless impulse $\bar{I}_S^+$ as a function of the scaled distance R_S/R_0 is shown in Fig. 2.28.

Fig. 2.28 Impulse changes with R_S/R_0

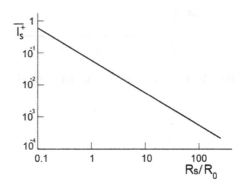

2.6.3   Cranz–Hopkinson Law for Scaling Impulse

An explosion releasing energy in the medium of explosion length λR_0 gave the same overpressure at a distance λR_S, as an explosion with explosion length R_0 at a distance R_S from the source (See Sect. 2.4.1). When the distance from a given source increases to λR_S from R_S with the explosion length increasing from R_0 to λR_0, the arrival time of the blast increases from t_a to λt_a. This is because the Mach number of the blast wave and hence its speed at distance λR_S for the explosion length λR_0 is the same as at distance R_S with explosion length R_0 (Sachs' scaling). The duration of impulse increases from t^+ to λt^+ since the distance increases by a factor λ and the duration of the pressure decay to the ambient value will increase by λ. With the magnitude of the peak overpressure remaining the same, the impulse from a source of explosion length λR_0 at a distance λR_S is λ times the impulse from a source of explosion length at a distance R_S. This is the Cranz–Hopkinson law for scaling of impulse.

The overpressure of the blast wave at a distance λR_S from an energy release of explosion length λR_0 was seen to be the same as the overpressure at a distance R_S from an energy release of explosion length R_0. The scaling law for overpressure and impulse is summarized in Fig. 2.29.

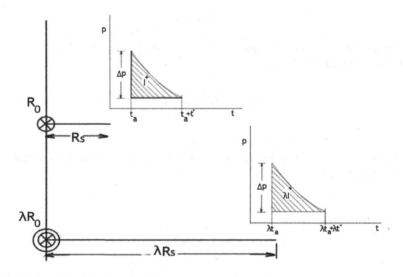

Fig. 2.29 Cranz–Hopkinson scaling for overpressure and impulse

2.7 Missiles, Shrapnels, and Fragments from Blast; Gurney Constant

The debris from the source and broken pieces from damaged structures could be propelled outward from the explosion due to the overpressure and impulse. The propelled objects are known as missiles. Shaped pieces of inert placed within the explosive are also propelled and are known as shrapnels. The distance traveled by the missiles and shrapnels depends on their mass, shape, and size. Following the earlier discussions, wherein overpressure and impulse were seen to increase with explosion length, the distance traveled by the missiles and shrapnels of a given mass and shape would also scale as the explosion length. They can lead to considerable damage at large distances from the source of the explosion.

Intentionally formed missiles in a warhead such as from its case are known as fragments.

The initial velocity of fragments from an explosion and their momentum can be determined assuming the following:

(a) Fragments are formed from the casing holding the explosive.
(b) Fragments fly at the velocity of gases generated by the explosive.
(c) Kinetic energy of the fragments and the energy of the expanding gases equal the internal energy from the explosion.
(d) Velocity of the gases increases linearly from zero at the center of the explosive to a maximum at the casing.

Assuming the internal energy per unit mass of the explosive as u J/kg, we have for a mass m_E of explosive and total mass of fragments m_M (corresponds to mass of casing):

$$\frac{1}{2}\left(\frac{1}{2}m_E V^2\right) + \frac{1}{2}m_M^2 V = m_E u \tag{2.92}$$

where V is the initial velocity of the fragments. Here the kinetic energy of the gases is taken as $\frac{1}{2}(\frac{1}{2}m_E V^2)$ since the velocity of the gas varies from zero at the center to the maximum at the front. Solving for the velocity, we get

$$V = \sqrt{2u}\left(\frac{m_E/m_M}{1 + m_E/2m_M}\right)^{1/2} \tag{2.93}$$

The value $\sqrt{2u}$, which relates the velocity with the mass of explosive and the casing forming the missiles, is known as Gurney constant and is a characteristic of an explosive. The unit of $\sqrt{2u}$ is $\sqrt{Jkg}$ = m/s and the Gurney constant is also referred to as Gurney velocity. The initial velocity of the fragments is the product of the Gurney velocity and the mass ratio term shown in brackets in Eq. 2.93. When the ratio $m_E/m_m < 2$, the velocity of the fragments would be less than the Gurney velocity. However, for $m_E/m_m > 2$, the fragment velocity exceeds the Gurney velocity.

The damage from fragments and missiles is due to its momentum. The total momentum for the missiles of mass m_M from Eq. 2.93 is

$$m_M V = m_M \sqrt{2u} \left(\frac{m_E/m_M}{1 + m_E/2m_M} \right)^{1/2} \tag{2.94}$$

2.8   Salient Features of Blast Waves

Blast waves, formed from spontaneous release of energy, are characterized by the overpressure at the wave front and the impulse. Missile effects are also present. The overpressure, impulse, and missiles cause damage to life and property. Blast-resistant walls, which comprise of heavy structure, are used to protect buildings, equipment, and personnel from the blast damages.

Only spherical blast waves were considered in this chapter. Cylindrical blast waves are formed in the near field of a linear source of energy such as the lightning discussed in Chap. 1. However, in the far field, it becomes spherical. The interaction of a shock front with blockages such as buildings and the ground form reflected shocks and the change in overpressures in the reflection process needs to be considered while evaluating damages from blast waves.

A strong blast wave has most of its mass concentrated very near its front. The energy released is characterized by an explosion length which is a very useful scaling parameter.

Examples

2.1 *Deceleration of Blast Wave from Dimensional Considerations*: From dimensional analysis, we obtained the blast wave trajectory for a given value of energy release in a given medium to be given by

$$R_S = At^{2/5}$$

Show that the velocity of the blast wave decays with increasing distance and the deceleration parameter defined by $(R_S \ddot{R}_S / \dot{R}_S^2)$ is a constant equal to $-3/2$.

Solution
We have from the blast wave trajectory:

$$R_S = At^{2/5}$$

On differentiating:

$$\frac{dR_S}{dt} = \dot{R}_S = \frac{2}{5} At^{-3/5}$$

Differentiating again with respect to time t gives

$$\frac{d^2 R_S}{dt^2} = \ddot{R}_S = -\frac{6}{25} A t^{-8/5}$$

The acceleration of the blast wave is negative. The blast wave therefore decelerates.

Combining the values of R_S, $\ddot{R}_S$ and $\dot{R}_S^2$ to give the parameter $R_S \times \ddot{R}_S / \dot{R}_S^2$, we get:

$$R_S \times \ddot{R}_S / \dot{R}_S^2 = \frac{A t^{2/5} \times (-6 A t^{-8/5}/25)}{(2 A t^{-3/5}/5)^2} = -3/2$$

The deceleration parameter $R_S \times \ddot{R}_S / \dot{R}_S^2$ is a constant equal to $-3/2$.

2.2 *Compression by Different Processes*: A compression process wherein the ambient density increases from a value of $1.16 \, \text{kg/m}^3$ to $1.25 \, \text{kg/m}^3$ takes place through either of the following three processes: (a) isothermal compression, (b) isentropic compression, and (c) shock wave compression. The value of the initial pressure is $100 \, \text{kPa}$. If the specific heat ratio γ of the medium is 1.4, determine the increase of pressure for the three processes and state the reasons for the changes obtained in the different compression processes.

Solution:

For isothermal compression $pV = C$ or $(p/\rho) = C$. Denoting the final pressure as p_2, we get

$$p_2 = 1000 \times \frac{1.25}{1.6} = 107.76 \, \text{kPa}$$

This gives the pressure rise in the isothermal compression as $\Delta p = 107.76 - 100 = 7.76 \, \text{KPa}$.

In the case of an isentropic process $pV^\gamma = C$ or $(p/\rho^\gamma) = C$. Substituting the values, we get

$$p_2 = 100 \times \left(\frac{1.25}{1.16}\right)^{1.4} = 111.03 \, \text{KPa}$$

The above gives the pressure rise in the isentropic process as $\Delta p = 111.03 - 100 = 11.03 \, \text{kPa}$.

The increase in pressure with a shock wave is obtained from the Rankine–Hugoniot relation given by Eq. 2.24 as

$$\frac{p_2}{p_1} = \frac{\frac{\gamma+1}{\gamma-1} \frac{\rho_2}{\rho_1} - 1}{\frac{\gamma+1}{\gamma-1} - \frac{\rho_2}{\rho_1}}$$

Substituting the value of

$$\frac{\rho_2}{\rho_1} = \frac{1.25}{1.16}$$

in the above, we get

$$\frac{p_2}{p_1} = 1.1292. \tag{2.95}$$

Hence, $\Delta p = 112.92 - 100 = 12.92$ kPa.

We find that shock compression gives the highest value of pressure increase of 12.92 kPa while the isentropic compression gives a lower value of 11.03 kPa. The isothermal compression gives the lowest value of 7.76 kPa.

The reason for difference between shock compression and isentropic compression is that though both of them are adiabatic and the shock compression is not reversible. The irreversibility causes an increase of its temperature and hence results in a higher value of pressure. The increase of pressure is by 12.92/11.03, i.e., by about 17%. The magnitude would increase as the pressure ratio would increase.

The isothermal and isentropic processes are reversible. However, in the isothermal process, the temperature is maintained constant. The increase of temperature in the isentropic process results in an increase in the pressure by a factor 11.03/7.76, i.e., 42%.

2.3 *Properties in a Shock Wave*: Determine the temperature, pressure, velocity, and density behind a shock wave traveling at a velocity of 1500 m/s in ambient air. The properties of ambient air are given by the following: temperature = 300 K; density = 1.1 kg/m^3; specific heat ratio =1.4, and sound velocity = 330 m/s. Would a strong shock solution reasonably predict these properties?

Solution:

Mach number of the shock wave M_S: $1500/330 = 4.55$. Using Eqs. 2.44, 2.45 and 2.46, we get

$$\frac{T}{T_0} = \left(\frac{2\gamma}{\gamma+1}M_S^2 - \frac{\gamma-1}{\gamma+1}\right)\left(\frac{\gamma-1+2/M_S^2}{\gamma+1}\right)$$

$$\frac{u_1}{\dot{R}_S} = \frac{\rho_0}{\rho} = \left(\frac{\gamma-1+2/M_S^2}{\gamma+1}\right)$$

and

$$\frac{p}{\rho_0 \dot{R}_S^2} = \frac{2}{\gamma+1} - \frac{\gamma-1}{\gamma(\gamma+1)}\frac{1}{M_S^2}$$

Substituting the values, we get:

$$T = 1482 \text{ K}$$
$$u_1 = 303 \text{ m/s}$$
$$\rho = 5.45 \text{ kg/m}^3$$
$$p = 2.05 \text{ MPa}$$

The velocity behind the shock is $\dot{R}_S - u_1 = 1500 - 303 = 1197$ m/s.

The temperature, pressure, velocity, and density behind the shock are 1482 K, 2.05 MPa, 1197 m/s, and 5.45 kg/m^3 respectively.
Under strong shock conditions for which $1/M_S^2 = 0$.

$$\frac{u_1}{\dot{R}_S} = \frac{\rho_0}{\rho} = \frac{\gamma - 1}{\gamma + 1}, \quad \frac{p}{\rho_0 \dot{R}_S^2} = \frac{2}{\gamma + 1}$$

and

$$\frac{T}{T_0} = \left(\frac{2\gamma(\gamma - 1)}{(\gamma + 1)^2} M_S^2\right)$$

we get

$$T = 1208 \text{ K}$$
$$u_1 = 250 \text{ m/s}$$
$$\rho = 6.6 \text{ kg/m}^3$$
$$p = 2.06 \text{ MPa}$$

The density is seen to be overpredicted. The predicted value of pressure is about the same while the temperature is under-predicted. The velocity behind the shock is $(1500 - 250) = 1250$ m/s, which is more than the value of 1197 m/s determined when the Mach number of the shock of 4.55 was considered.

2.4 *Energy of an Impacting Fragment*: Determine the impact energy when a missile of mass 2 kg hits a stationary target of very large mass at a speed of 300 m/s.

Solution
The impact energy is derived from the kinetic energy of the missile, which is

$$\frac{1}{2} m V^2 = \frac{1}{2} \times 2 \times 300^2 = 90 \text{ kJ}.$$

2.5 *Scaling of Overpressure*: An explosive device is found to generate a strong overpressure at a distance of 100 m and a weak overpressure at 2000 m from an explosive device in ground tests at sea-level conditions. Would the magnitude of the overpressure generated by the same explosive device in a high altitude region be more, less, or remain the same when compared to those determined at sea level at the same distances of 100 m and 2000 m from the explosive device?

Solution
The ambient pressure decreases with increase of altitude. The dimensionless pressure ratio (p_S/p_0) varies with the scaled distance (R_S/R_0). For a strong blast wave, the variation of overpressure $\Delta p \approx p_S$ and is given by the relation

$$\frac{\Delta p}{p_0} = \frac{C}{(R_S/R_0)^3}$$

while for a weak blast wave the variation has the form

$$\frac{\Delta p}{p_0} = \frac{D}{(R_S/R_0)}$$

C and D in the above expressions are constants and R_S is the distance from the explosive device and R_0 is the explosion length. p_0 is the ambient pressure.

(a) *Strong blast region*: The overpressure in the strong blast wave region (100 m) is

$$\frac{\Delta p}{p_0} = \frac{C}{(R_S/R_0)^3} = \frac{C}{R_S^3}\frac{E_0}{p_0}$$

This gives the overpressure Δp to be

$$\Delta p = \frac{CE_0}{R_S^3}$$

The overpressure is therefore independent of the ambient pressure and hence altitude. The overpressure would be the same at the high altitude as at sea level at the distance of 100 m from the energy release.

(b) *Weak blast region*: In the weak blast region ($R_S = 2000$ m), the overpressure is

$$\frac{\Delta p}{p_0} = \frac{D}{(R_S/R_0)}$$

Substituting the value of R_0, the overpressure Δp will be

$$\Delta p = \frac{D}{R_S}\frac{E_0^{1/3}}{p_0^{1/3}}p_0 = \frac{DE_0^{1/3}}{R_S}p_0^{2/3}$$

Since the ambient pressure p_0 is less at the high altitude, the overpressure for the weak blast region would be less.

2.6 *Overpressure in an explosion:* The rupture of a pressure vessel in ambient sea-level atmosphere releases 5 MJ of energy. The ambient pressure is 0.1 MPa. Assuming the energy release as sudden and at point, determine the overpressure and impulse per unit area at a distance of 20 m from the site of the energy release. Comment on the assumptions of point energy release.

Solution
A spherical blast wave is assumed to originate from the point source and to travel without interfering with the ground. The explosion length R_0 corresponding to the energy release of 5 MJ in a medium having pressure $p_0 = 0.1$ MPa is

$$R_0 = \left(\frac{E_0}{p_0}\right)^{1/3} = \left(\frac{5 \times 10^6}{0.1 \times 10^6}\right)^{1/3} = 3.684 \text{ m}$$

The value of the scaled distance corresponding to the distance $R_S = 20$ m is

$$\frac{R_S}{R_0} = \frac{20}{3.684} = 5.43$$

Based on the non-dimensional overpressure $(p - p_0)/p_0$ and non-dimensional impulse $\bar{I}_S^+$ charts obtained as a function of the scaled distance such as given in Figs. 2.24 and 2.28, we get at $R_S/R_0 = 5.43$:

$$\frac{(p - p_0)}{p_0} = 0.1$$
$$\bar{I}_S^+ = 0.001$$

The overpressure $p - p_0$ is $0.1 \times p_0 = 0.01$ MPa or 10 kPa.
The impulse per unit area

$$I_S^+ = \bar{I}_S^+ \frac{p_0 R_0}{a_0} = 0.001 \times \frac{0.1 \times 10^6 \times 3.684}{330} = 1.12 \frac{\text{Ns}}{\text{m}^2}$$

Since the value of R_S/R_0 is large, the strong blast wave formed at small values of R_S/R_0 would have decayed and the resulting blast wave is relatively weak as evidenced by the low value of overpressure of 10 kPa. It is stated in the problem that the source of energy release could be idealized as a point. The pressure vessel, however, has a finite volume. Since the explosion length R_0 is about 4 m and the dimensions of the pressure vessel characterized by its radius R_E is likely to be a fraction of a meter, $R_E \ll R_0$ and the decaying blast wave solution would be applicable.

2.7 *Scaling of Overpressure and Impulse*: If the explosion of a 15 kg charge of dynamite in the ambient atmosphere of 0.1 MPa produces an overpressure of 0.2 MPa and an impulse of 250 N-s/m² at a distance of 5 m from the explosion source, determine the overpressure and impulse at a distance of 10 m from the explosion of a geometrically similar dynamite charge having a mass of 120 kg and exploded in a similar manner in the same ambient atmosphere of 0.1 MPa.

Solution
If the energy release per unit mass of dynamite is E_0 J, the ratio of the explosion lengths (R_0) of the 15 kg and 120 kg charges are $(15E_0/p_0)^{1/3}$ and $(120E_0/p_0)^{1/3}$, i.e., in the ratio 1:2. The values of R_S in the two cases are 5 m and 10 m giving the ratio of the distances as 1:2. R_S/R_0 is therefore the same in the two cases. The overpressure will therefore be the same and equal to 0.2 MPa at the distance of 10 m with a charge of 120 kg.
The arrival time and the duration over which the impulse acts gets stretched as the distance increases even when the magnitude of the overpressure, and hence, the velocity of the shock front at 5 m and 10 m remains the same. The impulse at 10 m with a charge of 120 kg will therefore be $(10/5) \times 250 = 500$ N-s/m².

2.8 *Velocity of Fragments*: If the iron casing of a bomb enclosing 2 kg charge of RDX has a mass of 1 kg, determine the initial velocity of the fragments formed from the case. The change in internal energy associated with the combustion of 1 kg of RDX is estimated to be 3000 kJ/kg.

Solution

The Gurney velocity of RDX is $\sqrt{2u} = \sqrt{2 \times 3000 \times 1000} = 2450$ m/s
Initial velocity of the fragments: Assuming the internal energy is available for driving the fragments, the initial velocity of the fragments is determined using Eq. 2.93 as

$$V = \sqrt{2u} \left(\frac{m_E/m_M}{1 + m_E/2m_M} \right)^{1/2}$$

It is given in the problem $m_E = 1$ kg and $m_M = 2$ kg. This gives

$$V = \sqrt{2 \times 3000 \times 1000} \left(\frac{2}{1+1} \right)^{1/2} \text{ m/s} = 2450 \text{ m/s}$$

and is equal to the Gurney velocity. The initial velocity of the fragments is 2450 m/s.

The drag of the air will decrease the fragment velocity.

Nomenclature

A	Constant
a_0	Sound velocity in ambient medium (m/s)
B	Constant
C	Constant
C_P	Specific heat at constant pressure [J/(kg K)]
C_V	Specific heat at constant volume [J/(kg K)]
E_0	Energy release by the source (J)
$\dot{E}_0$	Rate of energy release by the source (J/s = W)
e	Specific internal energy of the shocked gas (J/kg)
e_0	Specific internal energy of the ambient medium (J/kg)
f	function; non-dimensional pressure given as $p/(\rho_0 \dot{R}_S^2)$
h	Enthalpy of the shocked gases [kJ/(kg K)]
h_0	Enthalpy of the ambient medium [kJ/(kg K)]
I	Energy integral defined by Eq. 2.63; Impulse (N-s)
I^+	Positive impulse(N-s)
I^-	Negative impulse(N-s)
I_S	Impulse per unit area (specific impulse) (N-s/m²)
$\bar{I}_S^+$	Non-dimensional positive impulse per unit area defined by Eq. 2.89
L	Length; distance from source
M	Mach number defined as velocity/sound speed
M_S	Mach number of shock/blast wave
$M_{S,L}$	Mach number of blast wave at distance L
m	Mass (kg)

m_E	Mass of explosive (kg)
m_M	Mass of casing housing explosive (kg)
p	Pressure of shocked gases (Pa)
p_E	Pressure of energy source (Pa)
p_0	Pressure of the ambient (Pa)
$p - p_0$	Overpressure (Pa)
$\bar{p}$	Dimensionless overpressure $[(p - p_0)/p_0]$
p_S	Pressure behind the shock (Pa)
q	Exponent in Eq. 2.19 for density change in a blast wave
R	Specific gas constant (kJ/(kg K))
R_0	Explosion length (Characteristic length of energy release) $[(E_0/p_0)^{1/3}]$
R_S	Distance of blast wave from source (m)
$\dot{R}_S$	Velocity of blast wave dR_S/dt (m/s)
$\ddot{R}_S$	Acceleration of blast wave (m/s^2)
r	Spatial location from source within the blast wave
T	Temperature of shocked gas (K)
T_0	Temperature of ambient (K)
t	Time (s)
t_a	Arrival time of blast wave (s)
t^+	Duration of pressure greater than ambient (s)
t^-	Duration of pressure less than ambient (s)
V	Velocity of missile (m/s)
u	Velocity of gases behind shock (m/s); change of specific internal energy (J/kg)
u_1	Velocity of gases behind shock in the frame of reference with shock stationary (m/s)
β	Term L/t_a in Eq. 2.82
φ	Non-dimensional velocity behind shock (u_1/R_S)
γ	Specific heat ratio of gases
η	Inverse of square of shock Mach number ($1/M_S^2$)
λ	Multiplication factor
π	Non-dimensional parameter
ρ	Density behind shock (kg/m^3)
ρ_0	Density of ambient medium (kg/m^3)
τ	Duration of energy release (s)
ξ	Non-dimensional distance from source referred to the distance of shock (r/R_S)
ψ	Non-dimensional density (ρ/ρ_0)

Further Reading

1. Bach, C.G., and Lee, J.H., Higher Order Perturbation Solution for Blast Waves, *AIAA Journal*, 7, 742–744, 1969.
2. Baker, W. E., *Explosions in Air*, University of Texas Press, Austin, TX, 1973.

3. Boddurtha, F. T., *Industrial Explosion Prevention and Protection*, McGraw Hill, New York, 1980.
4. Brode, H.L., *Numerical Solutions of Spherical Blast waves*, J. Applied Physics, 26(6), pp. 766–775 (1955)
5. Cooper, P. W and Kurowski, S. R., Introduction to the Technology of Explosives, Wiley-VCH, New York, 1996
6. Courant, R. and Freidricks, K. O., *Supersonic Flow and Shock Waves*, Interscience Publishers, New York, 1948.
7. Kinney, G.F. and Graham, K. J., *Explosive Shocks in Air*, Springer, Berlin, 1985.
8. Johansson C. H. and Persson, P. A., Detonics of High Explosives, Academic Press, London and New York, 1970
9. Landau, L. D. and Lifshitz, E. M., *Fluid Mechanics*, Pergamon Press, London, 1966.
10. Lee, J. H., *The Detonation Phenomenon*, Cambridge University Press, New York, 2008.
11. Shapiro, A.H., *The Dynamics and Thermodynamics of Compressible Flow*, Vol. 1, The Ronald Press, New York, 1953.
12. Shchelkin K. I. and Troshin, Y.K., *Gas Dynamics of Combustion*, Mono Book Corp., Baltimore, 1965.
13. Sedov, L. I., *Similarity and Dimensional Methods in Mechanics*, Academic Press, New York, 1961.
14. Strehlow, R. A. and Baker, W.E., The Characterization and Evaluation of Accidental Explosions, NASA CR-134779, National Aeronautics and Space Administration, Cleveland, OH, June 1965.
15. Stull, D. R., *Fundamentals of Fire and Explosion*, AIChE Monograph Series, Vol. 73, No. 10, 1977.
16. Taylor, G., The Formation of a Blast Wave by a Very Intense Explosion, *Proceedings of the Royal Society of London. Series A, Mathematical and Physical Sciences*, 201(1065), 159–174, 1950.

Exercises

2.1 (a) A blast wave generated from a concentrated energy release in the ambient reaches an observer positioned at a distance of 50 m from the source 10 ms after the energy release. If the Mach number of the blast wave at the observer's location is 4, determine the overpressure in kPa experienced by the observer. The pressure and temperature of the ambient air are 100 kPa and 27°C, respectively. The molecular mass of air is 28.9 kg/kmol. The gas constant is 8.314 kJ/(kmol K). The specific heat ratio of air (γ) is 1.4.

(b) What is the impulse per unit area (N-s)/m^2 felt by the above observer assuming that only positive impulse is experienced by the observer and the pressure (behind the blast wave) decays with time at the location of the observer as

$$p(t) = p - 2 \times 10^4 t^2$$

where $p(t)$ is the pressure in kPa at time t (seconds) after passage of blast wave at the location of the observer and p is the pressure behind the shock front of the blast wave in kPa.

(c) At what time after the release of the source energy does the observer stop feeling the impulse from the blast?

2.2 The accidental spill of combustible gases from a processing plant formed an explosive fuel–air cloud. The explosion of this cloud resulted in significant damage to buildings situated at a distance of 500 m from the site of the explosion. If the overpressure required to cause the observed damage to the buildings is 0.07 MPa, determine the mass of the combustible cloud. The energy content of 1 kg of the cloud is 2500 KJ/Kg. State any assumptions made.

2.3 The distance traveled (R_S) by a blast wave over a time t seconds from a source was derived from dimensional considerations as $R_S = C(E_0/\rho)^{1/5}t^{2/5}$, where energy E_0 Joules is released in a medium of density ρ_0 kg/m^3 and C is a constant. Use the above expression to:

(i) Determine the time taken for a blast wave generated in a particular explosion to travel from A to B which are at a distance of 30 and 90 m along a radius from the energy source at point O. The blast wave takes 1.5 ms (milliseconds) to arrive at point A after the energy is released at O.

(ii) The velocity of the blast wave at point A and at point B.

2.4 The rupture of a pressure vessel in ambient sea-level atmosphere releases 5 MJ of energy. The ambient pressure is 0.1 MPa. Assuming the energy release as sudden and at point determine the overpressure and impulse per unit area at a distance of 2 m from the site of the energy release. Comment on the assumptions of point energy release.

2.5 A tiny little asteroid of mean diameter of about 17 m and mass of about 10,000 tons entered the atmosphere at a height of 50 km over the city of Chelyaabinsk in Siberia on Friday February 15, 2013. The blast wave formed by the asteroid injured more than 1300 people and blew out windows of several multistory buildings. The tiny asteroid is reported to have entered the atmosphere with a velocity of 20 km/s. If the temperature of the atmosphere at 50 km height is $-65°C$ and the specific heat ratio of atmospheric air at the given altitude is 1.4, estimate the Mach number of the shock wave formed. State any assumptions made.

2.6 A spherical blast wave is formed in ambient atmosphere ($p_0 = 100$ kPa, $T_0 = 300$ K, $\rho_0 = 1$ kg/m^3) by the impulsive release of energy from an accidental explosion. The velocity of this blast wave is observed to be 2000 m/s when its distance from the source of energy release is 100 m. If the density ratio across the blast wave is given by the expression

$$\frac{\rho_S}{\rho_0} = \frac{\gamma + 1}{\gamma - 1 + 2/M_S^2}$$

where ρ_S is the density behind the wave, ρ_0 is the ambient density, γ is the specific heat ratio (= 1.4 for ambient air), and M_S is the Mach number of the blast wave, determine using the mass conservation equation across the shock front:

(a) Velocity of particles behind the blast wave in the frame of reference of the blast wave when the blast wave is at a distance of 100 m from the energy source

(b) Velocity of particles behind the blast wave in the frame of reference of a stationary observer on the ground

2.7 The internal energy of a solid explosive of mass 1.5 kg, contained in a rigid metal case is 1600 J/g. It is desired to achieve a fragment velocity of 2000 m/s. What should be the mass of the rigid metal casing housing the solid explosive?

2.8 (a) While photographing the fire in a fertilizer plant at the town of West in Texas, USA, on April 17, 2013, a photographer was knocked down by the explosion caused by the fire. It appears that while the fire fighters were dousing the fire with water, spontaneous chemical reaction of the fertilizer ammonium perchlorate (AP), stored in the plant, took place. The explosion caused severe damage to more than half the town and resulted in 14 people getting killed and about 160 being injured. You are required to determine the quantity of AP in kg, which would have exploded given the following data: You can assume that a spherical blast wave is generated in the explosion.

(i) The photographer was standing at a mean distance of 250 m from the center of the explosion when he was knocked down by the explosion.

(ii) The photographer is knocked down by an overpressure (Δp) of 70 kPa.

(iii) The ambient pressure (p_0) at the town of West is 100 kPa, and the ambient temperature at the time of the incident was 27 °C.

(iv) The non-dimensional overpressure (Δp)/p_0 in an ideal explosion is correlated with the non-dimensional distance [expressed as ratio of the dimensional distance (R_S) to the explosion length (R_0)] by the expression

$$\frac{\Delta p}{p_0} = \frac{0.16}{(R_S/R_0)^3} - 1$$

(v) The energy release per kg of AP is 1500 kJ/kg

(vi) The entire energy release may be assumed to drive the blast wave.

(b) If the non-dimensional specific positive impulse from the spherical blast wave generated by the explosion is given by the expression

$$\bar{I}_S^+ = 0.32/(R_S/R_0)$$

determine the positive impulse experienced by the photographer per unit surface area in (kg-m/s per m²) in the above problem. You may recall that the non-dimensional specific positive impulse is given by

$$\bar{I}_S^+ = I_S^+ a_0 / (p_0 R_0)$$

where I_S^+ is the impulse per unit area, a_0 is the speed of sound in the ambient atmosphere, and p_0 and R_0 are the ambient pressure and explosion length, respectively, in proper units. The universal gas constant is 8.314 J/(mole K), and the mean molecular mass of air is 28.8 g/mole. The specific heat ratio of air is 1.4.

2.9 If the overpressure produced by the blast wave at a distance of 0.1 km from the source of energy release in an explosion is 20 kPa and the non-dimensional overpressure is given by the expression

$$\frac{p - p_0}{p_0} = \frac{0.1568}{(R_S/R_0)^3} = 1$$

where R_S/R_0 denotes the distance R_S scaled by the explosion length R_0, determine the energy release in the explosion in MJ. The ambient pressure p_0 is 100 kPa.

2.10 The safety engineers in a chemical plant at Mumbai, handling explosive material, design, and erect a strong wall, known as a blast wall, around a hazardous chemical processing area. The distance of this wall from the source of the hazard is 100 m, and it is found to be adequate in protecting the remaining parts of the plant, and the neighboring township from an accidental explosion should it occur in the hazardous area. The management now wishes to construct a similar chemical plant at Lhasa in Tibet with a similar blast wall. The plant and the blast wall are therefore identically configured at Lhasa as the one at Mumbai with the same factor of safety and with the protective wall being at the same distance of 100 m from the source of the chemical hazard. Determine whether the explosive material to be handled in the chemical plant at Lhasa needs to be decreased/ increased/ kept the same as the one at Mumbai to maintain the same factor of safety in both the plants. In case the explosive material to be handled has to be increased or decreased to give the same factor of safety as at Mumbai, determine the percent increase or decrease. Mumbai is at sea level while Lhasa in Tibet is at an altitude of about 3.5 km for which the ambient pressure can be assumed to be 75 kPa as compared to 100 kPa at sea level at Mumbai.

2.11 A compression wave 15 kPa propagates in ambient atmosphere of 100 kPa and temperature of 27 °C. Calculate the temperature in the compression wave assuming it to be i. a weak shock and ii. an isentropic compression. The specific heat ratio of air is 1.4.

2.12 A bomb having a mass of 2 kg of a given explosive produces a blast wave of Mach number 4 at a distance 500 m from it. At what distance from a similar bomb but having a reduced mass of 0.5 kg of the same explosive will a blast wave of Mach number 4 be formed?

Chapter 3
Interaction of Blast Waves with Rigid and Non-rigid Bodies

Blast waves, formed from explosive energy release, would interact with bodies placed in their path of propagation. These could be the ground, water in the ocean, structures, humans, or for that matter any object in the path of the blast waves. When a blast wave strikes the surface of a rigid body, it partially rebounds into the medium in which it had been propagating and partly gets transmitted into the body. If the surface of the body is hard and unyielding, the blast wave fully rebounds back. However, if the surface of the body yields, the blast wave is partly transmitted into the body. The high particle velocities behind the blast wave, the high temperatures, and changes in the speed of sound of the medium processed by the blast wave cause the phenomenon of the interactions to be complex, and very often the observations of the interactions are opposite of what we would have intuitively expected. The scaling of the interactions with the geometry and size of the bodies and Mach number of the blast wave becomes difficult. The reflection and transmission of blast waves with the bodies placed in its path and the associated phenomena are examined in this chapter.

A shock wave is an integral part of the blast wave with the shock forming the front of the blast wave. We will, therefore, consider the interaction shock waves of different Mach numbers with surfaces before drawing conclusions on the blast wave interaction with bodies.

3.1 Reflection of Shock Waves from Non-yielding Surfaces

A planar shock wave could impact the surface of a body either head on normal to it or else impact it at an angle. These two cases are known as normal reflection and oblique reflection and are dealt with in the following.

© The Author(s), under exclusive license to Springer Nature Switzerland AG 2021
K. Ramamurthi, *Modeling Explosions and Blast Waves*,
https://doi.org/10.1007/978-3-030-74338-3_3

3.1.1 Normal Reflection

In the case of normal reflection, shown in Fig. 3.1, the planar shock wave incident to the surface called as incident shock propagates at a velocity $\dot{R}_S$ in a stationary medium at pressure p_0, density ρ_0, and temperature T_0 having a sound velocity a_0. The pressure, density, temperature, and sound speed behind the incident shock are p_1, ρ_1, T_1, and a_1, respectively. The medium follows the incident shock with a velocity. It strikes a surface normally, and the surface is considered to be planar and non-yielding. The incident shock gets reflected at the unyielding surface and is propagated back into the medium as a reflected shock. The pressure, density, and temperature behind the reflected shock increase beyond the values behind the incident shock to p_2, ρ_2, and T_2 (Fig. 3.1).

The velocity of the medium behind the incident shock and the associated sound speed a_1 change with the Mach number of the incident shock wave M_S ($M_S = \dot{R}_S/a_0$). This causes the ratio of the pressure behind the reflected shock p_2 to the pressure behind the indent shock p_1, viz. p_2/p_1 to vary in a highly nonlinear way with respect to the variations in the incident shock Mach number M_S. The value p_2/p_1, known as the reflection coefficient, is determined for different values of the incident shock Mach numbers in Appendix C. It is seen that as the incident shock Mach number M_S or equivalently as the pressure ratio p_1/p_0 across the incident shock wave increases, the ratio of the reflected to incident shock pressures p_2/p_1 reaches a limiting value of 8. For a very weak shock wave, the ratio p_2/p_1 is about 2. The value of the ratio of the reflected shock pressure of about 8 would be even higher in the case of non-ideal gases due to the locked in energy of the molecules. The high values result in a significant increase of blast overpressures on rigidly held bodies. As an example, at an incident shock number of Mach number $M_S = 5$, worked out in Appendix C, the value of p_1/p_0 across the incident shock is 29, while the ratio p_2/p_1 across the reflected shock wave is 7.09. The ratio of the reflected shock wave pressure to the ambient pressure p_2/p_0 is, therefore, $29 \times 7.09 = 205.6$. This implies that at an ambient pressure of 0.1 MPa, the pressure experienced at the surface of the rigid body is 20.56 MPa. This high pressure leads to intensive force at the surface wherein the incident shock collided head on.

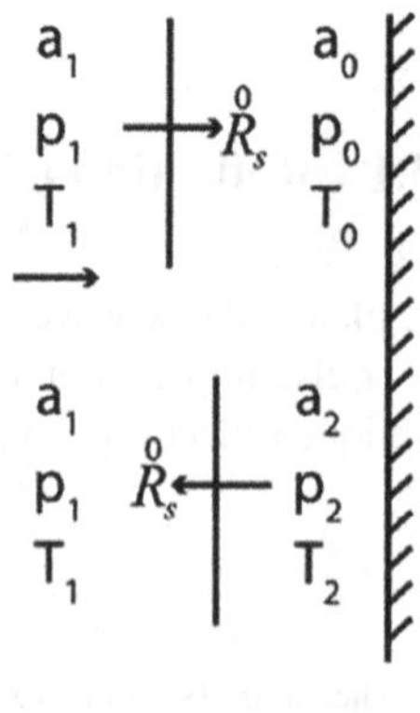

Fig. 3.1 Normal reflection of shock wave from a non-yielding surface

3.1.2 Oblique Reflection

Regular Reflection

When the incident shock wave propagates at a small value of angle of incidence α with respect to the non-yielding surface of a rigid body, it is reflected back into the medium with an angle of reflection β as shown in Fig. 3.2. However, unlike for the reflection of sound waves and light, the angle of incidence is not equal to the angle of reflection. The angle of reflection is such that the direction of flow at the surface is along it.

The properties behind the reflected shock can be readily determined using the shock relations. The component normal to the direction of propagation of incident shock component follows the normal shock relations, while the component of velocity along its direction of propagation remains invariant. For the reflected shock, the additional constraint is that the net flow velocity of the medium is along the surface. This is illustrated in Fig. 3.2.

Mach Reflection

When the angle of incidence α exceeds a threshold value, the reflected shock with regular reflection is unable to permit the flow of the medium behind the reflected wave to be along the surface. With medium of propagation of the shock being air, the threshold value of α is about $40°$. The incident shock spurts up with a normal shock being formed at the surface that can ensure the flow velocity behind it to be along the surface. This normal shock is called as Mach stem shock and is stronger than the incident and reflected shocks. The three shocks, viz. the incident shock, reflected shock, and Mach stem shock meet at point O. The oblique shock reflection forming Mach stem shock is known as Mach reflection. The Mach stem reflection is shown in Fig. 3.3.

Fig. 3.2 Regular reflection

Fig. 3.3 Mach reflection

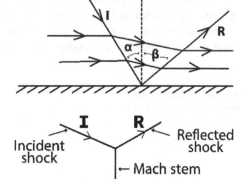

The Mach stem shock, being stronger, causes the velocity of the medium behind it to be smaller in the frame of reference of the Mach stem shock. However, in the inertial frame of reference, the velocity of the medium, which is the difference between the shock velocity and the above velocity in frame of reference of the shock, is very much higher. It results in strong winds being formed at the surface.

3.2 Reflection and Transmission of Shocks from Yielding Surfaces

When an incident shock wave (I) propagating in medium A impinges in a direction normal to the surface of a body B that is compressible, the shock reflects back as a reflected shock (R) in medium A, and a part is transmitted into the body B as a transmitted shock (T). The reflection and transmission are shown in Fig. 3.4. The reflection at the surface is no longer at the velocity of the incident shock as it happened for an unyielding surface. The surface between the medium A and the body B is characterized by density changes and is spoken of as an interface. We could, in general, state that at the interface separating media of two different densities, reflection and transmission of an incident shock normal to it take place. The magnitudes of the reflected and transmitted components are determined in the next section.

3.2.1 Mechanical Impedance of Medium and Determination of Reflected and Transmitted Waves

The magnitudes of the reflected and transmitted shocks formed by the normal impingement of a planar incident shock at a planar interface, wherein density changes are encountered, are determined using the mechanical impedances of the two media.

The mechanical impedance is analogous to electrical resistance and is defined as the ratio of the change in pressure to the corresponding change in velocity. If the change in pressure across the wave in a given medium is p' for a given velocity change u', the mechanical impedance Z of the medium is

Fig. 3.4 Reflection and transmission of an incident shock at an interface

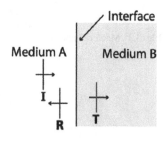

Fig. 3.5 Transmission and
reflection of blast waves
traveling from medium A to
medium B

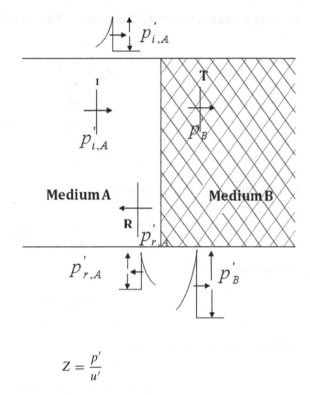

$$Z = \frac{p'}{u'}$$

With pressure and velocity having units of N/m^2 and m/s, respectively, mechanical impedance has units of N-s/m^3 or Pa-s/m.

Let an incident shock I travel from medium A into medium B. Let the mechanical impedance of medium A be Z_A and that of medium B be Z_B. When the shock wave I reaches the interface of the two media, it gets partially transmitted into medium B as T and is also reflected back into medium A as R. The incident, transmitted, and reflected shocks are shown in the two media as shown in Fig. 3.5. If the overpressure of the incident shock I in media A is denoted by $p'_{i,A}$ and the overpressures in the transmitted shock in medium B and reflected shock in medium A are denoted by p'_B and $p'_{r,A}$, respectively, we have from pressure balance at the interface:

$$p'_{i,A} + p'_{r,A} = p'_B \tag{3.1}$$

If the velocity induced in medium A due to the incident and reflected shocks are $u'_{i,A}$ and $u'_{r,A}$, respectively, and the velocity in medium B due to the transmitted shock is u'_B, the velocity balance at the interface can be expressed as follows:

$$u'_{i,A} - u'_{r,A} = u'_B \tag{3.2}$$

Denoting the value of mechanical impedance of medium A and B as Z_A and Z_B, we have

$$Z_A = \frac{p'_{i,A}}{u'_{i,A}} \tag{3.3}$$

$$Z_A = \frac{p'_{r,A}}{u'_{r,A}} \tag{3.4}$$

$$Z_B = \frac{p'_B}{u'_B} \tag{3.5}$$

From the above set of equations we have

$$u'_{i,A} = \frac{p'_{i,A}}{Z_A}, \ u'_{r,A} = \frac{p'_{r,A}}{Z_A} \ \text{and} \ u'_B = \frac{p'_B}{Z_B}$$

On substituting in Eq. 3.2, we get

$$\frac{p'_{i,A}}{Z_A} - \frac{p'_{r,A}}{Z_A} = \frac{p'_B}{Z_B} \tag{3.6}$$

Multiplying Eq. 3.6 by Z_A, we get

$$p'_{i,A} - p'_{r,A} = \frac{p'_B}{Z_B} \times Z_A \tag{3.7}$$

Solving Eqs. 3.1 and 3.7 for $p'_{i,A}$ and $p'_{r,A}$, we have

$$2p'_{i,A} = (1 + Z_A/Z_B)p'_B = \frac{Z_A + Z_B}{Z_B}p'_B \tag{3.8}$$

$$2p'_{r,A} = (1 - Z_A/Z_B)p'_B = \frac{Z_B - Z_A}{Z_B}p'_B \tag{3.9}$$

The values of overpressures of the transmitted shock wave and the reflected wave as a function of the overpressure of the incident shock wave, therefore, are

$$p'_B = \frac{2Z_B}{Z_A + Z_B}p'_{i,A} \tag{3.10}$$

$$p'_{r,A} = \frac{Z_B - Z_A}{2Z_B}p'_B = \frac{Z_B - Z_A}{Z_A + Z_B}p'_{i,A} \tag{3.11}$$

The magnitudes of the transmitted and reflected shock waves will therefore depend on the mechanical impedance of the two media. Once the mechanical impedance Z of the medium is obtained, the overpressure of the transmitted and reflected shocks can, therefore, be obtained from Eqs. 3.10 and 3.11.

Table 3.1 Acoustic impedance

S. No.	Medium	Acoustic impedance (Z) N-s/m^3 or Pa-s/m
1	Air	420
2	Water	1.5×10^6
4	Muscles in body	1.7×10^6
5	Bone	6.6×10^6
7	Pyrex glass	13×10^6
8	Steel	46×10^6
9	Tungsten	101×10^6

The determination of the mechanical impedance is given in Appendix D. It is obtained by balancing momentum across the wave. For a weak shock wave it is given by the value of the acoustic impedance Z, which equals the product of the density of the medium and the sound speed in the medium. For a strong shock wave the mechanical impedance equals the product of the density of the medium and the shock velocity. The acoustic impedance is a thermodynamic property of a medium, and the values for some substances are given in Appendix D. The use of acoustic impedance for moderate to strong shocks instead of shock impedance is also discussed in Appendix D. A few values of the acoustic impedance are given in Table 3.1.

3.2.2 Formation of Expansion Waves

Consider a shock wave of overpressure p_A traveling in a medium A and getting partially transmitted into medium B with an overpressure p_B and also being partially reflected at the interface back into medium A with an overpressure p_R (Fig. 3.6). The overpressure denotes excess pressure over the ambient, and we need not indicate it by the superscript 'prime' any longer. Denoting the acoustic impedances of media A and B as Z_A and Z_B, we have, from Eqs. 3.10 and 3.11, the magnitudes of the transmitted overpressure p_B and the reflected overpressure p_R as follows:

$$p_B = \frac{2Z_B}{Z_A + Z_B} p_A \quad \text{and} \quad p_R = \frac{Z_B - Z_A}{Z_A + Z_B} p_A$$

If the medium A is air and the medium B is steel, we have the overpressures in the reflected and transmitted waves using the values of acoustic impedances given in Table 3.1 as follows:

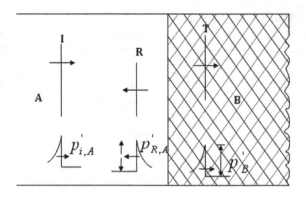

Fig. 3.6 Reflection and transmission of shock waves ($Z_B > Z_A$)

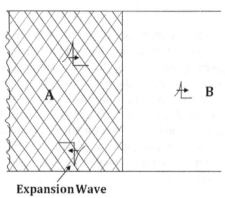

Fig. 3.7 Reflection from a lower impedance medium ($Z_B < Z_A$)

Expansion Wave

$$p_R = \frac{46 \times 10^6 - 420}{420 + 46 \times 10^6} p_A \approx p_A$$

$$p_B = \frac{2 \times 46 \times 10^6}{420 + 46 \times 10^6} p_A \approx 2p_A$$

While a shock with the same overpressure is reflected from the higher impedance medium, a higher overpressure of twice the magnitude of the incident shock is transmitted into the higher impedance medium.

If the two media are now reversed with the incident shock traveling in steel and getting transmitted into the air, the reflected wave now travels in the steel. The configuration is shown in Fig. 3.7.

The magnitudes of the transmitted and reflected waves are given by

$$p_B = \frac{2 \times 420}{420 + 46 \times 10^6} p_A = 0.18 \times 10^{-4} p_A$$

$$p_R = \frac{420 - 46 \times 10^6}{420 + 46 \times 10^6} p_A \approx -p_A$$

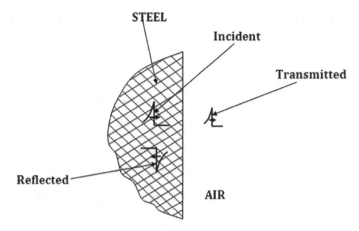

Fig. 3.8 Formation of expansion waves

This implies that a very weak compression wave gets transmitted into the air from the steel while a negative compression wave, i.e., an expansion wave is reflected back into the high impedance steel from the air. The expansion wave is significant in magnitude; however, the wave transmitted into the low impedance medium is very weak. The absence of material at the interface causes the expansion wave. The process of formation of the expansion waves and the transmission of the weak compression is schematically shown in Fig. 3.8.

The reflected wave becomes an expansion wave when the impedance of the second or subsequent medium in which the shock gets transmitted is less than the impedance of the first medium in which it is reflected back. The transmitted wave, however, is always a shock wave independent of the media involved and the mismatch of the acoustic impedance at the interface of the two media.

3.2.3 Spallation

The passage of a shock front from a high acoustic impedance medium to a medium having a low value of acoustic impedance was seen in the last section to result in a significant reflected expansion wave in the high impedance medium. The passage of the incident blast wave through the high acoustic impedance medium causes expansion behind the high pressure at the shock front and is shown in Fig. 3.9. The movement of the strong expansion wave back into the medium causes expansion in the opposite direction to the expansion behind the incident shock wave. These two expansion waves, as shown in Fig. 3.9, cause the particles in the medium to be pulled apart in opposite directions due to the expansion behind the incident shock and the reflected expansion wave. This sudden pull or tension at the meeting plane of the two expansion processes tears the material apart (at the plane) provided that the tension

Fig. 3.9 Spallation of a
material

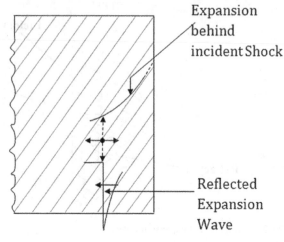

Fig. 3.10 Kidney stone

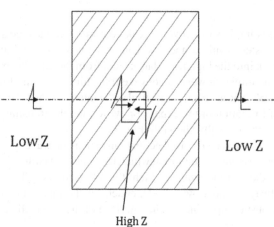

is greater than the dynamic strength of the material. This process of tearing apart of
the material in tension is known as spallation.

3.2.4 Crushing of Kidney Stones in Humans

The formation of expansion waves from the reflection of shocks at interfaces and
the associated spallation is used for breaking up of kidney stones. The kidney stones
have acoustic impedance of the order of a bone (about 6.6×10^6 N-s/m) while the
medium in which the stone is held is near to that of water and a muscle (acoustic
impedance 1.5×10^6 and 1.33×10^6 N-s/m). Figure 3.10 provides a sketch of the
stone of high acoustic impedance enclosed in a medium of lower acoustic impedance.

If a minute blast wave is generated and focused on the stone, it propagates from the lower acoustic impedance medium of water into the relatively higher impedance of the stone and thereafter into the lower impedance of the fatty tissue. The blast wave propagates with enhanced overpressure into the stone; however, it gets reflected as an expansion wave when it meets the lower impedance fatty tissue. The pulling apart of the expansion behind the incident blast wave and the expansion associated with the reflected expansion wave in the stone result in its spallation or stone getting broken into pieces.

A similar process causes the rupturing of eardrum and damage to lungs from blast waves.

3.3 Reflection of a Blast Wave from the Ground: Formation of Multiple Shocks and a Mushroom Cloud

Consider a blast wave generated from an explosion at a height H above the ground instead of the explosion occurring at the surface (Surface Burst). The progression of the blast wave is shown by the dotted lines in Fig. 3.11. This incident blast wave on striking the ground gets reflected, and the reflected waves propagating back into the air are shown by the full lines. The reflected waves will be stronger than the incident blast waves since it travels in a medium, which is hotter, and this has been discussed earlier. The reflected wave will, therefore, overtake the incident blast wave as shown in Fig. 3.11.

The expansion process behind the reflected waves and the associated reduction in pressure as it propagates higher up will cause upward air motion. The particles are, therefore, no longer directed toward the ground but are now reversed due to the reflected wave. The reflected wave also interacts with the ground surface. When the angle of impingement is lower than a threshold value, the flow along the surface of

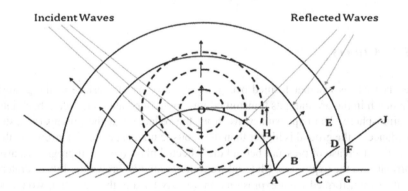

Fig. 3.11 Incident and reflected shocks from an explosion; regular expansion at *A* and *C* while Mach reflection at *G*

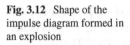

Fig. 3.12 Shape of the impulse diagram formed in an explosion

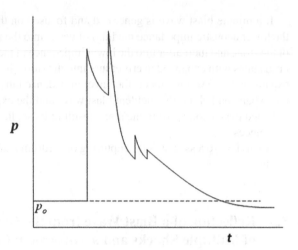

the ground can be maintained by a regular reflection shown by AB and CD. However, when the angle exceeds a threshold value of about 40°, regular reflection is no longer possible, and the reflected blast wave spurts from the surface to form a normal shock. The stand off shock is known as Mach stem shock, and the reflected shock is detached from the surface. This is shown by EF, FG and FJ and happens because the flow along the ground needs to be along it.

A number of shocks are, therefore, formed. An observer will experience not only the reflected shock, which does more damage than the incident blast wave but also be subject to multiple shock waves from the regular and Mach stem shock reflections at the ground. The impulse will consist of multiple spikes from the different shocks and the shape of a typical impulse diagram as shown in Fig. 3.12.

The reflected blast wave, which has overtaken the incident blast wave, carries the particulates from the expansion. The product cloud rolls up to give it the shape of a mushroom cloud as shown in Fig. 3.13.

3.3.1 Craters

The blast waves whether formed over the ground or on the surface of the ground or in depth in the ground gets transmitted into the ground and is reflected back into the atmosphere. The expansion wave from the reflection at the ground (acoustic impedance of the ground is higher than the acoustic impedance of air) throws up the mud. The decrease of pressure behind the reflected blast wave as it progresses also contributes to the expansion. Material from the ground is thrown up. Pits or craters are, therefore, formed from explosions. In a rocky terrain, the reflection would be stronger and the diameter of the crater formed will be larger.

Fig. 3.13 Mushroom cloud
in an explosion

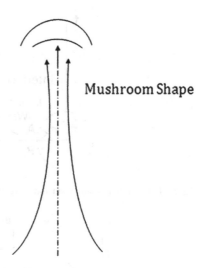

Mushroom Shape

The diameter of the craters is formed in an explosion scales with the strength of the blast wave and hence the explosion length. The conditions of the ground and the location of the energy release with respect to the ground surface influence the diameter and depth of the crater formed.

3.4 Multiple Pressure Spikes from a Finite Volume Explosion

The formation of expansion waves due to the change of impedance at the interface between the high pressure and hot gases of the explosion and the ambient results in multiple pressure spikes in the impulse diagram. Figure 3.14 shows an explosion originating from a finite volume of radius R_E. The X-axis shows the distance while the Y-axis shows the time. A blast wave A-1 is formed from the high pressure of the explosion at radius R_E, and the blast wave travels into the ambient medium outside it.

When a blast wave is formed at radius R_E, the gas within it expands at R_E and an expansion wave propagates inward into the volume of radius R_E from the interface that expands. The expansion wave, shown by the dotted line A-2 in Fig. 3.14, travels at high speeds since the sound speed of the high pressure and high temperature exploded gases is high. The blast wave propagates outward from radius R_E into the ambient, while the expansion wave propagates inward from R_E into the high-pressure exploded gas as shown in Fig. 3.14 and also in Fig. 3.15.

The expansion wave (A-2) gets reflected from the zone of symmetry and propagates out as an expansion wave shown by the dotted line 2-B-3 in Fig. 3.15. It meets the interface between the high-pressure expanding gases from the explosion (shown by dotted circle) at B in Fig. 3.15. It partly reflects back from the interface at B as

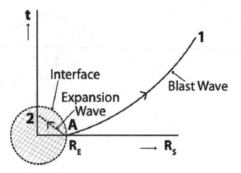

Fig. 3.14 Formation of an expansion wave at the interface of explosive energy release

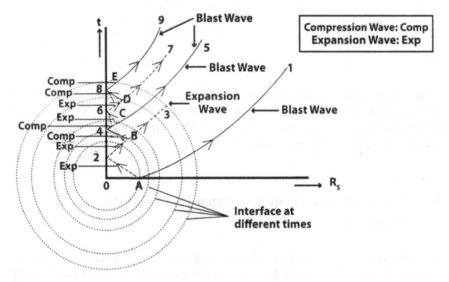

Fig. 3.15 Formation of shocks due to wave interactions at the expanding interface of the explosion products

a compression wave since the impedance drops at the interface. This is due to the impedance of the high-pressure medium being higher than the acoustic impedance of the ambient external to it. This compression wave travels inward and is shown by B-4. The compression wave B-4 within the expanded exploded gases is reflected from the symmetry line as a compression wave 4-C-5. At C, the compression wave gets transmitted at the interface into the lower impedance medium as a compression wave or a blast wave C-5. However, at the interface an expansion wave is sent back into the higher impedance medium, and in this way a series of shock compressions are generated.

An observer standing at a distance R_S from the explosion such that $R_S \gg R_E$ will encounter a series of pressure spikes starting from the pressure spike due to the

Fig. 3.16 Shape of the
impulse obtained in an
explosion

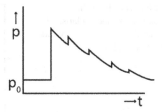

first blast wave A-1 are followed by pressure spike due to the blast wave C-5 and so
on. The pressure that is felt by the observer, located at R_S, as a function of time is
shown in Fig. 3.16. A continuous series of pressure spikes are observed starting with
the first blast wave, the second blast wave and then third wave and so on. Therefore,,
the impulse from an explosion need not correspond to a monotonically decreasing
pressure behind a single lead blast wave. It contains a series of shocks, and this
phenomenon is seen in all explosions.

3.5   Blast Waves in Water

3.5.1   Underwater Explosions and Associated Blast Wave

When an explosion takes place in water, the release of the gaseous products of the
explosion and the high energy in the explosion cause the formation of a high-pressure
gas bubble in the water. A shock front is formed at the surface of the bubble. The
blast wave propagates into the water at higher values of pressure than in the bubble,
since water has higher value of acoustic impedance compared to the medium of the
gas bubble. If the energy release is at depth in the water, the blast wave continues to
travel in the water until it either meets the surface of water at the top or at the bottom
or side surfaces of the containment, where there is an impedance mismatch.

The gas bubble formed by the explosion also expands in the water from the high
pressure of the explosion. The gas–liquid interface travels like an accelerating piston
behind the decaying shock wave. The outward flow of water causes the expanding
gas bubble to continue expanding even when the pressure within it has dropped below
the ambient hydrostatic pressure.

As the gas bubble expands, the pressure within it reduces till it becomes much
lower than the surrounding pressure. The higher ambient pressure now causes it
to collapse inward, i.e., implode. During the implosion process, pressure builds up
inside the bubble to high values. The process of the implosion continues even after
the pressure in the bubble exceeds the hydrostatic pressure of water at the particular
depth of the bubble. When the pressure within it is much greater than the hydrostatic
pressure, the bubble begins to expand again, and shock waves or pressure pulses are

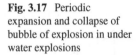

Fig. 3.17 Periodic expansion and collapse of bubble of explosion in under water explosions

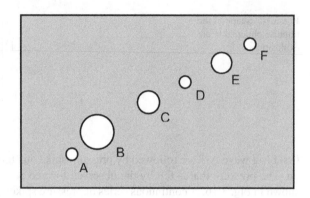

generated in the water again. The process continues with a series of expanding and imploding bubble with the result that series pressure pulses are generated.

Figure 3.17 shows the high-pressure water bubble at A formed by the explosion driving the initial shock with an overpressure $(p_s - p_h)$, where p_s is the pressure behind the shock and p_h is the hydrostatic pressure. The bubble expands to larger diameter at B and thereafter collapses to a lower diameter at C and D. The attendant rise in pressure within the bubble at D causes its pressure to increase beyond the hydrostatic pressure. The bubble expands then to E and collapses again to F.

The period of the pulses due to the alternate growth and collapse of the bubbles would be very much larger than the time of travel of the shock waves in water. Impulse is generated from the pressure pulses due to the expansion of the imploding bubbles in addition to the impulse from the initial value of overpressure. The impulse depends on the depth at which the explosion occurs in water and the nature of confinement of the water. The blast wave from the explosion is followed by a series of overpressures as sketched in pressure time diagram of Fig. 3.18. The variations in pressure at the different instants of time corresponding to the expanded bubbles at A, D and F and expanded bubbles at B, C and E are shown in Fig. 3.17.

Underwater explosions can lead to much greater damage than in explosions in air. The larger acoustic impedance of water compared to air would contribute to high overpressure damage. The periodic collapse of bubbles leads to an increase of impulse.

The buoyancy of the gas bubbles would lead to an upward motion in water during the process of their expansion and collapse.

3.5.2 Explosions over Surface of Water

The blast wave could be transmitted in the water from the energy release at the surface. The overpressure at the shock front causes the damage. The peak pressure continually decays as the blast wave progresses in the water. The pressure evolution would change as the reflected wave from the confinement interacts with the decaying shock wave.

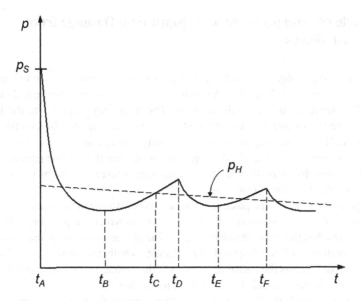

Fig. 3.18 Impulse from the periodic pressure spikes

The impulse from a surface explosion would be very much lower than in an underwater explosion since the collapse of the gas bubbles no longer takes place. If the pressure during the region of negative impulse is lower than the vapor pressure of water, cavitation could occur giving rise to small pressure pulses. It is the overpressure at the shock front, which causes the damage to the structures in water from such surface explosions.

3.6 Absorption of Blast Wave Energy in Layered Structures

The principle of spallation and the associated energy absorbed is used in practice to mitigate the destructive effects of blast waves in several applications. Layered structures with each layer having distinctly different acoustic impedances are used. The blast wave transmitted into a lower acoustic impedance medium will have a lower strength while the reflected expansion wave in the higher acoustic impedance medium will absorb energy due to spall. Estimates of the rate at which energies are absorbed per unit area are given by the product of the overpressure and the velocity. The energy dissipated per unit area could be expressed as follows:

$$E = \int p'u'dt = \int (p'^2/Z)dt \ \text{J/m}^2 \tag{3.12}$$

Here the term $p' \times u'$ denotes the rate of work done per unit time.

3.7 Role of Overpressure and Impulse on Damage from Blast Waves

Damage to bodies subject to blast waves from explosions could be due to spall and the energy that can be absorbed by the body before failure. Overpressure and impulse of the blast wave contribute to the damage. The frequency response of the bodies subject to the blast wave decides on whether overpressure or impulse contributes to the damage. If the frequency response of the body is very high, such as the eardrum in humans, overpressure does the damage. This is because the varying pressures after the overpressure do not really matter as the damage is already done by overpressure in the available short time span.

The period over which the positive impulse acts is between the arrival time t_a of the blast wave at the given location to the time $t_a + t^+$ at which the pressure behind the blast wave reaches the ambient value, i.e., t^+. The characteristic response time of the body or structure τ when compared to t^+ decides whether overpressure or impulse in a blast wave governs the damage. If $\tau \ll t^+$, the damage is by overpressure while if $\tau > t^+$, the impulse could also contribute to the damage.

In the zone wherein the characteristic response time of the body τ and the duration of the impulse t^+ are of the same order of magnitude, both overpressure and impulse influence the damage. This region is known as the dynamic region of failure. The uneven load on objects such as over buildings and the drag force on objects will cause dynamic effects. If the body is large and massive, it will respond to the overpressure while if the drag from the dynamics of flow around the body is dominant, then the impulse plays a major role. The transient crushing from overpressure and the transient wind effects from pressure variations over the body, thus, influence the failure.

Lines of constant damage known as isodamage curves are plotted with the axes overpressure and impulse. A typical isodamage curve is shown in Fig. 3.19. The shaded region shows the region of no damage. The vertical part of the isodamage parallel to the overpressure axis (Δp) denotes damage when $\tau \ll t^+$. The horizontal asymptote is for $\tau > t^+$ corresponding to impulse (I). Here the process of damage is quasi-static. In between overpressure and impulse zones of failure, we have the dynamic region of failure wherein both overpressure and impulse contribute and caused zone of failure and $\tau \approx t^+$.

Examples

3.1 *Impact loading on a Structure*: A planar blast wave traveling in atmospheric air hits a rigid wall of a building normally at a Mach number of 3. The frontal area of the rigid wall is $30\,\mathrm{m}^2$. Determine the impact load on the building.
Atmospheric air has a pressure of 100 kPa and temperature of 27 °C. It has a specific heat ratio of 1.4.

Solution: The pressure p behind the shock front at Mach 3 is given by

$$\frac{p}{p_0} = \frac{2\gamma}{\gamma + 1} M_S^2 - \frac{\gamma - 1}{(\gamma + 1)}$$

Fig. 3.19 Isodamage curve

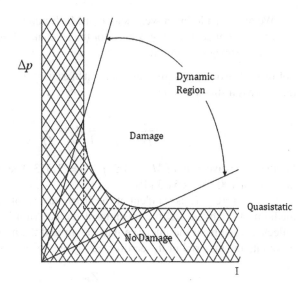

where p_0 is the ambient pressure. The pressure ratio p/p_0, therefore, for $M_S = 3$ is 10.33. This shock front gets reflected by the rigid unyielding wall to give the pressure ratio p_r/p, where p_r is the pressure behind the reflected shock front. The value of p_r/p from Appendix C is

$$\frac{p_r}{p} = \frac{(3\gamma - 1)(p/p_0) + (\gamma + 1)}{(\gamma - 1)(p/p_0) + (\gamma + 1)}$$

For the value of $p/p_0 = 10.33$, the above ratio works out to be $p_r/p = 5.428$. The wall experiences the pressure behind the reflected shock, and the value is $p_r = 5.428 \times 10.33 \times p_0 = 56.07 \times 100$ kPa $= 5.607$ MPa. This pressure is abnormally higher than the ambient pressure.

The overpressure is $5.607 - 0.1$ MPa $= 5.507$ MPa. The impact load on the wall is, therefore, $5.507 \times 30 = 165.2$ MN which is a significant load. The reflection magnifies the load.

3.2 *Blast Wave Interaction with Eardrum*: The human eardrum is approximately 0.1 mm in thickness, 8 to 10 mm in diameter, and has a mass of 14 mg. On both sides of the drum, air is present. If a blast wave at Mach number 1.2 strikes it, determine the compression wave transmitted into it from the front interface facing the blast wave and the nature and magnitude of the reflected wave transmitted into it from the rear interface. The acoustic impedance of air can be assumed as 415 Pa-s-m^{-1}, while the acoustic impedance of the material of the eardrum is 1.5×10^6 Pa-s-m^{-1}.

We are considering a weak shock at Mach number of 1.2. Hence, we can use the value of acoustic impedance for the mechanical impedance to determine the wave propagation into it.

Solution: The overpressure of the incident blast wave is determined from the pressure ratio across a shock given by

$$\frac{p}{p_0} = \frac{2\gamma}{\gamma+1} M_s^2 - \frac{\gamma-1}{(\gamma+1)}$$

For the blast wave in air of $M_S = 1.2$, $p/p_0 = 1.513$. The overpressure at the ambient pressure of 100 kPa is 51.3 kPa.

This overpressure is incident at the front surface of the eardrum. Denoting the medium of air by A and the medium of the eardrum by B, the magnitude of the reflected wave from the interface into the air and the magnitude of the transmitted wave into the material of the eardrum are given by

$$p'_{r,A} = \frac{Z_B - Z_A}{Z_B + Z_A} p'_{i,A}$$

$$p'_B = \frac{2Z_B}{Z_A + Z_B} p'_{i,A},$$

respectively. The value of the incident overpressure is 51.3 kPa. Hence substituting the values, we have

$$p'_{r,A} = \frac{1.5 \times 10^6 - 415}{1.5 \times 10^6 + 415} \times 51.3 \approx 51.3$$

and

$$p'_B = \frac{2 \times 1.5 \times 10^6}{1.5 \times 10^6 + 415} \times 51.3 \approx 102.6$$

Hence a compression wave of overpressure 51.3 kPa is reflected back into the air, while a compression wave of magnitude 102.6 kPa is transmitted into the eardrum.

The magnitude of the transmitted wave does not change over the very small distance of its propagation of 0.1 mm. At the edge of the eardrum, i.e., the interface between the drum and the air, this compression disturbance is reflected back in the eardrum and transmitted into the air as given below:

$$p'_{r,B} = \frac{Z_A - Z_B}{Z_A + Z_B} p'_{i,B}$$

and

$$p'_{i,A} = \frac{2Z_A}{Z_A + Z_B} \times p'_{i,B}$$

Substituting the value of the overpressure in the eardrum as 102.6 kPa, we get the reflected wave into the eardrum as -102.6 kPa which is an expansion wave. The pressure transmitted into the air is 0.057 kPa. Hence, a strong expansion wave is transmitted back into the fabric of the eardrum with the value of the peak expansion front being the same as the compression front which traversed in it. The expansion behind this initial compression and the expansion behind the expansion front cause the material to be pulled in tension, and the eardrum ruptures.

Nomenclature

a_0	Sound velocity in ambient medium (m/s)
i	Incident
E	Energy dissipated
M	Mach number defined as velocity/sound speed
M_S	Mach number of shock/blast wave
m	Mass (kg)
p	Pressure of shocked gases (Pa)
p_0	Pressure of the ambient (Pa)
$p - p_0$	Overpressure (Pa)
p'	Overpressure or change of pressure
$\bar{p}$	Dimensionless overpressure $[(p - p_0/p_0)]$
R_0	Explosion length (characteristic length of energy release) $[(E_0/p_0)^{1/3}]$
R_S	Distance of blast wave from source (m)
$\dot{R}_S$	Velocity of blast wave dR_S/dt (m/s)
R, r	Reflected
t	Time (s)
t_a	Arrival time of blast wave (s)
t^+	Duration of pressure greater than ambient (s)
t^-	Duration of pressure less than ambient (s)
Z	Mechanical impedance/acoustic impedance (N-s/m^3)
γ	Specific heat ratio of gases
τ	Characteristic time constant of the body

Further Reading

1. Boddurtha, F. T., *Industrial Explosion Prevention and Protection*, McGraw Hill, New York, 1980.
2. Cole, Robert H., Underwater Explosions, Dover Publications, New York, 1965.
3. Cooper, P. W and Kurowski, S. R., Introduction to the Technology of Explosives, Wiley-VCH, New York, 1996.
4. Courant, R. and Freidricks, K. O., *Supersonic Flow and Shock Waves*, Interscience Publishers, New York, 1948.
5. Holt, Maurice, Underwater Explosions, *Ann. Rev Fluid Mechanics*, 9, 187–211, 1977.
6. Kinney, G.F. and Graham, K. J., *Explosive Shocks in Air*, Springer, Berlin, 1985.

7. Johansson C. H. and Persson, P. A., Detonics of High Explosives, Academic Press, London and New York, 1970.
8. Liepman, H.W. and Roshko, A., *Elements of Gas Dynamics*, Wiley, New York, 1957.
9. Strehlow, R. A., Blast Waves Generated by Constant Velocity Flames - A Simplified Approach, *Combustion and Flame*, 24, 257–261, 1975.
10. Strehlow, R. A. and Baker, W.E., The Characterization and Evaluation of Accidental Explosions, NASA CR-134779, National Aeronautics and Space Administration, Cleveland, OH, June 1965.
11. Stull, D. R., *Fundamentals of Fire and Explosion*, AIChE Monograph Series, Vol. 73, No. 10, 1977.

Exercises

3.1 The accidental spill of combustible gases from a processing plant formed an explosive fuel–air cloud. The explosion of this cloud resulted in significant damage to buildings situated at a distance of 500 m from the site of the explosion. If the overpressure required to cause the observed damage to the buildings is 0.07 MPa, determine the mass of the combustible cloud. The energy content of 1 kg of the cloud is 2500 KJ/Kg. The blast wave may be considered to be strong. The ambient pressure is 0.1 MPa.

3.2 A blast wave traveling in atmospheric air strikes a wall of a building normally at a Mach number of 2. Determine the pressure and particle velocity behind the shock wave reflected from the wall in atmospheric air and the pressure and particle velocity behind the shock wave transmitted into the wall. The acoustic impedance of the material of construction of the wall is 7.5×10^6 N-s/m^3. The acoustic impedance of air is 430 N-s/m^3. The pressure p and velocity u_1 behind a shock wave in the frame of reference of the shock wave are given as

$$\frac{p}{p_0} = \frac{2\gamma}{\gamma + 1} M_S^2 - \frac{\gamma - 1}{\gamma + 1}$$

and

$$\frac{u}{\dot{R}_S} = \frac{\gamma - 1 + 2/M_S^2}{\gamma + 1}$$

p_0 and $\dot{R}_S$ are the ambient pressure (100 kPa) and the shock velocity in m/s. The sound speed in the ambient is 340 m/s. The value of specific heat ratio is 1.4 for air.

3.3 A blast wave progressing through atmospheric air at 950 m/s strikes a material having density 1700 kg/m^3 and sound speed of 600 m/s. Calculate the velocity of the blast wave transmitted into the material and velocity of the reflected blast wave? The temperature and pressure of atmospheric air are 300 K and 100 kPa. The molecular mass of air is 28.8 g/mole, and its specific heat ratio is 1.4. The universal gas constant is 8.314 J/(mol K).

3.4 (a) A blast wave traveling through air, shown by A, strikes a thin steel plate
with an overpressure of 3 MPa, as shown in the figure. The blast wave is
partially transmitted into the steel (shown by B) and is partially reflected
back into the air medium, shown by C. The wave transmitted into the steel
plate is further transmitted into the air (shown by D) and is also reflected
back into the steel plate as E. Determine the magnitude of the overpressure
of this reflected wave in the steel plate and the overpressure of the onward
transmitted wave in the air.

   (b) Are both these transmitted and reflected waves compression shock waves?
Determine the particle velocities behind the onward transmitted wave D and
the reflected wave E, respectively.

The mechanical impedance of steel is 46×10^6 Ns/m^3, while the mechanical
impedance of air is 430 Ns/m^3.

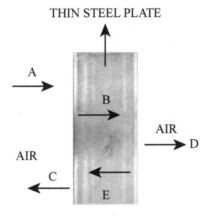

3.5 A blast wave traveling in air strikes the wall of a building normally at a Mach
number of 1.2. The acoustic impedance of the wall is 7.5×10^6 Ns/m^3 while
the acoustic impedance of air is 430 Ns/m^3. Determine the overpressure pressure
and velocity of the medium of

   (i) air behind the incident blast wave
   (ii) air behind the reflected blast wave
   (iii) wall behind the transmitted wave.

What would the velocities be in the frame of reference of an observer on the
ground viewing the blast wave interaction with the wall.

Chapter 4
Energy Release and Rate of Energy Release

Gaseous, liquid, and solid substances containing elements such as carbon and hydrogen can chemically react with oxygen in the air and release energy. The release of energy, if rapid, and when liberated in a restricted volume could generate a blast wave. The fire associated with the heat release and the pollutants formed in the chemical reaction also result in damage to environment, life, and property.

When a substance contains inbuilt oxidizing elements such as oxygen or chlorine in addition to the fuel elements such as carbon and hydrogen, reactions could take place between the fuel elements and oxidizer within the substance itself. Such substances having a composite of fuel and oxidizer elements were referred to as explosives in Chap. 1. In this chapter, we address a simple procedure of determining the energy release and the rate of energy release from the chemical reaction of fuels with ambient air and from chemical reactions in an explosive.

4.1 Energy Release

Chemical reactions between a fuel and oxidizer result in the formation of products of combustion. When the chemical structure of the product has lower energy than the reactant, the deficit in energy is released during the chemical reaction. The internal chemical energy associated with the reactants and products therefore decides the energy released in a chemical reaction and is dealt with in the following.

4.1.1 Heat of Formation

The heat of formation of a substance is defined as the heat released or absorbed while forming 1 mole of the substance at the standard state of 25 °C and 1 atmosphere pressure from its elements in the same standard state. It may be recalled that a mole of substance represents its amount in terms of the molecular mass. For example, one mole of hydrogen represents a mass of 2 g, while one mole of oxygen corresponds to a mass of 32 g. Moles do not represent the number of molecules. The number of molecules in one mole of a substance is the Avogadro number, which is equal to 6.023×10^{23}.

The naturally occurring elements at the standard ambient conditions of 1 atm pressure and 25 °C are used as a reference for specifying the heat of formation of a substance. These include molecular oxygen (O_2) as a gas, carbon $C(s)$ in solid form, and hydrogen (H_2) as a gas. The heat of formation is denoted by ΔH_f^0, where ΔH denotes the incremental enthalpy to form the substance from its elements while the subscript 'f' denotes the formation of the substance from its elements. The superscript '0' signifies that both the substance and the elements from which the substance is formed are at the same standard state of 1 atm pressure and 25°C. For solids and liquids, the standard state is defined as the temperature of 25 °C.

As an illustration, the heat of formation of carbon dioxide gas (CO_2) is represented by $\Delta H_{fCO_2}^0$ and is defined as the heat required to form 1 mole of CO_2 at 25 °C and 1 atm pressure from elements $C(s)$ at 25 °C and O_2 at 25 °C and 1 atm pressure. The reaction for forming CO_2 from elements $C(s)$ and O_2 is given by

$$C(s) + O_2 \rightarrow CO_2 \tag{4.1}$$

When 1 kg of carbon at 25 °C is burnt in O_2 at 1 atm pressure and 25 °C to CO_2, the measured value of the energy released is 32,800 kJ. The heat so released raises the temperature of CO_2. If the CO_2 were to be formed at 25 °C and 1 atm pressure, as is the condition of the elements $C(s)$ and O_2, an amount of heat corresponding to 32,800 kJ has to be removed for every kilogram of carbon burned.

Carbon has a molecular mass of 12 g/mole. According to Eq. 4.1, one mole of CO_2 is formed from one mole of carbon $C(s)$. The heat to be removed for the formation of 1 mole of CO_2 is $32,800 \times 0.012 = 394$ kJ. The heat of formation of CO_2 is therefore given as

$$\Delta H_{fCO_2}^0 = -394 \frac{kJ}{mole}$$

The negative sign signifies that heat has to be removed for the formation of CO_2 at the standard state from its elements at the standard state.

The heats of formation of different substances are determined in the above manner from simple reactions of their formation from the elements or from a combination of several chemical reactions involving intermediate substances. The heats of formation of the elements by the very definition are zero.

The heat of formation is representative of the chemical structure of a substance and hence its internal chemical energy. It can be determined as the sum of the energy of the bonds linking the atoms together in the substance. In case the linkages have additional energy like vibration or resonance between the different forms of its existence, the additional energy must also be taken into account.

The definition of heat of formation, it must be remembered, is at the standard state with reference to the naturally occurring elements at the standard state. It is therefore necessary to subtract the bond energy of the element from the bond energy of the substance for determining its heat of formation using bond energies.

The general practice is to determine the heats of formation by finding out the chemical reaction through which the substance is formed and using known values of heats of formation of the species involved in the reaction. Table 4.1 gives the standard heats of formation of some typical substances in gas, liquid, and solid phases.

It is shown in Table 4.1 that as the molecular structure of a substance becomes more complex, the value of the heat of formation becomes increasingly negative. This is readily observable in saturated hydrocarbons of the paraffin family, wherein the magnitude of heat of formation progressively increases from methane to kerosene (s. nos. 1 to 5 in Table 4.1). Similarly, if we compare the heat of formation of CO and CO_2 (s. nos. 15 and 16), we find the magnitude for CO_2 to be higher than CO as its molecular structure is more complex than CO.

Among the hydrocarbons, acetylene (C_2H_2) is seen to have a positive value of heat of formation. Acetylene is an alkyne with one triple bond and two single bonds. The positive value of the heat of formation indicates that heat has to be supplied to form acetylene from its elements C(s) and H_2 in the solid and gas phase, respectively.

Table 4.1 Heats of formation in kJ/mole

S. no.	Substance	Chemical formula	Phase	Heat of formation (kJ/mole)
1	Methane	CH_4	Gas	−74.9
2	Ethane	C_2H_6	Gas	−84.7
3	Propane	C_3H_8	Gas	−103.9
4	Butane	C_4H_{10}	Gas	−124.7
5	Kerosene	$C_{12}H_{26}$	Liquid	−292.9
6	Acetylene	C_2H_2	Gas	+226.7
7	Hydrazine	N_2H_4	Liquid	+50.3
8	Hydrogen peroxide	H_2O_2	Liquid	−187.8
9	Nitric acid	HNO_3	Liquid	−171.8
10	Nitroglycerine	$C_3H_5 (ONO_2)_3$	Liquid	−333.7
11	Ammonium nitrate	NH_4NO_3	Solid	−365.1
12	TNT	$C_7H_5N_3O_6$	Solid	−54.4
13	RDX	$C_3H_6N_6O_6$	Solid	+61.5
14	HMX	$C_4H_8N_8O_8$	Solid	+75
15	Carbon dioxide	CO_2	Gas	−393.5
16	Carbon monoxide	CO	Gas	−110.5
17	Water	H_2O	Liquid	−286.7

It therefore has more locked-up energy when compared to other hydrocarbons. As a result, it is very reactive and we shall deal with the acetylides later in the context of primary explosives. The other substances, shown in Table 4.1 having positive heats of formation, are hydrazine (+50.3 kJ/mole) and explosives RDX (+61.5 kJ/mole) and HMX (+75 kJ/mole). Substances having positive heats of formation tend to spontaneously react as compared to those having negative values of heats of formation.

4.1.2   Chemical Reactions and Energy Release

Consider the following two cases of chemical reactions. In the first one, a substance A undergoes a reaction to form products B and C as in an explosive

$$n_A A \rightarrow n_B B + n_C C \tag{4.2}$$

In the second one, the substance A reacts with another substance B, such as fuel and air, to form products C and D.

$$n_A A + n_B B \rightarrow n_C C + n_D D \tag{4.3}$$

Here n_A, n_B, n_C, and n_D denote the number of moles of the substances A, B, C, and D, respectively. If the heats of formation of n_A moles of A in Eq. 4.2 are greater than the combined values of heats of formation of n_B moles of B and n_C moles of C, energy is liberated during the chemical reaction. The energy liberated is called heat of combustion and is given by

$$\Delta H_C = - \left(n_B \times \Delta H_{fB}^0 + n_C \times \Delta H_{fC}^0 - n_A \times \Delta H_{fA}^0 \right) \tag{4.4}$$

Similarly, for reaction given by Eq. 4.3 the energy liberated is

$$\Delta H_C = - \left(n_C \times \Delta H_{fC}^0 + n_D \times \Delta H_{fD}^0 - \left\{ n_A \times \Delta H_{fA}^0 + n_B \times \Delta H_{fB}^0 \right\} \right) \tag{4.5}$$

The number of moles of the reactants and products and their heats of formation are therefore required to determine the energy release.

Let us illustrate the above with an example of butane gas reacting with oxygen to form combustion products, carbon dioxide and water. The chemical reaction is represented by

$$2C_4H_{10} + 13\,\Omega_2 \rightarrow 8CO_2 + 10\,H_2O \tag{4.6}$$

The energy released in the reaction is

$$\Delta H_C = -\left(n_{CO_2} \times \Delta H^0_{fCO_2} + n_{H_2O} \times \Delta H^0_{fH_2O} \right.$$
$$\left. - \left\{ n_{C_4H_{10}} \times \Delta H^0_{fC_4H_{10}} + n_{O_2} \times \Delta H^0_{fO_2} \right\} \right) \quad (4.7)$$

Substituting the number of moles from Eq. 4.6 and the values of the standard heats of formation from Table 4.1, we get the energy release as

$$\Delta H_C = 8 \times 393.5 + 10 \times 286.7 - 2 \times 124.7 = 5765.6 \text{ kJ}$$

The energy release in the reaction is therefore 5765.6 kJ. The mass of butane taking part in the reaction corresponds to 2 moles and is $2 \times (48 + 10) = 116$ g. The energy released per kg of butane is

$$\frac{5765.6}{0.116} = 49,703 \text{ kJ/kg}$$

The energy released per kg of the mixture of butane and oxygen is

$$\frac{5765.6}{0.116 + 0.416} = 10,838 \text{ kJ/kg}$$

Here 0.416 is the mass in kilogram of the 13 mol of oxygen taking part in the reaction.

4.1.3 Stoichiometry, Fuel-Rich and Fuel-Lean Compositions; Equivalence Ratio

The proportion of reacting substances, viz., the proportion of fuel and oxidizer, is referred to as stoichiometry. 'Stoichiometry' is derived from Greek words '*stoikheon*' meaning element and '*metria*' meaning amount. In practice, stoichiometry refers to the amount of fuel and oxidizer in a chemical reaction, which leads to completely oxidized products of combustion. By completely oxidized products of combustion, it is meant that no further oxidation of the products is possible. Carbon and hydrogen, when completely oxidized by oxygen, form carbon dioxide and water, respectively. The completely oxidized substances are very stable and have large negative values of heats of formation.

The reaction given by Eq. 4.6 is a stoichiometric reaction since completely oxidized products of combustion CO_2 and H_2O are formed.

A reaction between fuel and oxidizer need not always lead to the formation of completely oxidized products. If the amount of oxygen is less than that required for stoichiometric reaction, incompletely oxidized products such as CO or non-oxidized products such as C or H_2 could be formed. The composition is then said to be fuel-rich. However, if the amount of the oxidizer is more than what is required for stoichiometric reaction, completely oxidized products are formed and the excess unutilized oxygen is present in the combustion products. The composition is then said to be fuel-lean.

Most of the accidental explosions involving gaseous and volatile liquid fuels such as confined and unconfined fuel–vapor–air explosions and boiling liquid expanding explosions, considered in Chapter 1, involve the fuel mixing with air. The term equivalence ratio is defined as the ratio of the actual mass of fuel to air in the mixture formed to the mass of fuel to air required to form a stoichiometric mixture. The equivalence ratio is denoted by φ and is given as

$$\varphi = \left(\frac{\text{Mass of fuel} \div \text{Mass of air in the mixture}}{\text{Mass of fuel} \div \text{Mass of air for stoichiometric reaction}} \right)$$

An equivalence ratio $\varphi > 1$ implies a fuel-rich mixture, while $\varphi < 1$ denotes a fuel-lean or equivalently an oxygen-rich mixture. $\varphi = 1$ is a stoichiometric mixture.

Atmospheric air essentially contains 78.1% by volume of nitrogen and 21% oxygen and about 0.9% argon, CO_2, and other gases. Since argon and nitrogen are inert, air could be assumed to contain 79% by volume of nitrogen and 21% by volume of oxygen. From Avogadro's hypothesis that equal volume of all gases at the same temperature and pressure contains the same number of molecules, the number of moles of nitrogen per mole of oxygen in ambient air is 79/21 which equals 3.76. For all practical purposes, nitrogen is considered to be inert.

4.1.4 Energy Release in a Stoichiometric Mixture of Fuel Vapor and Air

The equation for the chemical reaction for a stoichiometric composition of butane–air mixture following Eq. 4.6 is:

$$2C_4H_{10} + 13\,O_2 + 3.76 \times 13 N_2 \rightarrow 8CO_2 + 10\,H_2O + 3.76 \times 13 N_2 \qquad (4.8)$$

The fuel–air ratio for the above reaction is

$$\left(\frac{F}{A} \right)_{\text{Stoich}} = \frac{2 \times (4 \times 12 + 10)}{13 \times 32 + 13 \times 3.76 \times 28} = 0.065$$

The heat released remains the same at 49,703 kJ/kg of butane (as obtained for the stoichiometric reaction of butane with oxygen) since the same number of moles of inert nitrogen is present in both the reactants and products. However, compared to the stoichiometric reaction with oxygen, the mass of the reacting mixture has gone up due to the presence of nitrogen (3.76×13 moles of nitrogen having mass of $3.76 \times 13 \times 0.028 = 1.368$ kg). The heat release per unit mass of the mixture therefore gets reduced and is $5765.5/(0.116 + 0.416 + 1.368) = 3033.5$ kJ/kg.

If we consider a stoichiometric cloud of $1\,m^3$ of butane–air mixture at 1 atm pressure and 25 °C, the energy release in the cloud can be determined either from the mass of butane or from the number of moles of butane in it. The partial volume of butane in $1\,m^3$ cloud from Eq. 4.8 is $2/(2 + 13 + 13 \times 3.76) = 0.031\,m^3$. Here we have used Amagat's law, which states that the volume of each of the species in a mixture of gases behaves as if it existed separately at the pressure and temperature of the mixture. In this particular example, the partial volume of butane in $1\,m^3$ of the mixture is $0.031\,m^3$. The number of moles of butane is determined from the partial volume of butane by using the equation of state for an ideal gas

$$pV = nR_0T$$

Here p and T are the pressure and temperature of the fuel–air mixture, R_0 is the universal gas constant $[= 8.314$ J/(mole K)$]$, and V is the volume of n moles of butane in the mixture. Taking atmospheric pressure as 10^5 Pa and temperature as 298 K, the number of moles of butane in the mixture of $1\,m^3$ is

$$n = \frac{10^5 \times 0.031}{8.314 \times 298} = 1.25 \text{ moles/m}^3 \tag{4.9}$$

Since for 2 moles of butane the energy released is 5765.6 kJ, the energy released in $1\,m^3$ of the butane–air mixture is

$$\frac{5765.5 \times 1.25}{2} = 3603.5 \text{ kJ/m}^3. \tag{4.10}$$

The energy released in $1\,m^3$ of the stoichiometric butane–air cloud is 3603.5 kJ.

4.1.5 Generalized Procedure for Determining Energy Release

In order to calculate the energy release, we need to know the reactants, the chemical species in the products of combustion, and the number of moles of the reactants and the products. If the chemical reaction is represented in a generalized form as

$$\sum_{i=1}^{N} n_i' M_i = \sum_{i=1}^{N} n_i'' M_i \tag{4.11}$$

where n_i' is the number of moles of species M_i appearing as reactants and n_i'' is the moles of species M_i in the products. The energy release in the reaction with the reactants being at the standard temperature and pressure is given from Eq. 4.7 as

$$\Delta H_C = - \left(\sum_{i=1}^{N} n_i'' \Delta H_{fM_i}^0 - \sum_{i=1}^{N} n_i' \Delta H_{fM_i}^0 \right) \qquad (4.12)$$

4.1.6   Influence of Variations in the Temperature And Pressure of the Reactants on Energy Release

The energy release ΔH_C determined in Eq. 4.12 raises the temperature of the products from the standard state of 25°C when the fuel and oxidizer are at the standard temperature of 25 °C. If the fuel and oxidizer are at a temperature different from 25 °C, it is necessary to first convert them to 25 °C so that the standard heats of formation can be used for determining ΔH_C. The change of temperature of the reactants to 25 °C requires heat to be either added or removed from it. If the temperature of the fuel and oxidizer is less than 25°C, part of the energy released at the standard state must contribute to enhance the temperature of the reactants to 25 °C. The net energy release therefore decreases by an amount equal to the enthalpy required to increase the temperature of the reactants to the standard temperature of 25 °C. Similarly, if the reactants are at temperatures in excess of 25 °C, the enthalpy of the reactants in excess of 25 °C is additionally available as energy release.

The standard heats of formation are given at a pressure of 1 atm. For an ideal gas, variations do not arise in the enthalpy from changes of pressure and the same values of the standard heats of formation can be used.

When chemical reactions take place at constant volume, changes of internal energy instead of enthalpy need to be considered. Considering that the enthalpy H is related to its internal energy U by the relation $H = U + pV$, only part of the enthalpy contributes to the change of internal energy. For constant volume combustion, it would be necessary to consider the internal energy changes and not the enthalpy changes. An internal energy of formation is evaluated as

$$\Delta U_f^0 = \Delta H_f^0 - R_0 T$$

and used instead of the heat of formation. The Gurney constant, discussed in Chap. 2, is considered the internal energy change, as the explosion was confined to the volume from which the fragments originated.

4.1.7 Energy Release for Fuel-Lean ($\varphi < 1$) And Fuel-Rich ($\varphi > 1$) Compositions

Fuel-lean Compositions

In the case of fuel-lean mixtures, the number of moles of oxygen available for the reaction is more than required for a stoichiometric reaction. The excess oxygen would therefore be available in the products along with the completely oxidized combustion products. This is illustrated in the following example, wherein instead of the stoichiometric reaction of butane with air, given in Eq. 4.8, two extra moles of oxygen and associated increase in nitrogen are considered. The chemical reaction is represented by

$$2C_4H_{10} + 15\,\Omega_2 + 3.76 \times 15N_2 \rightarrow 8CO_2 + 10\,H_2O + 2O_2 + 3.76 \times 15N_2 \tag{4.13}$$

Since only 13 mol of oxygen is required for complete combustion, the balance of 2 moles of oxygen is left over in the products. The fuel–air ratio corresponding to the above mixture is

$$\frac{F}{A} = \frac{2 \times 58}{15(32 + 3.76 \times 28)} = 0.0563 \tag{4.14}$$

The equivalence ratio φ is $0.0563/0.065 = 0.87$. Here 0.065 in the denominator is the fuel–air ratio for the stoichiometric composition.

Since oxygen gas is a reference element with the standard heat of formation of zero, the energy released in the reaction remains the same as for the stoichiometric reaction, viz., 5765.6 kJ for 2 moles of butane. The energy release per unit mass of the mixture is $5765.6/(2 \times 0.058 + 15 \times 0.032 + 15 \times 3.76 \times 0.028) = 2651$ kJ/kg of mixture. This is less than the value of 3033.5 kJ/kg of mixture for stoichiometric composition. Similarly, the heat released in a unit volume of the cloud is less than for a stoichiometric mixture. The energy released per unit volume or mass of mixture from the chemical reactions progressively decreases as the mixture becomes more fuel-lean; i.e., the equivalence ratio increasingly diminishes from the stoichiometric value of 1.

Fuel-rich Compositions

The amount of oxygen available is less than required to form completely oxidized products in fuel-rich mixtures. Under-oxidized products such as carbon monoxide and non-oxidized substances like carbon and hydrogen could result depending on the reduced quantity of oxygen that is available. It is difficult to determine the moles of the different substances in the products as the number of substances is more than the number of atom balance equations between the reactants and the products. The different substances and their amounts depend on whether they can coexist in equilibrium with each other. The chemical equilibrium of different species and their number of moles are to be determined at the temperature of the products and are addressed in Annexure E. However, the procedure is rather cumbersome and a simplified method based on the higher reactivity of hydrogen compared to carbon has been found to give fairly reasonable assessment of the moles of different species and the energy released in explosive mixtures comprising carbon, hydrogen, nitrogen, and oxygen.

In this simplified method, all nitrogen in the reactants is assumed to be inert and forms N_2 in the products. The oxygen first oxidizes the more reactive hydrogen to water. The leftover oxygen oxidizes carbon to carbon monoxide. If oxygen is still left, the balance oxygen converts part of the carbon monoxide to carbon dioxide. In case sufficient oxygen is not available for converting C(s) to CO, C(s) is left as soot in the products. The following illustrates the method for a fuel-rich butane–air mixture.

Consider the reaction of 2 moles of butane reacting with 12 moles of oxygen. The reactants consist of

$$2C_4H_{10} + 12O_2 + 3.76 \times 12N_2 \tag{4.15}$$

The fuel–air ratio of the mixture is: $2 \times 58/(12 \times 32 + 12 \times 3.76 \times 28) = 0.0704$. The equivalence ratio is $0.074/0.065 = 1.083$. For the reactants given by Eq. 4.15, the 20 atoms of H consume $5\,\Omega_2$ to form $10\,H_2O$. Four of the balance O_2 of $7\,\Omega_2$ react with 8C to form 8CO. The remaining $3O_2$ reacts with 6CO to form $6CO_2$. The balance 2CO cannot be oxidized. Nitrogen is left as it is in the products. The reaction is therefore written as

$$2C_4H_{10} + 12O_2 + 3.76 \times 12N_2 \rightarrow 10\,H_2O + 6\,CO_2 + 2CO + 3.76 \times 12N_2 \tag{4.16}$$

Under-oxidized CO is formed due to the reactant mixture being fuel-rich. If the mixture is considerably fuel-rich and is given as

$$2C_4H_{10} + 6\,\Omega_2 + 3.76 \times 6N_2 \tag{4.17}$$

the fuel–air ratio of the mixture is $2 \times 58/(6 \times 32 + 3.76 \times 6 \times 28) = 0.141$ and the equivalence ratio is $0.141/0.065 = 2.17$. In this case, 20 H combines with $5O_2$ to form $10\,H_2O$. The remaining $1O_2$ reacts with 2C (of the 8C available) to form 2CO. The remaining 6C is left in the product. The chemical reaction is therefore given as

$$2C_4H_{10} + 6\,\Omega_2 + 3.76 \times 6N_2 \rightarrow 10\,H_2O + 2CO + 6\,C(s) + 3.76 \times 5N_2 \tag{4.18}$$

The energy release in the reaction given by Eq. 4.16 ($\varphi = 1.083$) is

$$-[10 \times (-286.7) + 6 \times (-393.5) + 2 \times (-110.5) - 2 \times (-124.7)] = 5200 \text{ kJ}.$$

For the equivalence ratio $\varphi = 2.17$ corresponding to Eq. 4.18, the energy release is

$$-[10 \times (-286.7) + 2 \times (-110.5) - 2 \times (-124.7)] = 2838.6 \text{ kJ}.$$

In the above, the standard heat of formation of C(s) is taken as zero since it is a standard element.

Rather than forming carbon or soot, part of the hydrocarbon C_4H_{10} might as well remain unburned in practice. However, the above procedure gives reasonable estimates of energy released in fuel-rich reactions.

We observe that for fuel-rich mixtures, the heat release in the reaction is lower than for stoichiometric and fuel-lean compositions in view of the formation of under-oxidized and non-oxidized species. This is qualitatively indicated in Fig. 4.1, wherein variation of the energy release per unit mass of the fuel is plotted as a function of the equivalence ratio. The energy release decreases for fuel-rich mixtures per unit mass of the fuel; however, for fuel-lean mixtures, the energy release per unit mass of fuel does not change as completely oxidized products of combustion are formed. If the energy release per unit mass or per unit volume of the mixture is plotted, the energy is a maximum at stoichiometry ($\varphi = 1$) and decreases as the equivalence ratio drops below 1 or increases above 1. This is illustrated in Fig. 4.2. The small values of heat release in the mixture for very lean and very rich fuel–air mixture cannot sustain the combustion. We shall deal with this in Chap. 6.

Fig. 4.1 Variations of energy release per unit mass of fuel with equivalence ratio

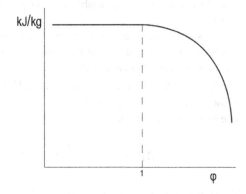

Fig. 4.2 Variations of energy release per unit mass of the mixture of fuel and air with equivalence ratio

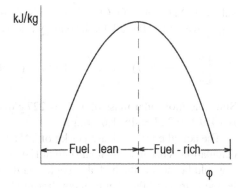

Heat Release from Condensed Phase Explosives

Substances in which the fuel and oxidizer are inbuilt as a single mass are known as explosives. The mixture of butane and air together, considered in the last section, constitutes an explosive, while the butane gas by itself is a fuel. An explosive could be a gas mixture, a liquid, or a solid. The energy release is determined for stoichiometric, fuel-lean, and fuel-rich conditions in the last section and was illustrated for the butane–air mixture.

Many solid and liquid explosives are fuel-rich, and we shall deal with them in subsequent chapters. We consider in the following the example of trinitrotoluene (TNT), which is a powerful explosive and is used as a standard to evaluate blast damages. TNT is derived from toluene (methyl benzene) in which three NO_2 radicals are substituted for three OH radicals to give $CH_3C_6H_2(NO_2)_3$. The chemical formula for TNT is therefore $C_7H_5(NO_2)_3$. The oxygen present in TNT reacts with H and C within it to release the chemical energy.

In order to determine the energy release, we need to know whether the composition is stoichiometric, fuel-lean, or fuel-rich. We have in one mole of TNT 6 atoms of O, 5 of H, and 7 of C. For stoichiometric composition, 5H requires $2\frac{1}{2}$O while 7C requires 14 of O calling for a total of $16\frac{1}{2}$O to form completely oxidized products H_2O and CO_2. However, the available O is only 6. The composition of TNT is therefore fuel-rich.

The simple procedure given in the last section for fuel-rich mixtures is followed to determine the moles of oxidized, under-oxidized, and non-oxidized products of combustion. The 3N in one mole of TNT becomes available in the products as $1\frac{1}{2}N_2$. The 6O oxidizes 5H to $2\frac{1}{2}$ H_2O leaving $3\frac{1}{2}$O. Of the 7C in one mole of TNT, $3\frac{1}{2}$CO is formed and $3\frac{1}{2}$C is left behind as carbon in the products. No CO_2 is generated. The reaction is given by

$$C_7H_5(NO_2)_3 = 2\frac{1}{2}H_2O + 3\frac{1}{2}CO + 3\frac{1}{2}C(s) + 1\frac{1}{2}N_2 \qquad (4.19)$$

The standard heat of formation ΔH_f^0 of TNT is -54.4 kJ/mole (Table 4.1). With the values of ΔH_f^0 for H_2O, CO, C(s), and N_2 being -286.7, -110.5, 0, and 0 kJ/mole, respectively (see Table 4.1), the energy release for Eq. 4.19 is

$$-\left[\frac{5}{2} \times (-286.7) + \frac{7}{2} \times (-110.5) - (-54.4)\right] = 1049.1 \text{ kJ/mole TNT.} \quad (4.20)$$

Since the molecular mass of TNT is 227 g/mole, the energy release per kg of TNT is $1049.1/0.227 = 4621.6$ kJ/kg.

The actual energy release per kg of TNT is 4520 kJ/kg. The over-estimate in the calculation is because the final state of H_2O is assumed as water, whereas in practice it would be steam. We had also neglected the dissociation of the products at higher temperatures and the chemical equilibrium between the products. The estimates are however seen to be reasonable.

For an accurate assessment of the number of moles of the different species in the products, their thermodynamic equilibrium at the temperature of the combustion products must be determined. A rigorous procedure for calculating the oxidized, under-oxidized, and non-oxidized species in the products is given in Appendix E. Since the temperature of the products of combustion of the fuel-rich explosives is not very high, significant dissociation of the species in the products is not to be expected.

4.2 Rate of Energy Release

4.2.1 Concentration, Law of Mass Action, And Activation Energy

While determining the energy release in the last section, we addressed the moles of substances formed in the products and their heats of formation. However, we have not addressed the rate at which a chemical reaction takes place, which would influence the rate of energy release. The rate of chemical reactions depends on the rate of collision of the individual molecules and their energy levels before the collision process, since the chemical structure of the molecules changes during the reaction. The momentum of the colliding molecules (reactants) must be high enough to break its structure upon collisions and form a different chemical structure corresponding to the products. The collisions must also be frequent so that higher rates of chemical reactions (and the associated high values of rate of energy release) are obtained as in an explosion.

As the number of molecules in a given volume increases, the possibility of collisions between molecules increases. Instead of considering moles of a reactant, as done for energy release, it becomes necessary to consider the moles per unit volume. The number of moles per unit volume is called as concentration and is denoted by C moles/cc.

The rate of chemical reactions is given by the law of mass action. The law is based on experimental observations and states that the rate of a chemical reaction varies directly as the active concentration of the reactants. If we consider the reactants as fuel F and oxidizer O and represent their concentration by C_F and C_O, respectively, the rate of consumption of the reactants (dC_F/dt) in moles/(cc-s) is

$$-\frac{dC_F}{dt} = kC_FC_O \tag{4.21}$$

Here k is a constant called as reaction rate coefficient or reaction rate constant.

In order that a reaction takes place, it is necessary to provide sufficient energy to the molecules. Consider, for example, a well-mixed mixture of gaseous hydrogen and gaseous oxygen in stoichiometric proportion ($H_2 + \frac{1}{2} O_2$) kept at room temperature and atmospheric pressure in a container. The mixture can be kept indefinitely long

Fig. 4.3 Progress of a
reaction

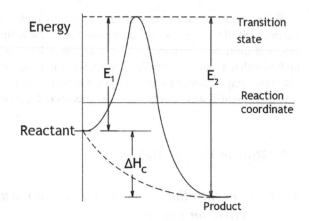

and will not combine to form water (H_2O) unless an electric spark, heat, or a catalyst
is added to it. Similarly, a solid explosive such as TNT can be stored for a long time at
the ambient temperature. The products are not produced automatically as indicated
by the dotted line in a plot of energy versus an arbitrary reaction coordinate in Fig. 4.3.
The energy released in the reaction is ΔH_C. Energy is however required to initiate
or activate the reaction. The energy for activation of the reaction, called as activation
energy, causes the formation of a transition state or an excited state for the reaction
to progress. The activation energy is shown by E_1 in Fig. 4.3. If the reaction were
to take place in the reverse direction and the products were to form the reactants,
the activation energy is seen to be E_2 from Fig. 4.3. The heat of reaction ΔH_C is the
difference between the activation energies E_2 and E_1 giving

$$\Delta H_C = E_2 - E_1 \quad \frac{kJ}{mole} \tag{4.22}$$

Smaller the value of the activation energy, the easier it is for a reaction to overcome
the energy barrier and form products.

4.2.2 Arrhenius Rate Equation

The reaction rate depends on the frequency of collisions between molecules and the
availability of the activation energy for the reaction to proceed. Geometric alignment
of the molecules provides better impact and is important. The rate constant for a
reaction (k in Eq. 4.21) is therefore of the form

$$k = A \exp(-E_a/(R_0 T)) \tag{4.23}$$

Here A is the pre-exponential factor, which considers the geometrical factor associated with the colliding molecules and the frequency of collisions. The exponential term contains the activation energy E_a. The activation energy is denoted by E_a instead of E_1 considered earlier for the forward reaction. The pre-exponential factor A has the same units as the rate constant k, whereas the activation energy has units of kJ/mole corresponding to the product $R_0 T$. The form of Eq. 4.23 is known as Arrhenius rate equation.

4.2.3 Rate of Chemical Reactions and Rate of Energy Release

The rate at which the concentration of the reactant decreases or equivalently, the rate at which the concentration of the products of combustion increases in kmoles/(m^{3}s) or moles/(cc-s) can be expressed by inserting the Arrhenius form of the rate equation (Eq. 4.23) in the law of mass action (Eq. 4.21). The reaction rate ω in moles/(cc-s) is given as

$$\omega = -\frac{dC_F}{dt} = AC^n \exp\left(-\frac{E_a}{R_0 T}\right) \qquad (4.24)$$

Here A is a new constant and C is the concentration. For most hydrocarbon–air mixture, the value of the exponent n is about 2. From ideal gas equation $pV = nR_0 T$ and the concentration $C = n/V = p/(R_0 T)$. The concentration is therefore proportional to pressure, and it is possible to write the reaction rate as

$$\omega = Bp^n \exp\left(-\frac{E_a}{R_0 T}\right) \quad \text{moles/(cc-s)} \qquad (4.25)$$

Here B is a constant.

Most of the substances that we shall deal with in explosions have activation energies between 80,000 and 160,000 J/mole. Since the gas constant $R_0 = 8.314$ J/(mole K), a change of temperature from say 500 to 1000 K will increase the overall reaction rate ω by four orders of magnitude if the activation energy is 80,000 J/mole and by about nine orders of magnitude for activation energy of 160,000 J/mole. The strong dependence of the reaction rate on the temperature is illustrated in Fig. 4.4. Temperature plays a crucial role in deciding the rate of the reaction when the activation energy is large.

The energy release from a chemical reaction was obtained per mole of reactant in Sect. 4.1 for fuel-lean, fuel-rich, and stoichiometric compositions. Denoting it by q kJ/mole, the rate of energy release per unit volume of the mixture is obtained using Eq. 4.24 as

$$\dot{q} = AqC^n \exp\left(-\frac{E_a}{R_0 T}\right) \left[\frac{kJ}{mole} \frac{mole}{cc-s} = \frac{kJ}{cc-s}\right] \qquad (4.26)$$

Fig. 4.4 Temperature
dependence of reaction
between fuel and oxidizer

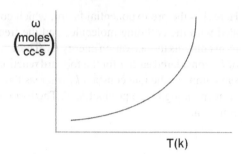

Fig. 4.5 Rate of energy
release per unit volume

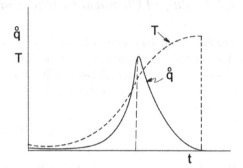

We shall use the above volumetric rate of heat release dependence on temperature to formulate the thermal theory of explosions in the next chapter.

Since the rate of reaction ω increases sharply with temperature (Fig. 4.4) and the temperature increases with time, the rate of reaction will increase sharply with time. Runaway or self-accelerating chemical reactions take place. However, the concentration of the fuel falls as the reaction progresses and this will lead to a drop in the rate of the reaction. When all the reactants are consumed, the rate of the reaction becomes zero. Therefore, the rate of heat release as a function of time t follows the shape given in Fig. 4.5. When all the fuel is consumed, the temperature is a maximum since all the energy is released. However, by then, the rate of energy release drops to zero since the chemical reaction ceases. The rate of energy released by the chemical reaction therefore has the characteristic feature discussed in Chap. 2 for explosions and is reproduced in Fig. 4.6.

Examples

4.1 *Hydrogen–Air Explosion:* Hydrogen is being increasingly used as a green fuel for different applications including automobiles. A major concern in using hydrogen is the hazard associated with fire and explosion. Assume a hydrogen car to be parked in a closed garage of size 4 m × 4 m × 6 m. If the hydrogen leaks from the hydrogen storage tank of the car and forms a stoichiometric mixture at the ambient pressure of 0.1 MPa and ambient temperature of 25 °C in the garage, determine the following:

Fig. 4.6 Energy release rate
as a function of time

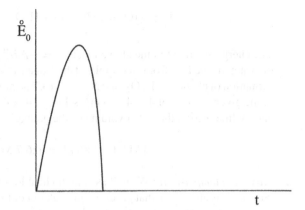

(a) Concentration of hydrogen in kmoles/m^3 in the garage
(b) Energy liberated on ignition of the mixture
(c) Maximum temperature reached
(d) Maximum pressure in the garage assuming that gases do not leak out. You
    can neglect the volume occupied by the car in the garage. The standard heat
    of formation of water is -286 kJ/mole. The boiling temperature of water is
    100 °C, and the latent heat of water is 40 kJ/mole at the ambient pressure.
    The specific heat of water is 90 J/(mole K), and specific heat of steam is
    58 J/(mole K). The specific heat of nitrogen is 36 J/(mole K).

Solution

(a) *Concentration of hydrogen:* It is given that a stoichiometric mixture of hydrogen
    and air is formed in the garage. For stoichiometric composition, fully oxidized
    water is formed as combustion product as per the reaction:

$$H_2 + \frac{1}{2}(O_2 + 3.76N_2) \rightarrow H_2O + 1.88N_2$$

Moles of oxygen and nitrogen in one mole of hydrogen $= 4.76/2 = 2.38$. Moles
of mixture: $2.38 + 1 = 3.38$. At the standard temperature of 25 °C and 0.1 MPa,
one mole of the ideal gas would occupy a volume given by the ideal gas equation
of state
$$v = \frac{1 \times R_0 \times T}{p} = \frac{8.314 \times 298}{10^5} = 0.0248 \text{ m}^3$$

Volume occupied by 3.38 moles of mixture $= 0.0248 \times 3.38$ m^3. In the above
volume, we have 1 mole of hydrogen. Concentration of hydrogen $= 1/(0.0248 \times 3.38)$ moles/m$^3 = 11.9$ mol/m$^3 = 0.0119$ kmoles/m^3

(b) *Energy liberated by the mixture in the garage:* The chemical reaction for the
    stoichiometric mixture is given by

$$H_2 + \frac{1}{2}(O_2 + 3.76N_2) \rightarrow H_2O + 1.88N_2$$

The energy liberated in the above reaction $= -\Delta H^0_{fH_2O} = 286$ kJ for one mole of water formed or from one mole of hydrogen consumed since the heats of formation of elements H_2, O_2, and N_2 are zero. The number of moles of hydrogen in the given volume of $4 \times 4 \times 6$ m^3 is $11.9 \times 4 \times 4 \times 6 = 1142.4$ moles. The energy liberated in the given volume of the garage:

$$1142.4 \times 286 \text{ kJ} = 326.7 \text{ MJ}.$$

(c) *Maximum temperature:* We will assume no heat loss from the products of combustion. It is also given that gases do not leak out of the garage so that the energy generated by combustion is conserved within the garage. Denoting the maximum temperature as T_f, we have from energy balance the energy required to raise the temperature to T_f as equal to the energy liberated in the reaction.

$$n_{H_2O}\left[C_{P,H_2O(l)}(T_B - 25) + L + C_{P,H_2O(g)}(T_f - T_B)\right]$$
$$+ n_{N_2} C_{P,N_2}(T_f - 25) = 286$$

Here n_{H_2O} and n_{N_2} refer to the moles of water/steam and nitrogen, T_B the boiling temperature of water ($= 100°C$), and C_P the specific heat. Subscripts '*l*' and '*g*' denote the liquid and gas phase, respectively. L is the latent heat. Substituting the values, we obtain

$$1 \times [0.090 \times 75 + 40 + 0.058 \times (T_f - 100)]$$
$$+ 1.88 \times 0.036(T_f - 25)] = 286$$

The above equation reduces to

$$0.126T_f + 39.26 = 286$$

The value of maximum temperature $= 246.74/0.126 = 1958\,°C$. We have neglected the dissociation of gases at high temperatures, and this value would therefore be an over-estimate.

(d) *Maximum pressure due to the combustion of hydrogen in the garage:* Assuming the ideal gas to hold for the products, the equation of state is $pV = nR_0T_f$. Here n represents the moles of combustion products in the garage, while T_f is the maximum temperature. R_0 is the universal gas constant $= 8.314$ J/(mole K). V is the volume of the garage, and p is the maximum pressure in the garage. The number of moles of combustion products consists of one mole of steam and 1.88 mol of nitrogen for every mole of hydrogen consumed. The number of moles of hydrogen in the garage is 1142.4, and this gives the total moles of combustion products to be $1142.4 \times (1 + 1.88) = 3290$. The maximum pressure is

$$p = \frac{nR_0 T_f}{V}$$

$$= \frac{3290 \times 8.314 \times (1958 + 273)}{(4 \times 4 \times 6)} = 635700 \text{ Pa} = 0.64 \text{ MPa}$$

which is seen to be about six and half times the initial ambient value of 0.1 MPa.

4.2 *Methane–Air Explosion:*

(a) A volume of 1 m^3 of methane gas at ambient temperature (25 °C) and pressure (0.1 MPa) escapes from a digester into the ambient atmosphere at 25 °C and 0.1 MPa and forms a cloud of methane–air mixture of volume 12 m^3. Determine the concentration of methane in the cloud in moles/cc and the equivalence ratio of the cloud.

(b) If the above cloud burns, determine the composition of burned gases in the cloud as volume and mass fractions.

(c) Determine the energy released in combustion of the cloud. The standard heat of formation of methane, carbon dioxide, and water is -75, -394, and -287 kJ/mole, respectively.

Solution

(a) • *Concentration*: From Avogadro's law, we know that equal volumes of all gases at the same temperature and pressure contain the same number of moles. Further, 1 mole of any gas has a volume of $(8.314 \times 298)/10^5 = 0.0248 \text{ m}^3$ at the standard temperature of 25 °C and 0.1 MPa pressure. The mixture of methane and air formed in the cloud consists of 1 m^3 of methane and 11 m^3 of air at 0.1 MPa and 25 °C. The number of moles of methane and air is therefore in the proportion 1:11. The composition of the methane–air mixture could be represented by

$$CH_4 + a(O_2 + 3.76N_2)$$

where a is the proportion of air per mole of methane. We have for the given molar ratio of 1:11

$$a(1 + 3.76) = 11$$

This gives $a = 11/4.76 = 2.31$. The molar composition of the reactive cloud is therefore given by

$$CH_4 + 2.31(O_2 + 3.76N_2)$$

1 mole of methane gas at 0.1 MPa and 25 °C has a volume of $= 0.0248 \text{ m}^3$. Therefore, the given volume of 1 m^3 of methane at 0.1 MPa and 25°C has $1/0.0248 = 40.32$ moles. These moles are contained in the cloud of volume 12 m^3. The concentration of methane is therefore $40.32/12 = 3.36 \text{ mole/m}^3$ $= 3.36 \times 10^{-06}$ moles/cc.

• *Equivalence ratio*: The stoichiometric reaction (corresponding to formation of completed products of combustion comprising CO_2 and H_2O) for methane–air is

$$CH_4 + 2(O_2 + 3.76N_2) \rightarrow CO_2 + 2H_2O + 2 \times 3.76N_2$$

The fuel–air ratio by mass for the above stoichiometric reaction is $(12 + 4)/[(2 \times 32) + (2 \times 3.76 \times 28)] = 0.0583$. The actual composition of the cloud gives the reaction:

$$CH_4 + 2.31(O_2 + 3.76N_2)$$

The fuel–air ratio for the above reaction is $(12 + 4)/[(2.31 \times 32) + (2.31 \times 3.76 \times 28)] = 0.0504$.

$$\text{The equivalence ratio } \varphi = \frac{(\text{Fuel}/\text{Air})_{\text{Actual}}}{(\text{Fuel}/\text{Air})_{\text{Stoichiometric}}}$$
$$= \frac{0.0504}{0.0583}$$
$$= 0.86$$

Since equivalence ratio is less than 1, the composition is fuel-lean.

(b) *Composition of combustion products*: The lean mixture results in completed combustion products with excess oxygen left over in the combustion products. The reaction is given as

$$CH_4 + 2.31(O_2 + 3.76N_2) \rightarrow$$
$$CO_2 + 2H_2O + 2.31 \times 3.76N_2 + 0.31O_2$$

Volumetric composition of the combustion products is the same as the molar composition and is

$$CO_2 : H_2O : N_2 : O_2 = 1 : 2 : 2.31 \times 3.76 : 0.31$$

Here we have assumed that H_2O exists in the vapor phase. The volumetric composition of the products is

$$CO_2 : H_2O : N_2 : O_2 = 0.083 : 0.167 : 0.724 : 0.026$$

The composition by mass is

$$CO_2 : H_2O : N_2 : O_2 =$$
$$(0.083 \times 44) : (0.167 \times 18) : (0.724 \times 28) : (0.026 \times 32)$$

The composition by mass of the products is therefore

$$CO_2 : H_2O : N_2 : O_2 = 0.13 : 0.11 : 0.73 : 0.03$$

(c) *Energy from the cloud*: The energy release in the reaction

$$CH_4 + 2.31(O_2 + 3.76N_2) \rightarrow$$
$$CO_2 + 2H_2O + 2.31 \times 3.76N_2 + 0.31O_2$$

is

$$\Delta H_C = - \left\{ n_{CO_2} \Delta H^0_{fCO_2} + n_{H_2O} \Delta H^0_{fH_2O} - n_{CH_4} \Delta H^0_{fCH_4} \right\}$$

Here n refers to the number of moles of the particular specie and ΔH^0_f is the standard heat of formation. The standard heats of formation of H_2 and O_2, being elements, are zero. Substituting the number of moles from the chemical reaction and the values of heats of formation given in the data, we get

$$\Delta H_C = - \{1 \times (-394) + 2 \times (-287) - 1 \times (-75)\}$$
$$= 893 \text{ kJ per mole of methane2}$$

The number of moles of methane in the cloud was determined earlier to be $= 40.32$. The energy generated by combustion of cloud $= 40.32 \times 893$ kJ $= 36{,}006$ kJ.

4.3 *Energy release in explosive RDX*:

(a) The explosive RDX (trimethylene trinitramine) has the molecular formula $(CH_2)_3(NNO_2)_3$. Determine through appropriate analysis whether the composition of RDX is fuel-rich, oxidizer-rich, or stoichiometric.

(b) If the standard heat of formation of RDX is $+61.5$ kJ/mole and the standard heats of formation of water $= -286.7$ kJ/mole, carbon monoxide $= -110.5$ kJ/mole, and carbon dioxide $= -393.5$ kJ/mole, determine the energy released in the explosion of 5 kg of RDX.

Solution

(a) *Fuel-rich or oxygen-rich?* RDX $[(CH_2)_3(NNO_2)_3]$ has the chemical formula $C_3H_6N_6O_6$. In order to obtain completed products of combustion comprising CO_2 and H_2O, we need $(3 + 6/4)$ of O_2, i.e., 9 atoms of O. However, we have only 6 atoms of O in the composition of RDX. Hence completed combustion products cannot be formed and the composition of RDX is fuel-rich.

(b) *Composition of products and energy release:* Of the 6Ω available in RDX $[C_3H_6N_6O_6]$, 3Ω is used in oxidizing the reactive 6H in the explosive to form $3H_2O$. The balance 3Ω oxidizes the 3C in RDX to form 3CO. There is no balance O to oxidize either part or total amount of CO to CO_2. The 6N is available in the products as $3N_2$. The fuel-rich reaction is

$$C_3H_6N_6O_6 \rightarrow 3H_2O + 3CO + 3N_2$$

The energy liberated in the reaction:

$$\Delta H_C = - \left\{ n_{CO}\Delta H_{fCO}^0 + n_{H_2O}\Delta H_{fH_2O}^0 - n_{RDX}\Delta H_{fRDX}^0 \right\}$$

Substituting the values, we get

$$\Delta H_C = -\{3 \times (-110.5) + 3 \times (-286.7) - 1 \times (+61.5)\}$$
$$= 1253.1 \text{ kJ per mole of RDX}$$

However, the molecular mass of RDX $= 3 \times 12 + 6 \times 1 + 6 \times 14 + 6 \times 16 = 222$ g/mole. Therefore, energy liberated in 5 kg of RDX $= (1253.1/0.222) \times 5 = 28,223$ kJ or 28.22 MJ. Energy released in explosion of 5 kg of RDX is 28.22 MJ.

4.4 *Hydrazine Explosion in a Spy Satellite:* 450 kg of explosive hydrazine (N_2H_4) was left behind in a US spy satellite orbiting the Earth at a height of 247 km above the Earth's surface. The entire 450 kg of hydrazine was detonated by impacting it with a missile. Determine the energy released in the explosion in MJ. You can assume that hydrazine explodes as per the reaction

$$3N_2H_4 \rightarrow 4NH_3 + N_2$$

The heat of formation under standard conditions for hydrazine and ammonia is 50 kJ/mol and -81.17 kJ/mol, respectively. **Solution:** Energy release for the chemical reaction

$$3N_2H_4 \rightarrow 4NH_3 + N_2$$

is given by

$$\Delta H_C = - \left\{ n_{NH_3}\Delta H_{fNH_3}^0 - n_{N_2H_4}\Delta H_{fN_2H_4}^0 \right\}$$
$$\Delta H_C = -\{4 \times (-81.17) - 3 \times (+50)\} - 474.68 \text{ kJ}$$

for 3 mol of hydrazine. Molecular mass of hydrazine $M = (2 \times 14 + 4 \times 1) = 32$ g/mole. Therefore,

$$\text{Heat released for 450 kg of hydrazine} = \frac{474.68}{3 \times 0.032} \times 450 = 2225 \text{ MJ}$$

The energy released in the explosion is 2225 MJ.

4.5 *Energy Release in a Lean Propane–Air Cloud*: A lean propane (C_3H_8)–air cloud is formed in ambient atmosphere on a cold winter night when the temperature is 2 °C. The cloud consists of 2% by volume of propane. Determine

(a) equivalence ratio of the mixture in the cloud and
(b) energy released per unit volume of the cloud.

The standard heats of formation of propane, carbon dioxide, and water are -104, -394, and -287 kJ/mole, respectively. The specific heat at constant pressure for propane, oxygen, and nitrogen can be assumed to be independent of temperature and to be given as 75, 290, and 290 J/(mole K), respectively.

Solution

(a) *Equivalence ratio of the mixture*: The volume fraction of propane in the fuel-lean cloud is given as 0.02. Since we are considering a fuel-lean mixture, completed products of combustion comprising CO_2 and H_2O will be formed and some O_2 along with N_2 will be left in the combustion products. The reaction of 1 mole of propane can therefore be written as

$$C_3H_8 + a(O_2 + 3.76N_2) \rightarrow 3CO_2 + 4H_2O + (2a - 5)O_2 + 3.76aN_2$$

The value of a in the above equation is determined from the given value of volume fraction of propane in the cloud. The volume fraction is the same as the molar fraction ($= 0.02$). Hence:

$$\frac{1}{(1 + a + 3.76a)} = 0.02$$

The value of a works out to be $49/4.76 = 10.3$. The reaction between propane and air is therefore

$$C_3H_8 + 10.3O_2 + 38.7N_2 \rightarrow 3CO_2 + 4H_2O + 5.3O_2 + 38.7N_2$$

The fuel–air ratio (by mass) $= (3 \times 12 + 8 \times 1)/[(10.3 \times 32) + (38.7 \times 28) = 44/1413.2 = 0.031$. The stoichiometric reaction between propane and air is

$$C_3H_8 + 5O_2 + 18.8N_2 \rightarrow 3CO_2 + 4H_2O + 18.8N_2$$

The fuel–air ratio (by mass) for the above stoichiometric reaction is

$$\frac{(3 \times 12 + 8 \times 1)}{[(5 \times 32) + (18.8 \times 28)]} = \frac{44}{1413.2} = 0.064$$

The equivalence ratio

$$\varphi = \frac{(\text{Fuel/Air})_{\text{Actual}}}{(\text{Fuel/Air})_{\text{Stoichiometric}}}$$
$$= \frac{0.031}{0.064}$$
$$= 0.48$$

The equivalence ratio of the mixture is 0.48.

(b) *Energy released per unit volume of the cloud*: The standard heats of formation can be used only when the reactants and products are at the standard temperature of 25°C. If the reactants comprising $C_3H_8 + 10.3O_2 + 38.7N_2$ were to be heated up by the combustion products to 25°C, the energy released at the standard temperature of 25°C is given by

$$\Delta H_C = -\left\{ n_{CO_2} \Delta H^0_{fCO_2} + n_{H_2O} \Delta H^0_{fH_2O} - n_{C_3H_8} \Delta H^0_{fC_3H_8} \right\}$$

Hence for the reaction

$$C_3H_8 + 10.3O_2 + 38.7N_2 \rightarrow 3CO_2 + 4H_2O + 5.3O_2 + 38.7N_2$$

$\Delta H_C = -\{3 \times (-394) + 4 \times (-287) - 1 \times (-104)\} = 2226$ kJ/mole of propane. However a part of this energy contributes to increase the temperature of 1 mole of propane and the associated air from 2 °C to 25 °C $= 1 \times 0.075 \times (25 - 2) + 10.3 \times 0.29 \times (25 - 2) + 38.7 \times 0.29 \times (25 - 2) = 328.6$ kJ. This energy is used from the combustion energy release of 2226 kJ per mole of propane. Therefore, the energy release from combustion of the low-temperature cloud $= 2226 - 328.6 = 1897.4$ kJ per mole of propane. The partial volume of propane in $1\,m^3$ of cloud at 0.1 MPa pressure and 2°C is given to be $0.02\,m^3$. The moles of propane n are determined from the ideal gas equation $pV = nR_0T$. Here R_0 is the universal gas constant $= 8.314$ J/(mole K). Substituting the value of $V = 0.02, p = 10^5$ Pa, and $T = 273 + 2 = 275$ K, we get the number of moles of propane as $n = 10^5 \times 0.02/(8.314 \times 275) = 0.875$ moles. Energy released from $1\,m^3$ of the fuel-lean cloud of propane $= 0.875 \times 1897.4 = 1660$ kJ.

4.6 *Energy release from a fuel-rich mixture of butane and air:* Butane gas (C_4H_{10}) is used for cooking purposes in our homes. Determine the maximum energy release per unit mass of butane and the energy release for a fuel-rich mixture containing 4% by volume of butane in the butane–air mixture. The standard heats of formation of butane, water, carbon dioxide, and carbon monoxide are given as -124.7, -286.7, -393.5, and -110.5 kJ/mole, respectively.

Solution

(a) *Maximum energy release:* Maximum energy is released when completely oxidized products of combustion are formed. The stoichiometric reaction for butane with air, which gives completely oxidized products of combustion, is

$$2C_4H_{10} + 13(O_2 + 3.76N_2) \rightarrow 8CO_2 + 10H_2O + 13 \times 3.76\,N_2$$

The energy released in the above reaction is

$$\Delta H_C = -\left\{ n_{CO_2}\Delta H^0_{fCO_2} + n_{H_2O}\Delta H^0_{fH_2O} - n_{C_4H_{10}}\Delta H^0_{fC_4H_{10}} \right\}$$

Substituting the number of moles of the different species and their heats of formation, we get

$$\Delta H_C = -\{8 \times (-393.5) + 10 \times (-286.7) - 2 \times (-124.7)\}$$
$$= 5765.6\,\text{kJ}$$

The above energy is released from 2 moles of butane. The corresponding mass of butane is $2 \times (4 \times 12 + 1 \times 10) = 116$ g. Therefore, maximum energy liberated per kg of butane $= 5765.6/0.116 = 49,703$ kJ or 49.703 MJ.

(b) *Energy release for fuel-rich mixture:* When the volume fraction of butane in the butane–air mixture is 0.04, the composition of the reactants can be written as

$$C_4H_{10} + a(O_2 + 3.76N_2)$$

giving

$$\frac{1}{(1 + 4.76a)} = 0.04$$

or

$$a = \frac{24}{4.76} = 5.042$$

The chemical reaction becomes

$$C_4H_{10} + 5.042(O_2 + 3.76N_2) \rightarrow$$
$$5H_2O + 1.084CO_2 + 2.916CO + 5.042 \times 3.76\,N_2$$

While determining the above, we consider 10 hydrogen atoms (in butane) to first react with 5Ω from $5.042 \times 2 = 10.084$ of O available in the air. The balance $(10.084 - 5 = 5.084)$ of O reacts with 4 of C to form 4 CO. We are still left with $5.084 - 4 = 1.084$ of O. Thus, of the 4 CO formed,

1.084 of CO gets oxidized to 1.084 CO_2 leaving $4 - 1.084 = 2.916$ of CO. Substituting the moles and heats of formation for the above reaction, we get the energy release as

$$\Delta H_C = -\left\{ n_{H_2O}\Delta H^0_{fH_2O} + n_{CO_2}\Delta H^0_{fCO_2} + n_{CO}\Delta H^0_{fCO} \right.$$
$$\left. -n_{C_4H_{10}}\Delta H^0_{fC_4H_{10}} \right\}$$

$$\Delta H_C = -\{5 \times (-286.7) + 1.084 \times (-393.5)$$
$$+2.916 \times (-110.5) - 1 \times (-124.7)\}$$
$$= 2061 \text{ kJ}$$

The above energy is released from 1 mole of butane. The molecular mass of butane being $(4 \times 12 + 1 \times 10) = 58$ g/mole, the energy released per kg of butane is $2061/0.058 = 35{,}534$ kJ or 35.534 MJ. It is seen that the energy release per kg of butane is grossly reduced for fuel-rich mixture compared to stoichiometric composition by about 14.2 MJ, i.e., by about 26%.

4.7 *Explosion of a firecracker*: A firecracker is made by tightly packing 100 g of a composition consisting of potassium nitrate (KNO_3), carbon (C), and sulfur (S) in several layers of paper glued together. The density of the compressed charge is 1.5 g/cc. The composition of KNO_3:C:S $= 0.69:0.20:0.11$ by mass. Determine the energy released in kJ when the cracker is fired, the maximum temperature of the products of combustion, and the maximum pressure when the cracker explodes. You can assume the following combustion products are formed: potassium sulfide (K_2S), carbon dioxide (CO_2), carbon monoxide (CO), and nitrogen (N_2). The atomic mass of potassium is 39. The heats of formation of KNO_3, K_2S, CO_2, and CO are $-492, -411, -394,$ and -110 kJ/mole, respectively. The mean molar specific heat at constant volume of the products of the combustion can be assumed as 60 J/(mole K). The perfect gas equation may be assumed to be valid for the products formed in the reaction.

Solution

(a) *Energy released by the cracker*: The molecular mass of $KNO_3 = 39 + 14 + 48 = 101$ g. The molecular mass of C and S is 12 g and 32 g, respectively. Based on the mass proportion of KNO_3:C:S, which is $0.69:0.20:0.11$, the molar composition is in the proportion $(0.69/101):(0.20/12):(0.11/32)$ $= 0.00683 : 0.0167 : 0.00344$. Therefore, number of moles of C and S in 1 mole of KNO_3 is $0.0167/ 0.00683 = 2.45$ and $0.00344/0.00683 = 0.5$. The molar composition of the reactants is therefore $KNO_3 + 2.45C + 0.5S$ for one mole of KNO_3. We first determine the products of combustion: It is given that K combines with S to form K_2S. We have 3 of O and 2.45 of C. The 2.45 of C combines with 2.45 of O to give 2.45 CO. We are now left with 0.55 Ω. Of the 2.45 CO formed, 0.55 forms CO_2 leaving 1.9 CO. The nitrogen in the reactant forms N_2 in the product. The chemical reaction is therefore given as

$$KNO_3 + 2.45\,C + 0.5\,S \rightarrow \frac{1}{2}K_2S + 0.55CO_2 + 1.9CO + \frac{1}{2}N_2$$

The heat released in the above reaction is

$$\Delta H_C = - \left\{ n_{K_2S}\Delta H_{fK_2S}^0 + n_{CO_2}\Delta H_{fCO_2}^0 + n_{CO}\Delta H_{fCO}^0 \right.$$
$$\left. -n_{KNO_3}\Delta H_{fKNO_3}^0 \right\}$$

The elements N_2, C, and S have the standard heats of formation of zero.

$$\Delta H_C = -\{0.5 \times (-411) + 0.55 \times (-394)$$
$$+1.9 \times (-110) - 1 \times (-492)\}$$
$$= 139.2 \text{ kJ}$$

Since the molecular mass of $KNO_3 = 101$ g, energy released by 100 g of the composition corresponds to 69 g of KNO_3. The energy release is therefore

$$\frac{139.2}{101} \times 69 = 95.08 \text{ kJ}.$$

(b) *Maximum temperature of the combustion products:* We assume no heat loss to the ambient. The mean molar specific heat of the combustion products at constant volume is given as 0.06 kJ/(mole K), and the total number of moles of products is $0.5 + 0.55 + 1.9 + 0.5 = 3.45$ moles. Taking the initial temperature as 25°C, the maximum temperature reached by the combustion products is obtained from the energy balance:

$$3.45 \times 0.06 \times (T - 25) = 139.2$$

Therefore, $T = 697.5\,°C$.

(c) *Pressure when cracker explodes*: We shall assume the entire charge of 100 g gets burnt when the cracker explodes.

$$\text{Volume of the charge} = \frac{\text{Mass of charge}}{\text{density of charge}}$$
$$= \frac{100}{1.5} = 66.7 \text{ cc.}$$

From ideal gas equation corresponding to the products of combustion we have

$$pV = mRT$$

where R is the specific gas constant of the combustion products, V the volume of the products, m the mass, and p the pressure. The specific gas constant of the mixture of the combustion products is determined from the universal gas constant R_0 in J/(mole K) by dividing it by the mean molecular mass M of the combustion products $(R = R_0/M)$. The molecular mass M is given as

$$M = \frac{\{n_{K_2S}M_{K_2S} + n_{CO_2}M_{CO_2} + n_{CO}M_{CO} + n_{N_2}M_{N_2}\}}{\{n_{K_2S} + n_{CO_2} + n_{CO} + n_{N_2}\}}$$

$$= \frac{(0.5 \times 110 + 0.55 \times 44 + 1.9 \times 28 + 0.5 \times 28)}{(0.5 + 0.55 + 1.9 + 0.5)}$$

$$= 42.43 \text{ g/mole}$$

$$R = \frac{R_0}{M} = \frac{8.314}{42.43} = 0.196 \text{ J/(gK)} = 196 \text{ J/(kgK)}$$

Substituting the values of volume of 66.7×10^{-6} m³, mass of 0.1 kg, and $T = 697.5 + 273 = 970.5$ K, we get

$$p = \frac{mRT}{V} = \frac{0.1 \times 196 \times 970.5}{(66.7 \times 10^{-6})} \left\{ kg\frac{J}{kgK}\frac{K}{m^3} = \frac{N}{m^2} \right\}$$

$$= 285 \text{ MPa}.$$

A very high pressure of about 285 MPa is seen to be reached. At higher values of pressure, the ideal gas equation should be used with caution. The products, if not in the gas phase, cannot contribute to pressure. In practice, the several layers of glued paper, holding the charge, would give way at much lower values of pressure. We can also determine the change of internal energy rather than enthalpy if the combustion takes place at constant volume, such as in the following Problem 4.8, and use this change of internal energy to determine the final pressure.

4.8 *Change in internal energy in the burning of RDX:* Determine the change of internal energy associated with the combustion of 1 kg RDX given the following data: The standard heats of formation of RDX, water, and CO are +61.5, −286.7, and −110.5 kJ/mole, respectively. The specific heats at constant pressure of $H_2O(g)$, CO, and N_2 are 58, 37, and 37 J/(mole K). The specific heat of water is 90 J/(mole K). The boiling temperature of water at 1 atm. pressure is 100°C, and the latent heat of vaporization is 40 kJ/mole.

Solution

Internal energy associated with combustion of RDX: The heat of combustion ΔH_C has been derived for constant pressure combustion process and given as

$$\Delta H_C = -\left\{ \sum_{\text{Products}} n_i \Delta H^0_{fi} - n_{\text{RDX}} \Delta H^0_{f\text{RDX}} \right\}$$

The pressure at 1 bar at which the heats of formation are considered does not influence the value of energy release at different pressures, since for an ideal gas the internal energy and enthalpy are functions of temperatures only [$u = u(T)$ and $h = h(T)$]. For a perfect gas the values of the specific heats are constants. The heat release is therefore taken to be independent of pressure. The heat of explosion of RDX is determined for the chemical reaction:

$$C_3H_6N_6O_6 \rightarrow 3H_2O + 3CO + 3N_2$$

$$\Delta H_C = -\{3 \times (-286.7) + 3 \times (-110.5) - 1 \times (+61.5)\} \text{ kJ}$$
$$= 1253.1 \text{kJ per mole of RDX}$$

The above heat release contributes to increase in the temperature of the combustion products H_2O, CO, and N_2 in the reaction $C_3H_6N_6O_6 \rightarrow$ products from 25°C to T_f. The energy balance gives

$$\Delta H_C = n_{H_2O} \left\{ C_{P,H_2O(l)}(100 - 25) + L + C_{P,H_2O(g)}(T_f - 100) \right\}$$
$$+ n_{CO} C_{P,CO}(T_f - 25) + n_{N_2} C_{P,N_2}(T_f - 25)$$

Here the first term in curly brackets on the right represents the increase in the enthalpy of water from the standard temperature of 25 °C to its boiling temperature of 100 °C (sensible heat), the latent heat of vaporization, and the heating of steam from the boiling temperature of 100 °C to the final temperature T_f. The next two terms denote the heating of CO and N_2 from 25 °C to T_f. Substituting the values of the specific heats and latent heat, we get

$$1253.1 = 3\left\{0.09(75) + 40 + 0.058(T_f - 100)\right\} + 3 \times 0.037(T_f - 25)$$
$$+ 3 \times 0.037(T_f - 25)$$
$$1253.1 = 48.9 + 0.396T_f$$

The value of $T_f = 3040.9°C = 3314$ K. The dissociation of gases, which would bring down the temperature, is not considered. The change of internal energy corresponding to the above change of enthalpy is found using the relation between enthalpy and internal energy:
$$u = h - pv$$

where u and h are the specific internal energy and specific enthalpy and v is the specific volume. For the 9 mol of the products formed in the reaction

$$U = H - pV$$

Here H, U, and V denote the enthalpy, internal energy, and volume of the products comprising the 9 mol. The value of pV for an ideal gas is written as

$$pV = nR_0T$$

Substituting the above in the equation for internal energy U, we get

$$U = H - nR_0T$$

Substituting the values of $H = 1253.1$ kJ obtained earlier, $n = 9$ moles, $R_0 = 8.314$ kJ/(mole K), and $T = 3314$ K, we get

$$U = 1005 \text{ kJ}$$

This internal energy corresponds to 1 mole of $C_3H_6N_6O_6$, i.e., 222 g of RDX. Hence the specific internal energy released from 1 kg of RDX (u) is $1005/0.222 = 4528$ kJ/kg. This could have been determined directly if the values of the standard internal energy of formation of RDX, H_2O, and CO were made available.

Nomenclature

A	Constant; mass of air (kg)
B	Constant
C	Concentration (kmoles/m^3, moles/cc)
E_0	Energy released by source (kJ)
$\dot{E}_0$	Rate of energy release by source (J/s, W)
E_a	Activation energy (J/mole)
F	Mass of fuel (kg)
H	Enthalpy (J)
h	Specific enthalpy (J/kg)
ΔH_C	Energy release in a reaction (kJ)
ΔH_f^0	Standard heat of formation (kJ/mole)
k	Rate constant for a chemical reaction defined by Eq. 4.23
M	Species in a chemical reaction
M	Molecular mass (g/mole)
n	Moles of a substance
N	Number of species in Eq. 4.12
p	Pressure (Pa)
q	Energy released by a chemical reaction per mole of the reactant (kJ/mole)
$\dot{q}$	Rate of energy release per unit volume (kJ/m^3)
R_0	Universal gas constant [J/(mole K)]
T	Temperature (K)
t	Time (s)

V	Volume (m^3)
U	Internal energy (J)
u	Specific internal energy (kJ/kg)
φ	Equivalence ratio
ω	Rate of chemical reaction (moles/(m^3s))

Superscripts

$'$ Reactant
$''$ Product

Subscripts

f Formation

Further Readings

1. Emmons, H. W., Scientific progress on fire, *Annual Review in Fluid Mechanics*, 12, 223, 1980.
2. Kubota, N., *Propellants and Explosives: Thermochemical Aspects of Combustion*, Wiley-VCH, Weinheim, 2002.
3. Kuo, K. K., *Principles of Combustion*, Wiley, New York, 1968.
4. Mader, C. L., *Numerical Modeling of Explosives and Propellants*, 3rd ed., CRC Press, Boca Raton, FL, 2008.
5. Penner, S.S., *Chemistry Problems in Jet Propulsion*, Pergamon Press, London, 1957.
6. Stevans, B., *Chemical Kinetics*, Chapman and Hall, London, 1970.
7. Strehlow, R. A., *Fundamentals of Combustion*, 2nd Ed., McGraw-Hill Book Company, New York, 1984.
8. Stull, D. R., Fundamentals of fire and explosion, *AIChE Monograph Series*. 73, 10, 1977.
9. Turns, S. R., *An Introduction to Combustion: Concepts and Applications*, 2nd ed., McGraw Hill, Boston, MA, 2006

Exercises

4.1 Propane (C_3H_8) leaks from a pipeline and forms a $10\,m^3$ cloud of propane–air mixture at atmospheric pressure and 25 °C. The volumetric ratio of gaseous propane to air in the cloud is 0.025. Determine

   (i) Equivalence ratio of the cloud
   (ii) Energy released by the cloud in MJ assuming that the products of combustion are CO_2, $H_2O(l)$, O_2, and N_2.

You can assume the air and propane to be mixed well in the cloud. You can also assume the values of the standard heats of formation of C_3H_8, CO_2, and $H_2O(l)$ as -104, -394, and -286 kJ/mole, respectively.

4.2 Liquified petroleum gas (LPG) consists mainly of propane and butane. Different countries have different specifications for the volumetric ratio of propane to butane in LPG. In an accidental spill of LPG containing 60% by volume of butane and 40% propane, a cloud of $20\,m^3$ of LPG air mixture at the ambient pressure of 0.1 MPa and 25 °C is formed. The equivalence ratio of the cloud is 0.75. If this cloud gets ignited, determine the energy liberated in kJ. The heats of formation are:

$$\Delta H^0_{f C_3H_8} = -104 \; \frac{kJ}{mole},$$

$$\Delta H^0_{f C_4H_{10}} = -125 \; \frac{kJ}{mole},$$

$$\Delta H^0_{f CO_2} = -394 \; \frac{kJ}{mole}, \quad and$$

$$\Delta H^0_{f H_2O(l)} = -286 \; \frac{kJ}{mole}.$$

4.3 A lean methane–air cloud with 6% volume of methane in the cloud is formed on a cold day when the ambient temperature is 5 °C. Determine

(a) Equivalence ratio of the mixture
(b) Heat released per kg of the mixture
(c) Heat released per m^3 of the mixture
(d) If the ambient temperature was 35 °C, what would be the change in the energy released per kg of the cloud?

You can take the standard heat of formation of methane as -75 kJ/mole. The values of the standard heats of formation of carbon dioxide and water are -394 and -286 kJ/mole. The mean specific heat of the methane–air mixture can be assumed as 40 J/(mole K).

4.4 A sealed can of volume $2\,m^3$ contains air (at ambient pressure of 0.1 MPa and 25 °C) and 100 g of liquid kerosene. The kerosene–air mixture is mixed well by shaking it vigorously and then ignited. Determine the following:

(a) Fuel–air ratio in the can
(b) Equivalence ratio
(c) Whether the mixture is fuel-lean or fuel-rich
(d) Energy liberated during the combustion.

You can assume the molecular formula for kerosene as $C_{12}H_{24}$. The density of kerosene is $800\,kg/m^3$. The standard heats of formation in kJ/mole for kerosene $= -159$, $CO_2 = -390$, $H_2O\,(l) = -286$.

4.5 A spherical container of volume $1\,m^3$ contains a mixture of hydrogen and air at a pressure of 0.12 MPa and 300 K. The percentage volume of hydrogen in the mixture of hydrogen and air is 10%. If the mixture gets ignited, determine the following:

(a) Energy released in kJ
(b) Temperature of the combustion products
(c) Maximum pressure in the container. The container is strong and can hold a maximum pressure of 50 MPa.

You can assume the following data:

$$\text{Standard heat of formation of water} = -286 \frac{\text{kg}}{\text{mole}}$$

$$\text{Latent heat of steam} = 40 \frac{\text{kg}}{\text{mole}}$$

$$\text{Specific heat of water} = 90 \frac{\text{J}}{\text{mole K}}$$

Specific heat at constant pressure of steam, nitrogen, oxygen, and hydrogen is 58, 37, 38, and 35 (J/mole K).

$$\text{Universal gas constant} = 8.314 \frac{\text{K}}{\text{mole K}}$$

$$\text{Boiling temperature of water} = 373 \text{ K}$$

4.6 2 kg of fine sugar powder gets ignited when mixed with $1\,m^3$ of air at the atmospheric pressure of 0.1 MPa and 25 °C. If the chemical formula of sugar can be assumed as glucose $C_6H_{12}O_6$, determine the products of combustion and the energy released during the combustion. The standard heat of formation of glucose can be taken as -1270 kJ/mole. The standard heats of formation of CO, CO_2, and H_2O are -110, -394, and -296 kJ/mole, respectively.

4.7 (a) If the heat of formation of ammonium nitrate (NH_4NO_3) is -365 kJ/kg and the heat of formation of the fuel oil $C_{18}H_{38}$ is -310 kJ/kg, determine the energy released in the explosion of 1 kg of the above stoichiometric mixture of ammonium nitrate and fuel oil (ANFO). The heats of formation of water, carbon monoxide, and carbon dioxide are -287, -110, and -394 kJ/kg, respectively.

(b) The mass of ammonium nitrate has to be limited in practice to a maximum of 60% of the above mass of ammonium nitrate per kg of the fuel oil (required for a stoichiometric mixture) in order to form a slurry of the liquid fuel and solid oxidizer. Determine the products of combustion of this explosive slurry and the ratio of the mass of carbon monoxide to carbon dioxide in the products assuming hydrogen to be much more reactive than carbon.

Chapter 5
Thermal Theory of Explosions

The rate of energy release from the chemical reaction of a given mass or volume of an explosive was seen to increase as its temperature increases. The increase was particularly intense at the higher values of temperatures when the activation energy of the reaction is large. If the temperature of a chemically reacting substance could increase to large values during interaction with the surroundings, an explosion due to rapid energy release is possible. We examine the possibilities of such explosions from an excursion in the temperature and explore the conditions for which explosions could occur.

5.1 Formulation of Theory

5.1.1 Lumped Mass Assumption

We consider a reactive substance of volume V at an initial temperature T_i kept in the surroundings that are maintained at temperature T_∞. If the substance has very high values of thermal conductivity and diffusion coefficient, its temperature and composition would be reasonably uniform throughout its volume. All spatial locations in it will always have the same values of temperature and concentration. Single values of temperature and concentration are adequate to describe it at a given time, and such an assumption is known as 'lumped mass assumption.'

Figure 5.1 shows a lumped mass of reactive substance having arbitrary volume V with initial temperature T_i and concentration C_0 with a boundary corresponding to surface area S. The boundary separates it from the ambient at temperature T_∞.

When heat is released by chemical reactions in the given volume V, its temperature increases above the initial value T_i. Let us presume that at a given time t, its

© The Author(s), under exclusive license to Springer Nature Switzerland AG 2021
K. Ramamurthi, *Modeling Explosions and Blast Waves*,
https://doi.org/10.1007/978-3-030-74338-3_5

Fig. 5.1 Lumped volume of
reactive substance at
temperature T_i and
concentration C_0

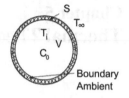

Fig. 5.2 Schematic of
temperature at time t

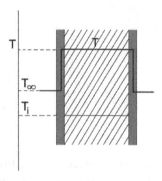

Fig. 5.3 Dependence of heat
release on temperature,
concentration, and volume

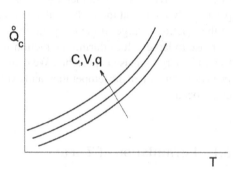

temperature rises to a value T that is higher than the ambient value T_∞, as shown in
Fig. 5.2.

The rate of energy release due to chemical reactions in the volume V can be
written following Eq. 4.26 in Chap. 4 as

$$\dot{Q}_C = AqC^n \exp\left(-\frac{E_a}{R_0 T}\right) \times V \text{ kJ/s} \tag{5.1}$$

The energy release rate is available as heat and is denoted by $\dot{Q}_C$. The magnitude of
$\dot{Q}_C$ varies with temperature T and fuel concentration C, and the variation, based on
Eq. 5.1, is shown in Fig. 5.3. As the temperature increases, the rate of heat release
goes up exponentially. The heat release also increases with increase in concentration
C, volume V, and energy release per mole of the substance q (Fig. 5.3).

Fig. 5.4 Rate of heat transfer from volume V

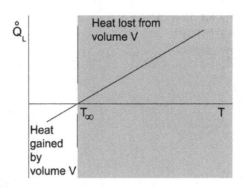

The magnitude of the temperature T in the given volume V, in relation to the ambient temperature T_∞, would govern whether heat is lost to the ambient from the lumped volume or gained by it. Initially when $T_i < T_\infty$, heat gets conducted into the volume V and its temperature increases. A rise in temperature would enhance the heat release $\dot{Q}_C$, and the temperature in the volume increases. When the temperature is greater than the ambient temperature T_∞, heat loss to the ambient takes place. Denoting the heat loss by $\dot{Q}_L$ and the heat transfer coefficient at the surface S by h, we have:

$$\dot{Q}_L = hS(T - T_\infty) \tag{5.2}$$

The dependence of the heat loss rate with temperature T is a straight line since the heat transfer coefficient is about a constant (Fig. 5.4).

5.1.2 Variations of Heat Release and Heat Loss Rates

Figure 5.5 shows two arbitrary variations of heat release rates with temperature corresponding to two different values of concentrations C_1 and C_2 of a substance in the volume V and bounded by surface area S. The heat release variations corresponding to these two concentrations are shown as $\dot{Q}_{C1}$ and $\dot{Q}_{C2}$, respectively, as a function of the temperature. In the same figure, the variation in the rate of heat loss from surface S, given by $\dot{Q}_L$ (Eq. 5.2), is shown by the dotted line. When the temperature equals the ambient value, the heat loss is zero. Heat is gained by the volume for temperatures below the ambient value. The slope of the heat loss $\dot{Q}_L$ line represents the product of the heat transfer coefficient and the surface area S.

It is shown in Fig. 5.5 that the heat release $\dot{Q}_{C2}$ is greater than the heat loss rate $\dot{Q}_L$ over the entire range of temperature. The accumulation of heat within the volume from the heat release increases the temperature, which in turn leads to a further enhancement of the heat release rate. The process of heat release enhancing the temperature and the temperature enhancing the heat release rate are bootstrapped, leading to a runaway or acceleration in heat release rate and a spontaneous explosion.

Fig. 5.5 Heat release and
heat loss rates

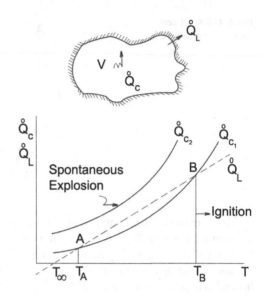

When the heat release rate is about the same as the heat loss rate $\dot{Q}_L$ (shown by
$\dot{Q}_{C1}$ in Fig. 5.5), the curves for heat release and heat loss intersect at two points A
and B as shown. In the regions of temperatures less than T_A and greater than T_B,
the rate of heat release $\dot{Q}_{C1}$ is greater than the heat loss rate $\dot{Q}_L$. In the range of
temperatures $T_A < T < T_B$, the rate of heat release is less than the heat loss rate.

5.1.3 Stable Temperature and Ignition Temperature

The temperatures T_A and T_B corresponding to points A and B in Fig. 5.5, for which
the heat generation rate $\dot{Q}_{C1}$ and heat loss rate $\dot{Q}_L$ are equal, have special significance. When the substance is at a temperature less than T_A, the heat generation rate
$\dot{Q}_{C1}$ is shown in Fig. 5.5 to be greater than the heat loss rate $\dot{Q}_L$ and therefore the
temperature of the substance will increase until it reaches a temperature T_A for which
heat generation equals the heat loss rate ($\dot{Q}_{C1} = \dot{Q}_L$). If the temperature exceeds T_A,
the heat loss rate $\dot{Q}_L$ becomes greater than the heat generation rate $\dot{Q}_{C1}$ and therefore
the temperature falls until the temperature corresponding to point A is reached. Initial
value of temperature less than T_A or greater than T_A (but less than T_B) results in the
temperature increasing or decreasing, respectively, to reach the equilibrium value T_A.
The temperature at A is therefore called as stable temperature since any perturbation
about it causes it to come back to T_A. A reactive substance at this temperature T_A
exists in equilibrium in the given surrounding. No explosion is possible.

An increase in temperature beyond B causes the heat release $\dot{Q}_{C1}$ to be greater
than the heat loss rate $\dot{Q}_L$ (see Fig. 5.5). If by some means, such as by supply of
energy from an external source or otherwise, the temperature is increased beyond

T_B, the excess rate of heat accumulation over the loss will cause the temperature to increase further. This leads to further enhancement of heat release rates, and a runaway in temperature occurs. If the temperature falls below T_B, the heat loss rate being higher (Fig. 5.5) brings down the temperature to T_A. B is therefore an unstable point. If temperature is increased beyond it, runaway reactions or explosion of the substance takes place. T_B is called as ignition temperature. The external energy to be supplied to the substance, to increase its temperature to T_B and thus bring about ignition, is the ignition energy.

5.1.4 Critical Temperature and Auto-ignition

Between the heat release rate curves $\dot{Q}_{C1}$ and $\dot{Q}_{C2}$ shown in Fig. 5.5, it is possible to have a heat release rate curve $\dot{Q}_{C3}$, which is tangent to the heat loss curve $\dot{Q}_L$. This is shown in Fig. 5.6. The curve $\dot{Q}_{C3}$ defines the critical conditions for spontaneous explosion, since for heat release rate indicated by it, the rate of heat generation is just above the rate of heat loss $\dot{Q}_L$ at temperatures both greater and less than the value of T_C at the point of tangency C.

The limiting curve $\dot{Q}_{C3}$, which is tangent to the heat loss curve, represents the threshold value of the heat release rate corresponding to the heat loss condition specified by $\dot{Q}_L$ for which the substance can get ignited without the addition of external energy. The self-heating leads to ignition and an explosion. This critical condition corresponds to the threshold for auto-ignition, and the temperature T_C is known as the critical temperature.

5.1.5 Changes of Ambient Temperature

For a specified rate of heat release, the condition of spontaneous or auto-ignition can also be achieved by varying the heat loss. Figure 5.7 shows a reduction in the heat loss

Fig. 5.6 Heat release rate curve tangent to heat loss rate curve

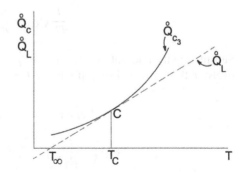

Fig. 5.7 Variations in heat
loss rate

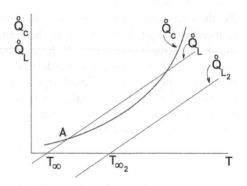

Fig. 5.7 Variations in heat loss rate

by increasing the ambient temperature from T_∞ to $T_{\infty 2}$ (curve $\dot{Q}_{L2}$ instead of $\dot{Q}_L$) keeping the heat transfer coefficient and surface area to be the same. The substance with the heat release rate characterized by $\dot{Q}_C$ which is stable at the temperature T_A, corresponding to heat loss rate $\dot{Q}_L$, spontaneously ignites for the heat loss curve specified by $\dot{Q}_{L2}$. The increase in ambient temperature causes the auto-ignition.

A reduction of heat transfer coefficient and the surface area for a given volume of the substance would likewise lead to conditions favorable for auto-ignition. Auto-ignition conditions would accordingly vary with the characteristics of the system containing the reactants and its surroundings.

5.2   Critical Conditions and Preheat

The critical conditions for which the substance can auto-ignite by self-heating are obtained when the heat loss rate curve is tangent to the heat release rate curve. At the tangency point C (Fig. 5.6), we have the magnitude of heat release rate to be the same as the heat loss rate and the slope of the release and loss rate curves are equal. We can therefore write

$$\dot{Q}_{C3} = \dot{Q}_L \tag{5.3}$$

$$\frac{d}{dT}\dot{Q}_{C3} = \frac{d}{dT}\dot{Q}_L \tag{5.4}$$

Substituting the relevant expressions for $\dot{Q}_{C3}$ and $\dot{Q}_L$ from Eqs. 5.1 and 5.2, respectively, in the above and simplifying for $T = T_C$, we get

$$AqC^n \exp\left(-\frac{E_a}{R_0 T}\right) \times V = hS(T_C - T_\infty) \tag{5.5}$$

$$AqC^n \exp\left(-\frac{E_a}{R_0 T}\right)\frac{E_a}{R_0 T^2} \times V = hS \quad \text{at } T = T_C \tag{5.6}$$

Dividing Eq. 5.6 by Eq. 5.5, we have

$$\frac{E_a}{R_0 T_C^2} = \frac{1}{T_C - T_\infty} \tag{5.7}$$

The following quadratic equation is therefore obtained for T_C:

$$T_C^2 - \frac{E_a}{R_0} T_C + \frac{E_a}{R_0} T_\infty = 0 \tag{5.8}$$

Solving the above equation for T_C gives

$$T_C = \frac{E}{2R_0} \left(1 \pm \sqrt{1 - \frac{4R_0 T_\infty}{E_a}} \right) \tag{5.9}$$

The value of activation energy E_a for most substances of interest in explosions is large with values around 120,000 J/mole. The term $4R_0 T_\infty / E_a$ is therefore a small number less than about 0.1. Expanding $\sqrt{1 - 4R_0 T_\infty / E_a}$ as a binomial series $[(1 + x)^{1/2} = 1 + x/2 - x^2/8 + 3x^3/16, \ldots]$, we get, on neglecting the third- and higher-order terms:

$$\left(1 - \frac{4R_0 T_\infty}{E_a} \right)^{1/2} = 1 - \frac{1}{2} \left(\frac{4R_0 T_\infty}{E_a} \right) - \frac{1}{8} \left(\frac{4R_0 T_\infty}{E_a} \right)^2 \tag{5.10}$$

Substituting in Eq. 5.9, T_C is obtained as:

$$T_C = \frac{E_a}{2R_0} \pm \frac{E_a}{2R_0} \left[1 - \frac{2R_0 T_\infty}{E_a} - 2 \left(\frac{R_0 T_\infty}{E_a} \right)^2 \right] \tag{5.11}$$

The use of the positive sign in the above expression gives the value of T_C in excess of about 10,000 °C which is a fictitious number. This is due to the large value of the activation energy. The value of $T_C - T_\infty$ is obtained as

$$T_C - T_\infty = \frac{E_a}{2R_0} - \frac{E_a}{2R_0} \left[1 - 2 \left(\frac{R_0 T_\infty}{E_a} \right)^2 \right] \tag{5.12}$$

or

$$T_C - T_\infty = \frac{R_0 T_\infty^2}{E_a} \tag{5.13}$$

The heat required to increase the temperature from the ambient value of T_∞ to the critical value T_C is known as preheat, and $(T_C - T_\infty)$ is defined as preheat temperature. Considering that the activation energy E_a of interest is about 120,000 J/mole, we get the temperature corresponding to preheat to be about 6 °C. The preheat temperature required for auto-ignition is generally small. From Eq. 5.13, the preheat temperature can be expressed in a non-dimensional form as

$$\frac{T_C - T_\infty}{T_\infty} = \frac{R_0 T_\infty}{E_a} \tag{5.14}$$

Equation 5.7, derived at the point of tangency of the rate of heat release and heat loss curves can be rewritten as

$$\frac{T_C - T_\infty}{T_C} = \frac{R_0 T_C}{E_a} \tag{5.15}$$

since in the limit of $E_a \to \infty$, gives $T_C \to T_\infty$ and is compatible with the conclusion drawn that the preheat temperature is small for large values of activation energy of interest in explosions.

5.3 Characteristic Times of Heat Generation and Heat Loss

5.3.1 Characteristics of Heat Release From Chemical Reaction

The rate of heat generation per unit volume of a combustible is given by the product of the rate of reaction ω in moles/(cc-s) and the heat generated per mole of the reactant q in kJ/mole. If we denote the initial concentration of the reactant as C_0 moles/cc, the amount of energy released in volume V in cc of the unreacted medium is $q \times C_0 \times V$ in kJ. The energy release leads to an increase in temperature of the combustion products. The maximum temperature is reached when all of the reactant gets consumed. A plot denoting the depletion of concentration of the reactant starting from the initial value C_0 and the increase in temperature of the combustion products with time is illustrated in Fig. 5.8.

The maximum temperature is reached when the concentration of the reactant drops to zero. Denoting the maximum temperature as T_m, we have from energy balance:

$$\rho V C_V (T_m - T_0) = q C_0 V \tag{5.16}$$

Here $q \times C_0 \times V$ represents the energy liberated in the reaction and $\rho V C_V (T_m - T_0)$ is the increase in energy of the combustion products. T_0 is the initial temperature. ρV is the mass in the volume, while C_V is the mean specific heat of the combustion products. When chemical reactions have not progressed to completion, and the concentration of the reactant has reached an intermediate value of C moles/cc, the heat liberated corresponds to the decrease in concentration from C_0 to C and is given by

Fig. 5.8 Depletion of reactant concentration and evolution of temperature

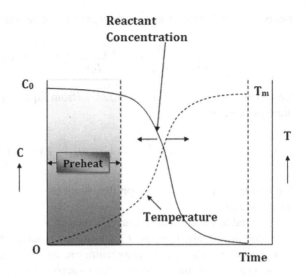

$$q(C_0 - C)V \tag{5.17}$$

If the temperature is T when the concentration is C, we have following Eq. 5.16

$$\rho V C_V (T - T_0) = q(C_0 - C)V \tag{5.18}$$

The temperature and concentration are related using Eqs. 5.16 and 5.18 as

$$\frac{C_0 - C}{C_0} = \frac{T - T_0}{T_m - T_0} \tag{5.19}$$

On simplification, we get

$$C = C_0 \left(\frac{T_m - T}{T_m - T_0} \right) \tag{5.20}$$

The rate of energy release in the given volume $\dot{Q}_C$ is given as the product of the rate of consumption of the reactant $\omega (= dC/dt)$, the energy release per mole of the reactant q, and the volume V and is given in Eq. 5.1. Substituting the value of C from Eq. 5.20 in Eq. 5.1 and simplifying, we have the rate of energy release as

$$\dot{Q}_C = Aq C_0^n \left(\frac{T_m - T}{T_m - T_0} \right)^n \exp\left(-\frac{E_a}{R_0 T} \right) V \tag{5.21}$$

This equals the rate of energy increase in the volume and is given by

$$\frac{d}{dt}(\rho V C_V T) = \dot{Q}_C \tag{5.22}$$

Simplifying the above equation after substituting the value of $\dot{Q}_C$ from Eq. 5.21, we get

$$\frac{dT}{dt} = \frac{A}{\rho C_V} q \frac{C_0^n}{(T_m - T_0)^n} (T_m - T)^n \exp\left(-\frac{E_a}{R_0 T}\right) \qquad (5.23)$$

Considering that $(T_m - T_0) = q C_0 / \rho C_V$ (from Eq. 5.16), we get the rate of temperature increase as

$$\frac{dT}{dt} = A \left(\frac{\rho C_V}{q}\right)^{n-1} (T_m - T)^n \exp\left(-\frac{E_a}{R_0 T}\right) \qquad (5.24)$$

We find from the above equation that when the temperature $T = T_m$, the value of dT/dt is zero.

At initial times, the temperature T would be near the ambient value T_∞ and since the value of activation energy E_a considered is large, the rate of temperature rise would be small. The rate of temperature rise therefore starts off slowly, accelerates as the temperature increases but falls to zero when the maximum temperature is reached and the reaction gets completed. The trend is illustrated by the dotted line in Fig. 5.9.

The rate of energy release in the volume follows the rate of temperature increase as shown in Fig. 5.9. The rate of energy release is initially very slow. It increases sharply after a certain time, which is called the induction time. During the induction time, the concentration of active intermediate species, which help to progress the reaction, is built up. A typical variation of the heat release along with the evolution of temperature is shown in Fig. 5.10. The induction time is denoted by t_d. The characteristic time over which the bulk of the energy is released is indicated by t_c.

The spurt in energy release after the induction time t_d would be true only if the activation energy for the chemical reaction is large. When the activation energy is small, the reaction takes place readily (even at the lower range of temperatures) as there is hardly any resistance to be overcome by the reactants to form the products. Typical examples are the chemical reactions of white phosphorus in ambient air, the combination of vapor of sodium and chlorine, and the reaction of hydrogen bromide

Fig. 5.9 Trend of temperature increase and rate of temperature change with time

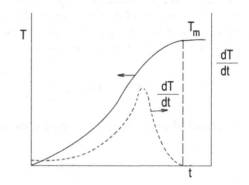

Fig. 5.10 Rate of energy
release

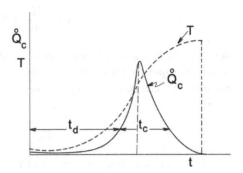

with ozone. There is no distinct spurt in energy release. In these cases, the reaction
starts off at the low ambient temperature itself.

The spurt in energy release after an induction delay, obtained for reactions with
large values of activation energies, gives rise to sudden pressure rise of gaseous fuel–
air mixtures in open surroundings and in solid and liquid explosives when not kept
in confined conditions. We shall consider this aspect while dealing with unconfined
fuel–air explosions and explosions of condensed phase explosives.

5.3.2   Characteristic Time for Energy Release

The characteristic time for the energy release is the ratio of the initial concentration
to the rate of the reaction and is given as

$$\frac{C_0}{dC/dt} \tag{5.25}$$

As the rate of reaction dC/dt increases, the characteristic time for heat release
decreases.

5.3.3   Characteristic Heat Loss Time

The heat loss time represents the time required for cooling the lumped mass. An
explosive having a large thermal mass (mass × specific heat) cools slowly unlike
one with a small thermal mass in a given surrounding. The temperature of the ambient
and the heat transfer coefficient at the surface would additionally influence the time
required for cooling. If we consider an explosive of volume V having density ρ at
a temperature T, and if the temperature T is higher than the ambient value T_∞, the
rate of heat loss from it is given by

$$-\rho V C_V \frac{dT}{dt} = hS(T - T_\infty) \tag{5.26}$$

Here h is the heat transfer coefficient at the surface of area S and C_V is the specific heat of the explosive for which the lumped mass assumption is presumed to be valid. Equation 5.26 can be rearranged to give

$$\frac{dT}{T - T_\infty} = -\frac{hS}{\rho V C_V} dt \tag{5.27}$$

If T_h is the initial high temperature, and T is its temperature at time t, Eq. 5.27 can be integrated between time $t = 0$ when the temperature is T_h and time t when the temperature is T to give

$$-\frac{hS}{\rho V C_V} t = \int_{T_h}^{T} \frac{dT}{T - T_\infty}$$

$$\ln \frac{T - T_\infty}{T_h - T_\infty} = -\frac{hS}{\rho V C_V} t \tag{5.28}$$

The above expression is simplified to give

$$\frac{T - T_\infty}{T_h - T_\infty} = \exp\left(-\frac{hS}{\rho V C_V}\right) t \tag{5.29}$$

A trend of the variation of temperature T with time t from the above equation is shown in Fig. 5.11. The temperature approaches the ambient value T_∞ at very large times.

It is difficult to identify a characteristic time for the exponential decrease in temperature shown in Fig. 5.11. We, however, observe that most of the temperature change takes place in the initial period whereas for later times, the temperature drop is very slow and asymptotically approaches the ambient temperature T_∞. We define a characteristic time as the time taken for the temperature difference $T - T_\infty$ to reach $1/e$

Fig. 5.11 Gradual cooling of a hot substance initially at temperature T_h

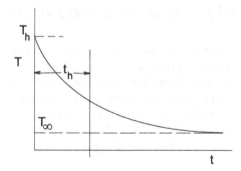

Fig. 5.12 Characteristic heat loss time

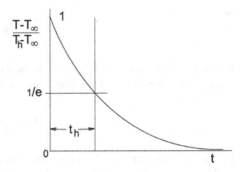

of the initial difference $T_h - T_\infty$. This is shown in the plot of $(T - T_\infty)/(T_h - T_\infty)$ versus t by t_h in Fig. 5.12.

The characteristic heat loss time t_h from Eq. 5.29 becomes

$$\frac{1}{e} = \exp\left(-\frac{hS}{\rho V C_V}\right) t_h$$

giving

$$t_h = \frac{\rho V C_V}{h S} \tag{5.30}$$

This characteristic time is also referred to as thermal relaxation time. Equation 5.29 is written as

$$\frac{T - T_\infty}{T_h - T_\infty} = e^{-t/t_h} \tag{5.31}$$

5.4 Conditions for Explosion to Occur

If the rate of heat release from chemical reactions of a substance of volume V is much greater than the rate of heat loss from it, energy accumulates within the volume leading to runaway reactions and an explosion. A small value of the characteristic time of energy release implies a larger value of energy release rate. When the characteristic time for heat generation t_C is very much smaller than the characteristic time for heat loss t_h, the heat generated within the volume gets accumulated within the volume, before it can be dissipated out and the temperature increases. The rapid increase of reaction rate with temperature leads to further enhancement of energy release rates and hence leads to an explosion. The condition to be satisfied for an explosion to occur is therefore

$$t_C < t_h \tag{5.32}$$

or

$$\frac{t_h}{t_C} \geq K \tag{5.33}$$

where K denotes the threshold value to be exceeded. The ratio of the characteristic times t_h/t_C is known as Damkohler number and is denoted by Da.

$$\text{Da} = \frac{t_h}{t_C} \tag{5.34}$$

Da has to be greater than the threshold value of K for explosion to occur.

The conditions for which Da $> K$ can be derived from the lumped mass thermal theory considered so far. From Eq. 5.6, for the tangency condition between the heat release curve and the heat loss curve when the heat generated and heat loss rates are equal (corresponding to critical conditions of auto-ignition), we had

$$Aq\,C^n \exp\left(-\frac{E_a}{R_0 T_C}\right) \frac{E_a}{R_0 T_C^2} \times V = hS \tag{5.35}$$

The characteristic time for energy release t_C is given in Eq. 5.25 as

$$\frac{C_0}{dC/dt}$$

Since the rate of reaction is given by $dC/dt = AC^n \exp(-E_a/R_0 T)$ and the effective energy release takes place starting at $T = T_C$ when the concentration is still very near the initial value C_0, the value of t_C is given as

$$t_C = \frac{C_0}{AC^n \exp[-E_a/(R_0 T_C)]}$$

Substituting the value of $AC^n \exp[-E_a/(R_0 T_C)]$ from the above in Eq. 5.35, we get

$$\frac{C_0}{t_C} \frac{qVE_a}{R_0 T_C^2} = hS \tag{5.36}$$

The characteristic heat loss time t_h is shown in Eq. 5.30 to be $\rho V C_V/hS$. Equation 5.36 is therefore expressed as

$$\frac{C_0}{t_C} \frac{qVE_a}{R_0 T_C^2} = \frac{\rho V C_V}{t_h}$$

Simplifying, we have

$$\frac{t_h}{t_C} \frac{qE_a}{R_0 T_C^2} \frac{C_0}{\rho C_V} = 1 \tag{5.37}$$

Noting that the value of the critical temperature T_C is given for large activation energy by

$$T_C = T_\infty + \frac{R_0 T_\infty^2}{E_a}$$

giving $T_C \approx T_\infty$, we obtain on substituting in Eq. 5.37:

$$\frac{t_h}{t_C} \frac{q E_a}{R_0 T_\infty^2} \frac{C_0}{\rho C_V} = 1 \tag{5.38}$$

From the above equation, we get the ratio of the characteristic times t_C / t_h as

$$\frac{t_C}{t_h} = \frac{q E_a}{R_0 T_\infty^2} \frac{C_0}{\rho C_V} \tag{5.39}$$

The above condition corresponds to the critical or threshold condition of auto-ignition. If

$$\frac{t_C}{t_h} \gg \frac{q E_a}{R_0 T_\infty^2} \frac{C_0}{\rho C_V},$$

the heat loss rate is more than the rate of heat generation (characteristic heat generation time is greater than the heat loss time) and auto-ignition is not possible. On the contrary, if

$$\frac{t_C}{t_h} \ll \frac{q E_a}{R_0 T_\infty^2} \frac{C_0}{\rho C_V},$$

the condition of auto-ignition is met and explosion occurs. The heat release q in the above equation is in kJ/mole. The density ρ and the specific heat C_V have units of kg/m^3 and kJ/(kg K). If the concentration is expressed in mole/m^3 and activation energy in kJ/mole, the right side is dimensionless corresponding to the non-dimensional ratio of the heat release time and heat loss time given on the left side. The condition for explosion therefore requires the value of the parameter $(q E_a C_0)/(\rho C_V R_0 T_\infty^2)$ to be large. This implies that large values of heat release q and activation energy E_a and small values of specific heat C_V and initial temperature T_∞ would be desirable for a reactive substance to be susceptible for an explosion. The requirement of a large value q (energy release per mole) for an explosion to occur is understandable from energy considerations. A high value of activation energy and a low initial temperature will result in the parameter $E_a/(R_0 T_\infty)$ being large, which provides a long induction time. A spurt in energy release therefore takes place after the long induction time. If the activation energy were small and the ambient temperature high, the rate of heat release would start gradually at the initial time itself when $T = T_\infty$ instead of obtaining a spurt in energy release after the induction period. A small value of specific heat gives rise to higher temperature and a larger volume of the combustion products.

The non-dimensional parameter $qE_aC_0/(\rho C_V R_0 T_\infty^2)$ in Eq. 5.39 can be written as a product of two non-dimensional parameters $E_a/(R_0 T_\infty)$ and $q/(\tilde{C}_V T_\infty)$, where $\tilde{C}_V$ represents the specific heat in kJ/(mole K) with $\tilde{C}_V = \rho C_V/C_0$ [J/(mole K) = (kg/cm^3)(J/kgK)/(mole/cm^3)]. While $E_a/(R_0 T_\infty)$ indicates the peaking of the rate of energy release after an induction time, $q/(\tilde{C}_V T_\infty)$ represents the ratio of energy released from chemical reactions and the initial thermal energy stored in the substance. The product of these two non-dimensional numbers, representing a spurt of energy release and the chemical energy release to the initial thermal energy, is required to be large for an explosion to take place.

A small value of t_C implies a spurt in the reaction and is associated with a long value of induction time t_d (see Fig. 5.10) giving $t_d \gg t_C$.

5.5 Ignition and Auto-ignition

The ignition temperature and the critical temperature corresponding to auto-ignition depend on the heat loss and would therefore be expected to depend not only on the concentration of the reactant and its volume but also on the containment, heat transfer coefficient, and the ambient conditions. For reactive substances having high values of activation energy, we observe that the heat release effectively takes place after an induction time. Since the heat generation rate at the end of the induction time is very much higher than the heat loss rate, the auto-ignition conditions depend on the properties of the reactive substance. The auto-ignition temperatures tend to be useful to characterize a reactive substance. Measured values of auto-ignition temperatures of stoichiometric mixtures of some gaseous fuels and air at 1 atmosphere pressure, as available in the literature, are given in Table 5.1. It is to be noted that the auto-ignition temperatures would depend on the experimental procedure employed for its measurement.

The auto-ignition temperatures of fuels containing a mixture of hydrocarbons such as kerosene and JP-4 with air are also given in Table 5.1. The chemical structure of kerosene is close to that of n-dodecane, while JP-4 denotes an aviation fuel, which is a blend of kerosene and gasoline fuel along with a few additives to improve its physical properties. These have lower values of auto-ignition temperatures than hydrocarbons, methane, ethane, and propane. It is shown in Table 5.1 that the auto-ignition temperature decreases as the number of carbon bonds increases for the saturated hydrocarbons or the ratio of the carbon atoms to hydrogen atoms in the hydrocarbon increases. The trend of changes in auto-ignition temperature follows the heat release for saturated hydrocarbons. Significant changes of the auto-ignition temperature, however, are seen to take place when the molecular structure of the substance changes such as for methanol and ethanol. This is due to the change in activation energy for the overall chemical reaction, leading to different rates of energy release and hence significant changes to the auto-ignition temperature.

Table 5.1 Auto-ignition temperatures for stoichiometric mixture with air

Fuel/vapor	Chemical formula	Auto-ignition temperature (°C)
Methane	CH_4	540
Ethane	C_2H_6	515
Propane	C_3H_8	450
n-Hexane	C_6H_{14}	225
n-Heptane	C_7H_{16}	215
Kerosene	$\sim C_{12}H_{24}$	210
JP-4	–	240
Methanol	CH_3OH	385
Ethanol	C_2H_5OH	365
Hydrogen	H_2	406

5.6 Induction Times and Nature Of Chemical Reactions

The heat release $\dot{Q}$ is derived on the basis of the overall reaction of the form

$$\omega = AC^n \exp\left(-\frac{E_a}{R_0 T}\right) \frac{moles}{(cc\text{-}s)}$$

A number of reactions such as decomposition of fuel and oxidizer, formation of active intermediates, and further reactions leading to the formation of products take place when a fuel reacts with an oxidizer to form products of combustion. It is these different chemical reactions that are responsible for the existence of the induction time. This is illustrated in the following with the example of the chemical reaction between hydrogen and oxygen.

Hydrogen and oxygen, when heated, dissociate to form atoms and free radicals. The atoms and free radicals are inherently unstable and very reactive unlike the stable molecules of hydrogen and oxygen. They bring about further reactions and are referred to as chain carriers. The formation of chain carriers from a mixture of hydrogen and oxygen is given below by the following reactions:

$$H_2 \rightarrow H + H \tag{5.40}$$

$$O_2 \rightarrow O + O \tag{5.41}$$

$$H_2 + O_2 \rightarrow OH + OH \tag{5.42}$$

The above reactions are called as chain initiation reactions since the active radicals or chain carriers are formed in them. These chain carriers are very reactive and generate more chain carriers by reacting with H_2 and O_2; i.e., the chain carriers get propagated. A chain propagation reaction is given by

$$OH + H_2 \rightarrow H_2O + H \tag{5.43}$$

Here for each OH radical consumed in the reaction, one H radical is formed, thus maintaining the propagation of the chain carriers.

When the chain carriers increase during the chain propagation reaction, they are known as chain branching reactions. The chain branching reactions for hydrogen–oxygen reaction are

$$H + O_2 \rightarrow OH + O \tag{5.44}$$
$$O + H_2 \rightarrow OH + H \tag{5.45}$$

The chain carriers, when generated in plenty, combine among themselves to form relatively stable substances. The combination of the chain carriers is called as recombination reaction. Examples of recombination reaction are the following:

$$OH + H + M \rightarrow H_2O + M \tag{5.46}$$
$$H + O_2 + M \rightarrow HO_2 + M \tag{5.47}$$

M in the above refers to the molecules, which promote the reaction. They allow for the collisions to occur and hence aid in progress of the reaction between the respective chains and molecules. Some molecules are more effective than others.

The reaction between H_2 and O_2 in a simplified scheme can be represented by the sequence of reactions given by Eqs. 5.40–5.47. They are summarized below along with the rate ω of each reaction in moles/(cc-s).

1. Chain initiation reactions

a. $H_2 + M \rightarrow H + H + M$; $\omega = 2.08 \times 10^{15} T^{0.07} \exp\left[\dfrac{-434{,}000}{(R_0T)}\right] C_{H_2} C_M$

b. $O_2 + M \rightarrow O + O + M$; $\omega = 1.85 \times 10^{11} T^{0.5} \exp\left[\dfrac{-399{,}500}{(R_0T)}\right] C_{O_2} C_M$

c. $H_2 + O_2 \rightarrow OH + OH$; $\omega = 1.70 \times 10^{13} \exp\left[\dfrac{-199{,}700}{(R_0T)}\right] C_{H_2} C_{O_2}$

2. Chain branching reaction

d. $H + O_2 \rightarrow OH + O;$ $\omega = 1.22 \times 10^{17} T^{-0.907} \exp\left[\dfrac{-69,470}{(R_0 T)}\right] C_H C_{O_2}$

e. $O + H_2 \rightarrow OH + H;$ $\omega = 4.76 \times 10^{17} T^{-0.907} \exp\left[\dfrac{-69,470}{(R_0 T)}\right] C_O C_{H_2}$

3. Chain propagation reaction

f. $OH + H_2 \rightarrow H + H_2O;$ $\omega = 5.2 \times 10^{13} \exp\left[\dfrac{-27,170}{(R_0 T)}\right] C_{OH} C_{H_2}$

4. Chain termination reaction

g. $H + O_2 + M \rightarrow HO_2 + M;$ $\omega = 2 \times 10^{15} \exp\left[\dfrac{-3,640}{(R_0 T)}\right] C_H C_{H_2} C_M$

h. $H + OH + M \rightarrow H_2O + M;$ $\omega = 7.5 \times 10^{23} T^{-2.6} \exp\left[\dfrac{-434,000}{(R_0 T)}\right]$

$$\times C_H C_{OH} C_M$$

The value of R_0 (universal gas constant) is 8.314 J/(mole K). The chain initiation reactions a, b, and c absorb heat (endothermic). This can be readily seen since the standard heats of formation of the chain carriers are positive while for the standard elements H_2 and O_2, the heats of formation are zero. The chain branching reactions and chain propagation reactions are also either endothermic or mildly exothermic. Only the chain termination reactions, for which completed reaction products such as H_2O (having large negative values of heat of formation) are generated, can liberate significant amounts of energy. The initial phase of progress of a reaction, when chain carriers are predominantly getting formed, does not lead to any significant release of energy and corresponds to the induction time. This is shown in Fig. 5.13.

Fig. 5.13 Induction time from chain propagation and chain branching reactions

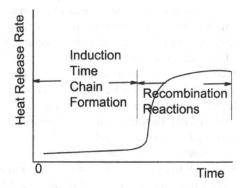

Substances forming products through chain propagation and chain branching mechanism are characterized by an induction time. Empirical expressions have been obtained for the induction times in gaseous mixtures such as hydrogen–oxygen and hydrocarbon–oxygen. For hydrogen–oxygen and acetylene–oxygen mixtures, the induction time, shown as τ in seconds, is given as function of temperature T in K and concentration C in moles/cc by the following expressions:

$$\log_{10}[\tau C_{O_2}] = -10.162 + 16,328/(4.58T) \qquad (5.48)$$

$$\log_{10}[\tau C_{O_2}^{1/3} C_{C_2H_2}^{2/3}] = -10.81 + 17,300/(4.58T) \qquad (5.49)$$

At high pressures, the recombination reactions begin to occur in the initial phase of reaction itself and a clear demarcation of the branching and propagation of chain carriers followed by their recombination is difficult.

5.7 Branched Chain Explosions in Closed Vessels

When chain carriers multiply rapidly, the resultant recombination reactions would provide large energy release rates leading to explosions. The initiation, propagation, branching, and termination reactions for the chains can be used to infer the explosion limits.

As an example, based on the chain initiation, propagation, and termination reactions for the reaction between hydrogen and oxygen in the last section, it appears possible that the formation of explosions could depend on the initial value of its pressure. At very low pressures, the concentration of hydrogen and oxygen is small, hence the rate of formation of chain carriers is negligible, and there is hardly any energy release from the recombination reactions. As the pressure increases, chain carriers are formed in greater amounts in the chain initiation, branching, and propagation reactions and this leads to higher rates of recombination reactions. Hence the rate of energy release rate increases. This is shown by the exponential increase in energy release rate (line AB) in Fig. 5.14 and leads to an explosion.

However, as pressure increases beyond a certain threshold of pressure, shown in Fig. 5.14 by p_B, the recombination reactions consume the chain carriers faster than they get formed in the initiation, branching, and propagation reactions. The net rate of production of chain carriers gets arrested. This leads to a net decrease in the rate of recombination reactions and the energy release could be reduced to a level, wherein no explosion is possible. But if the pressure exceeds a value greater than p_C, the higher rate of collisions between the molecules causes the H_2O and HO_2 formed in the recombination reactions to further break up into chains and the rate of chain formation accelerates rapidly. The increased chain carriers bring about more chain termination reactions and higher energy release rates as shown by CD in Fig. 5.14. This leads to an explosion. However, the temperature would also influence the energy release rates and is considered while formulating the explosion limit diagram.

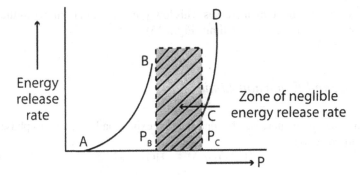

Fig. 5.14 Energy release rates based on chain reactions at the different pressures

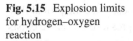

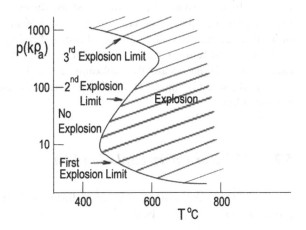

Fig. 5.15 Explosion limits for hydrogen–oxygen reaction

An explosion diagram, in the form of a reversed S curve in a plot of pressure versus temperature, is shown in Fig. 5.15 for the hydrogen–oxygen reaction. Three limiting regions, viz., the first, second, and third explosion limits, are encountered as shown in the figure. The existence of the explosion limits is discussed in the following.

5.7.1 First Explosion Limit

At low pressures, typically less than about 10 kPa, the concentration of H_2 and O_2 is small and the small number of chain carriers, generated during the reaction, is removed at the walls. An increase in pressure leads to the production of more chain carriers that accumulate, leading to recombination reactions and an explosion. This corresponds to the first explosion limit.

A higher value of temperature results in the generation of sufficient chain carriers even at low pressures. The pressure at which the chain carriers increase therefore decreases as the temperature increases. This low-pressure limit is known as the lower explosion limit or first explosion limit (Fig. 5.15).

5.7.2 Second Explosion Limit

When the pressure is increased beyond the first explosion limit, the gas phase chain termination reaction

$$H + O_2 + M \rightarrow HO_2 + M \tag{5.50}$$

removes the chain carrier H. HO_2 (hydroperoxide) molecules are produced. HO_2 is relatively unreactive and effectively destroys the chain branching process. As the pressure increases, the reaction rate given by Eq. 5.50 increases, since the probability of the three molecules colliding gets enhanced. A value of pressure therefore exists above which the removal rate of the chain carriers by the chain termination reaction exceeds their formation and the explosion ceases. This is shown in Fig. 5.15 and is called as the second explosion limit. The pressure increases with temperature for this explosion limit.

5.7.3 Third or Upper Explosion Limit

At values of pressure above the second explosion limit, HO_2 participates in the chain propagation process given below:

$$HO_2 + H_2 \rightarrow H_2O_2 + H \tag{5.51}$$

$$H_2O_2 \rightarrow OH + OH \tag{5.52}$$

This process leads to a rapid build-up of chain carriers and an explosion. The value of pressure corresponding to the above is called as third or upper explosion limit.

An increase of temperature beyond about 600°C results in breakup of HO_2, and explosion is observed at all values of pressure as shown in Fig. 5.15.

5.8 Limitations of Lumped Mass Assumption

The conditions for an explosion to occur are derived in this chapter based on the premise that the chemical reactions occurred at the same time uniformly over the entire spatial location in the given volume of the reactant. A small value of preheat is observed to be adequate to bring about an explosion when the overall activation energy of the chemical reaction is large. The large activation energy also provides an induction time during which no perceptible heat release takes place and is followed by a spurt in heat release. This feature leads to the existence of an auto-ignition temperature.

Ignition temperature and critical temperature are defined based on heat generated and heat lost from the volume of the combustible substance. A non-dimensional Damkohler number, defined as the ratio of t_h/t_C, is required to be large for an explosion to occur. This implies that the product of the ratio of the chemical energy release to the thermal energy stored in the substance $[q/(C_V T_\infty)]$ and a parameter representing the peaking of the reaction after an induction time $[E_a/(R_0 T_\infty)]$ must be high for an explosion to occur.

In order to reduce the spurt in chemical reaction after an induction time and thus reduce the susceptibility of a substance to an explosion, the inhibition of chain carriers, formed during the induction time, is a suitable option. Halogenated compounds containing bromine and iodine are very effective in arresting the multiplication of chain carriers. The removal of chain carriers (say H) by bromine and iodine compounds is given by the reaction

$$RX + H \rightarrow R + HX \tag{5.53}$$

where RX is the bromine/iodine compound. The sequence of reactions leading to the depletion of chain carrier H by hydrogen bromide, as an example, is given as

$$H + HBr \rightarrow H_2 + Br$$

$$Br + HBr \rightarrow H + Br_2$$

$$Br + Br + M \rightarrow Br_2 + M$$

$$Br + HO_2 \rightarrow HBr + O_2$$

The regeneration of HBr during the reactions effectively reduces the rate of formation of H. Methyl bromide (CH_3Br) and methyl chloride (CH_3Cl) are used for inhibiting the rate of chemical reactions and preventing the spurt of energy release. Phosphorus compounds are also effective in suppressing the formation of chain carriers. Monoammonium phosphate ($NH_4H_2PO_4$) and trisodium phosphate (Na_3PO_4) are also used as suppressants for fire and explosions.

While many features of an explosion could be discerned from the lumped mass model in this chapter, the zone of reactions is known to propagate from the place of ignition at a finite rate. The rate of heat release would depend on the spatial rate of propagation of the chemical reactions. We consider the propagation of the reaction front in the next chapter.

Examples

5.1 *Explosion of Fertilizer Grade Ammonium Nitrate in the Texas City Disaster:*

(a) The major explosion involving 7700 tons of fertilizer grade ammonium nitrate—FGAN (NH_4NO_3) stored in a hull of a ship at the Texas City port on April 16, 1947—resulted in extensive damage to property and loss of lives. The ammonium nitrate was contained in paper bags at the ambient temperature of 30 °C. It decomposed spontaneously leading to the explosion. Given that the rate of decomposition of ammonium nitrate has activation energy = 90 kJ/mole, estimate the increase in temperature that should have taken place in the stored FGAN for the explosion to occur.

(b) What is the value of critical temperature in the above explosion? Is it a function of amount of ammonium nitrate and the storage conditions?

(c) If the decomposition reaction of ammonium nitrate can be represented as: $NH_4NO_3 \rightarrow N_2O + 2H_2O$, determine the energy released during the explosion in MJ. You can assume the standard heats of formation of NH_4NO_3, N_2O, and H_2O as -369 KJ/mole, $+82$ KJ/mole, and -286 KJ/mole, respectively.

(d) The rate of decomposition of AN measured in small samples in the laboratory gave the decomposition rate in fraction of moles of AN decomposed per second as $10^{13.8} \exp[-80, 200/(R_0 T)]$ s^{-1}. Assuming the data to be valid over the range of temperatures of interest, determine the rate of energy release in the explosion.

(e) What is the characteristic time of the energy release?

Solution

(a) *Increase in temperature*: The increase in temperature at which FGAN would explode corresponds to the preheat temperature derived in Eq. 5.13. This is given as

$$T_C - T_\infty = \frac{R_0 T_\infty^2}{E_a}$$

Substituting the activation energy $E_a = 90,000$ J/mole, the universal gas constant $R_0 = 8.314$ J/(mole K), and the ambient temperature as 303 K, we get

$$T_C - T_\infty = \frac{8.314 \times 303^2}{90,000} = 8.48°C$$

A small temperature increase of about 8.5 °C above the ambient temperature of 30 °C is seen to be adequate to initiate the runaway reaction and the explosion.

(b) *Value of critical temperature*: From the preheat temperature, calculated in (a.), the critical temperature T_C is obtained as

$$T_C = 30 + 8.48 = 38.48°C.$$

Whether this temperature is reached in practice depends on the rate of heat loss rate from FGAN. The storage conditions and the quantity of the FGAN stored would be expected to influence the critical temperature. The hull of the ship would not allow any significant heat loss. As the quantity of FGAN increases, the relative proportion of heat loss to heat generation would decrease.

(c) *Energy released in the explosion*: The decomposition reaction of FGAN is given as

$$NH_4NO_3 \rightarrow N_2O + 2H_2O$$

The energy released in the reaction is

$$\Delta H_C = -\left\{ n_{N_2O}\Delta H^0_{fN_2O} + n_{H_2O}\Delta H^0_{fH_2O} - n_{AN}\Delta H^0_{fAN} \right\}$$

Substituting the number of moles and the corresponding values of the standard heats of formation:

$$\Delta H_C = -\{1 \times (+82) + 2 \times (-286) - 1 \times (-369)\} = 121 \text{ kJ}$$

The above energy release is from 1 mol of NH_4NO_3. The molecular mass of NH_4NO_3 is $(2 \times 14 + 4 \times 1 + 3 \times 16) = 80$ g/mole. Hence energy liberated from $7700t$ is

$$\frac{121 \times (7700 \times 1000)}{0.080} \text{ kJ} = 11.65 \times 10^6 \text{ MJ or } 11.65 \text{ million MJ}$$

(d) *Rate of energy release*: The rate of decomposition at the critical temperature of $38.48\,°C$ is:

$$10^{13.8} \exp\left[\frac{-80,200}{R_0 T}\right] s^{-1} = 10^{13.8} \exp\left[\frac{-80,200}{8.314 \times 311.48}\right]$$

$$= 0.631 \text{ s}^{-1}$$

$$\text{The rate of energy release} = 11.65 \times 10^6 \times 0.631 \text{ MJ/s}$$

$$= 7.35 \times 10^6 \text{ MW}$$

(e) *Characteristic time of energy release*: The characteristic time of energy release corresponds to the rate of decomposition and is $1/(0.631) = 1.58$ s. The heat loss time considering the near-insulated conditions in which the FGAN was stored would be very much larger. This would have led to accumulation of heat and caused an accelerated rate of heat release resulting in the explosion.

5.2 *Ammonium Nitrate Explosion at Beirut City*: The explosion of $2750t$ of ammonium nitrate (AN) stored in a warehouse at Beirut port on August 4, 2020, has been reported to have been initiated by the incendiaries from firecrackers that got ignited in an adjacent warehouse. Would it be necessary for the fireworks from the crackers to start the massive explosion or could the explosion have originated even without the firecrackers?

Solution Ammonium nitrate was stored in a large number of bags in an uncontrolled atmosphere in the warehouse. The humidity of the air at the port would be high, and AN being hygroscopic absorbs the moisture. Over the prolonged storage of about six years some amount of moisture would have been absorbed especially at the surface. However, the reaction of AN with water is endothermic and cannot cause an increase of temperature though small quantities of ammonia (NH_3) and NO_2 gets released. The bags of AN are not kept in a ventilated atmosphere. The small amount of incipient reactions between the ammonia and AN along with natural incipient decomposition of AN at the slightly elevated temperatures of summer in Beirut can slowly and gradually lead to the preheat corresponding to a degree or a fraction of a degree Celsius. But the heat transfer from the bulk of AN is almost impossible and is further aggravated by the insulated lumps of the dissociated AN from moisture absorption at the surface. The characteristic heat transfer times become even larger due to the reduced heat transfer from AN. Hence the very small or negligible excursions in the heat generated within the AN get accumulated, and over a significant duration of time, preheating in some relatively insulated regions of the AN to a value about 8 °C determined in the previous example is possible. The preheat results in AN decomposing and formation of the white smoke that preceded the first minor explosion before the massive explosion at Beirut.

5.3 *Characteristic Heat Loss Time*: A cloud of a gas of spherical volume of diameter 0.25 m has a density of $1.2\,kg/m^3$, molecular mass 30 g/ mole, and specific heat ratio of 1.25. The heat transfer coefficient at the surface of the cloud is $30\,W/m^2$ °C. Calculate the characteristic heat loss time.

Solution: The characteristic heat loss time is derived from the lumped mass assumption in Eq. 5.29:

$$t_h = \frac{\rho V C_V}{hS}$$

For a spherical volume of radius r, the ratio of volume to surface V/S is

$$\frac{V}{S} = \frac{4\pi^2 r^3/3}{4\pi r^2}$$
$$= \frac{r}{3}$$

The specific heat of the gas at constant volume is from its specific heat ratio as

$$C_V = \frac{R_0}{\gamma - 1}$$
$$= \frac{8.314}{0.25}$$
$$= 33.256 \text{ J/(mole K)}$$

The value in J/(kg K) is

$$\frac{33.256}{0.03} = 1108.53 \text{ J/(kg K)}$$

Here 0.03 is the molecular mass of the gas in kg/mole.

Substituting the values of density ρ, specific heat C_V, heat transfer coefficient h, and the volume-to-surface ratio V/S in the equation for heat loss time, we get

$$t_h = \frac{1.2 \ (\text{kg/m}^3) \times 1108.53 \ [\text{J}/(\text{kgK})] \times 0.25/6 \ (\text{m})}{30 \ [\text{J}/(\text{m}^2\text{Ks})]}$$

The characteristic heat loss time is 1.85 s.

Nomenclature

A	Constant
C	Concentration (moles/cc)
C_0	Initial concentration of the reactant (moles/cc)
C_V	Specific heat at constant volume [kJ/(kg K)]
C_V'	Molar specific heat at constant volume [J/(mole K)]
E_a	Activation energy (J/mole)
h	Heat transfer coefficient [kJ/(m^2 s)]
n	Exponent of concentration in Arrhenius reaction rate law
$\dot{Q}_C$	Rate of heat release from chemical reactions (kJ/s)
$\dot{Q}_L$	Rate of heat loss from the given volume (kJ/s)
q	Energy release per mole (kJ/mole)
R_0	Universal gas constant (8.314 J/(mole K)
S	Surface area (m^2)
T	Temperature (°C)
T_C	Temperature corresponding to critical conditions (°C)
T_h	Hot gas temperature (°C)
T_m	Maximum temperature (°C)
t_C	Characteristic time for chemical heat release (s)
t_d	Induction time for chemical reactions (s)
t_h	Characteristic heat loss time (s)
V	Volume (m^3)
ρ	Density (kg/m^3)
ω	Rate of chemical reactions [moles/(cc-s)]

Subscripts

c	Critical conditions; chemical reaction
i, o	Initial conditions
w	Wall
∞	Ambient conditions

Further Reading

1. Hainer, R. M., *The Application of Kinetics to the Hazardous Behavior of Ammonium Nitrate*, 5th International Symposium on Combustion, Reinhold, New York, Chapman and Hall, London, 1955, pp. 224–230.

2. Emmons, H. W., Scientific progress on fire, *Annual Review in Fluid Mechanics*, Vol. 12, 223, 1980.
3. Hermance, C.E., What do we Know About Ignition? Presented at 25th Canadian Chemical Engineering Congress, Montreal, November, 1975.
4. Lewis, B. and von Elbe, G., *Combustion, Flames and Explosion of Gases*, Academic Press, New York, 1961.
5. Penner, S.S., *Chemistry Problems in Jet Propulsion*, Pergamon Press, London, 1957.
6. Quintiere, J.G., *Fundamentals of Fire Phenomenon*, Wiley, Sussex, 2006.
7. Semenov, N.N., *Some Problems in Chemical Kinetics and Reactivity*, Princeton University Press, Princeton, NJ, 1959.
8. Sokolik, A.S., *Self Ignition, Flame and Detonation in Gases*, Akad. Nauk. SSR, English Translation, Jerusalem, 1963.
9. Strehlow, R. A., *Fundamentals of Combustion*, 2nd ed., McGraw-Hill, New York, 1984.
10. Stull, D. R., *Fundamentals of Fire and Explosion*, AIChE Monograph Series, Vol. 73, No. 10, 1977.
11. Turns, S. R., *An Introduction to Combustion: Concepts and Applications*, 2nd ed., McGraw Hill, Boston, MA, 2006.

Exercise

5.1 A number of buildings stocking firecrackers have reported accidental explosions and fires in recent years. Based on the thermal theory of explosions dealt with in this chapter, suggest suitable safety aspects that need to be incorporated in the storing of the firecrackers.

5.2 (a) The activation energy of a given stoichiometric reaction of propane and air is given as 90,000 J/mole. If the given mixture is at 25 °C and 100 kPa pressure, determine the preheat temperature.

(b) What is the concentration of propane in moles/m^3 in the above stoichiometric mixture?

(c) If the above mixture is contained in a spherical volume of diameter 0.5 m and the value of the heat transfer coefficient at the surface is 800 W/(m^2 °C), calculate the characteristic heat transfer time. The specific heat of the mixture is 700 J/(kg °C).

(d) If the characteristic chemical reaction time for the above mixture is 1.9 ms, calculate the Damkohler number. Also determine whether the mixture would explode in the context of the thermal theory.

5.3 The preheat required for a mass of 2 kg of a lumped mass of reactive gas of specific heat $C_P = 1200$ J/(kg K) to undergo spontaneous chemical reactions is 162 kJ. If the ambient temperature is 30°C, determine the critical temperature.

Chapter 6
Propagation of Reaction Front: Detonation, Deflagration and Quasi-Detonation

When the temperature of a reactive medium is increased, chemical reactions take place in it and energy is released. The region of energy release from chemical reactions constitutes the combustion zone, and temperature increases significantly across it. The chemical reaction is initiated in a small region of the reactive medium by an ignition source, and the zone of chemical reaction spreads or propagates over the entire volume of the reactive medium. The zone of chemical reaction is generally of very small thickness and is referred to as a combustion wave. We shall, in this chapter, determine the characteristics of a combustion wave and define three types of combustion waves assuming them to propagate as one-dimensional waves. The structure of the waves and their properties are also discussed.

6.1 Propagation of One-Dimensional Combustion Waves: Reaction Hugoniot and Rayleigh Line

Consider a one-dimensional combustion wave propagating in a stationary medium whose initial density is ρ_0, pressure is p_0 and temperature is T_0. The scheme of propagation is similar to the wave propagation considered in Chap. 2 except that heat is additionally released in the wave. Denoting the heat release per unit mass of the medium as Q and the velocity of propagation of the wave as v_0, instead of the symbol $\dot{R}_S$ used in Chaps. 2 and 3 on blast waves, the energy balance gets modified as

$$h_0 + \frac{v_0^2}{2} + Q = h + \frac{v_1^2}{2} \qquad (6.1)$$

© The Author(s), under exclusive license to Springer Nature Switzerland AG 2021
K. Ramamurthi, *Modeling Explosions and Blast Waves*,
https://doi.org/10.1007/978-3-030-74338-3_6

Fig. 6.1 Change of
parameters across the
combustion wave

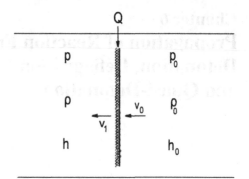

Fig. 6.2 Reaction Hugoniot
as distinct from shock
Hugoniot

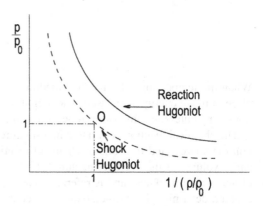

The schematic of the changes with the wave fixed in position (frame of reference
is with respect to the combustion wave) is shown in Fig. 6.1. Here the unburned
medium approaches the combustion wave with a velocity v_0 while the burned gas
medium leaves the wave with a velocity v_1. The enthalpy of the unburned gas is h_0
while that of the burnt gas is h.

Further, assuming the medium as a perfect gas, the above equation, following the
derivation of the Hugoniot in Chap. 2, reduces to the form

$$\frac{\gamma}{\gamma - 1}\left(\frac{p}{\rho} - \frac{p_0}{\rho_0}\right) - \frac{1}{2}(p - p_0)\left(\frac{1}{\rho_0} + \frac{1}{\rho}\right) = Q \tag{6.2}$$

Here, the heat release term Q gets added on the right side of Eq. 6.2 when compared
to the shock Hugoniot. The equation is known as reaction Hugoniot as distinct from
the shock Hugoniot for a shock wave. Figure 6.2 shows the reaction Hugoniot to be
above the shock Hugoniot for a given positive value of Q. The shock Hugoniot is
shown dotted in Fig. 6.2. Unlike the shock Hugoniot, the reaction Hugoniot does not
pass through the initial density ρ_0 and initial pressure p_0. The final density ρ and
pressure p of the combustion wave would be on the reaction Hugoniot.

Fig. 6.3 Different combustion wave velocities from reaction Hugoniot

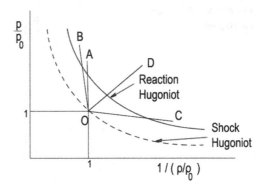

The final state of the medium is determined as in Chap. 2 by solving the reaction Hugoniot and the Rayleigh line for given values of velocity of the wave. The Rayleigh line is derived from momentum equation and is

$$v_0^2 = \frac{p_0}{\rho_0} \frac{p/p_0 - 1}{1 - \rho_0/\rho} \tag{6.3}$$

Different values of the slope of the lines connecting the initial point O with properties (p_0, ρ_0) to the reaction Hugoniot, illustrated in Fig. 6.3 by OA, OB, OC and OD, give different values of velocities v_0 for the combustion wave. However, all the different lines are not physically realizable in practice, and all the points on the reaction Hugoniot cannot be achieved. There are restrictions on possible velocities of combustion waves.

6.2 Physically Realizable States on Reaction Hugoniot: Detonations and Deflagrations

The realizable states on the reaction Hugoniot curve are determined by subdividing it into five regions by drawing tangents to it from the initial state 'O' (Fig. 6.4) and drawing vertical and horizontal lines from 'O'. Five regions are thus formed on the reaction Hugoniot, and these are indexed from I to V as shown in Fig. 6.4. The upper and lower tangent points on the reactive Hugoniot are indicated by U and L, respectively, while the intersection points of the vertical and horizontal lines with the Hugoniot are shown by A and B, respectively in Fig. 6.4.

The region 'V' between A and B on the reaction Hugoniot has $p > p_0$ and $\rho < \rho_0$. While the increase in pressure shows it as a compression, the decrease in density indicates it to be an expansion. This is therefore a physically impossible region. The value of v_0^2 from Eq. 6.3 would be negative, leading to imaginary value of the combustion wave velocity v_0. Regions I and II in Fig. 6.4 correspond to $p > p_0$ and

Fig. 6.4 Realizable states on
the Hugoniot

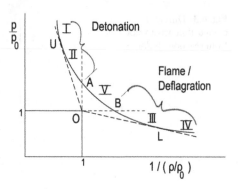

$\rho > \rho_0$, while regions III and IV correspond to $p < p_0$ and $\rho < \rho_0$. Regions I and II imply compression of the medium behind the combustion wave (such as shock wave) as against regions III and IV for which expansion takes place in the combustion wave.

We therefore have two possibilities for the combustion waves—one compression (regions I and II) and the other (regions III and IV) expansion. The compression is realized from a shock wave compressing the gases and heating it for the burning to occur. This form of combustion wave is known as detonation. Since the speed of a shock wave is supersonic, the velocity of detonation would also be supersonic.

The expansion region (III and IV) is obtained from diffusion and heating up of the reactive medium from the adjacent burned layers. The pressure and density decrease across the wave. This form of combustion wave is known as deflagration or flame.

The points U and L, which are tangent to the reaction Hugoniot, are called as upper and lower Chapman–Jouguet points, respectively.

6.2.1 Chapman–Jouguet (CJ) Points

The points in the upper branch of the Hugoniot (detonation) have higher values of slope of the Rayleigh line and therefore higher combustion velocities than those corresponding to the lower deflagration branch. The velocities of the combustion wave in the upper branch realized by shock compression are supersonic, while for the lower deflagration branch the wave speed is subsonic.

Among the points in the upper branch, the CJ point 'U' in the upper branch of the Hugoniot is seen to have minimum velocity based on the slope of the Rayleigh line. In the lower deflagration branch, however, the CJ point L has the maximum velocity. At the points of tangency, the slopes of the line joining the CJ points U and L with the initial state 'O' are given by

$$\frac{p - p_0}{1/\rho - 1/\rho_0}$$

and the slope of the Hugoniot curve $dp/(d(1/\rho))$ is the same. The slope of the Hugoniot is obtained from the energy equation (Eq. 2.2) by differentiating it. We get on differentiation

$$\frac{d}{d(1/\rho)}\left[\frac{\gamma}{\gamma-1}\left(\frac{p}{\rho}-\frac{p_0}{\rho_0}\right)-\frac{1}{2}(p-p_0)\left(\frac{1}{\rho_0}+\frac{1}{\rho}\right)-Q\right]=0 \qquad (6.4)$$

which on simplification gives

$$\left[\frac{\gamma}{\gamma-1}\left(p+\frac{1}{\rho}\frac{dp}{d(1/\rho)}\right)-\frac{1}{2}(p-p_0)-\frac{1}{2}\left(\frac{1}{\rho_0}+\frac{1}{\rho}\right)\frac{dp}{d(1/\rho)}\right]=0$$

The slope $dp/d(1/\rho)$ from the above equation is

$$\frac{dp}{d(1/\rho)}=\frac{(p-p_0)-\frac{2\gamma}{\gamma-1}p}{\frac{2\gamma}{\gamma-1}\frac{1}{\rho}-\left(\frac{1}{\rho_0}+\frac{1}{\rho}\right)} \qquad (6.5)$$

Equating the above slope with the slope of the Rayleigh line joining the tangency point with the initial state, we get

$$\frac{(p-p_0)-\frac{2\gamma}{\gamma-1}p}{\frac{2\gamma}{\gamma-1}\frac{1}{\rho}-\left(\frac{1}{\rho_0}+\frac{1}{\rho}\right)}=\frac{p-p_0}{\frac{1}{\rho}-\frac{1}{\rho_0}} \qquad (6.6)$$

On simplifying

$$2(p-p_0)\frac{1}{\rho}\left(\frac{\gamma}{\gamma-1}-1\right)=-\frac{2\gamma}{\gamma-1}p\left(\frac{1}{\rho}-\frac{1}{\rho_0}\right)$$

or

$$\frac{p-p_0}{\frac{1}{\rho_0}-\frac{1}{\rho}}=-\gamma p\rho \qquad (6.7)$$

From Eq. 6.3 of the Rayleigh line, we have

$$v_0^2=\frac{1}{\rho_0^2}\left(\frac{p-p_0}{\frac{1}{\rho_0}-\frac{1}{\rho}}\right) \qquad (6.8)$$

The velocity v_1 behind the combustion wave is in the frame of reference of the combustion wave itself (Fig. 6.1); using the mass conservation equation $\rho_0 v_0 = \rho v_1$, Eq. 6.8 becomes

$$v_1^2=\frac{1}{\rho^2}\left(\frac{p-p_0}{\frac{1}{\rho_0}-\frac{1}{\rho}}\right) \qquad (6.9)$$

From Eqs. 6.7 and 6.9, we get

$$v_1^2 = \frac{\gamma p}{\rho} \tag{6.10}$$

But $\gamma p/\rho$ is the square of the sound velocity behind the combustion wave. At the CJ points, therefore, the velocity of the medium behind the combustion wave is sonic ($M = 1$).

6.2.2 Detonation Branch of Hugoniot

a. Chapman–Jouguet (CJ) Velocity The velocity of burned gas corresponding to the CJ point U (see Fig. 6.5) is at sonic velocity with respect to the wave and will therefore prevent any rarefaction disturbance from reaching and interacting with the combustion wave. The velocity of propagation of the combustion wave will therefore be a constant. The constant or steady value of velocity achieved at the point U is called as the Chapman–Jouguet (CJ) detonation velocity.

b. Overdriven Detonation; Region I of Detonation Branch

In the region I above the CJ point U on the reaction Hugoniot (Figs. 6.4 and 6.5), the value of pressure is higher than at the CJ point. Hence, the gas is compressed more in the region I than at point U. The velocity v_0 would be higher considering the larger slope of the Rayleigh line. This region having higher velocity and higher pressure than the CJ point is known as strong detonation or overdriven detonation region.

Being more compressed than at the CJ point U, the velocity v_1 in the medium behind the wave in the overdriven region is less than at U where it is sonic. The subsonic speed allows the rarefaction disturbances to catch up with the combustion wave and slow it down. The region, though having a higher speed of the wave than a CJ detonation, corresponds to decaying detonation waves. A detonation exists in this overdriven region only as a decaying wave.

Fig. 6.5 Detonation branch of Hugoniot

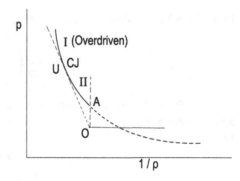

c. Weak Detonations; Region II of Detonation Branch

Here the pressure behind the wave is less than at the CJ point. The higher expansion compared to point U results in its velocity v_1 being higher than the speed of sound. Expansion of the gases to supersonic speeds in a constant flow area is not possible from subsonic velocities such as obtained behind the shock wave. It is therefore not possible to have a detonation in this region. The region is known as weak detonation region and cannot physically exist.

6.2.3   Deflagration Branch

The deflagration branch of the Hugoniot wherein the pressure and density decreases across the wave is shown in Fig. 6.6. The points in branch III (portion BL) and branch IV (portion beyond L) have smaller values of velocity than at the CJ point L.

In branch III between B and L (Fig. 6.6) density ρ and pressure p are higher than at the CJ point L. The gases behind the wave are not sufficiently accelerated as at L, and the region is called as weak deflagration.

Branch IV of the Hugoniot has density and pressure lower than at L so that the velocity behind the wave would be higher than the sonic velocity. It is known as strong deflagration region. However, it is not possible to expand the gases to supersonic speeds from subsonic speeds in a constant area flow as cited earlier. The strong deflagration region cannot therefore exist in practice.

6.2.4   Realizable Combustion Waves

The only realizable combustion waves, based on discussions in the last section, are CJ detonation and CJ deflagration (U and L in Fig. 6.4), decaying overdriven detonation

Fig. 6.6 Deflagration branch of Hugoniot

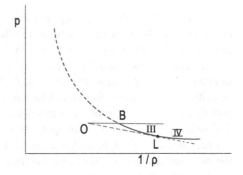

Fig. 6.7 Realizable
combustion waves

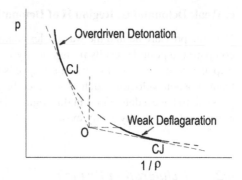

(region I in Fig. 6.4) and weak deflagration (region III in Fig. 6.6). These are shown
by the CJ points and the darkened portions of the reaction Hugoniot in Fig. 6.7. We
shall consider these realizable states of combustion waves in the following sections.

6.3 Detonations

6.3.1 Detonation Velocity V_{CJ} and Pressure p_{CJ} at the CJ Point U

Values of pressure p_S, temperature T_S and density ρ_S behind a shock were determined
in Chap. 2 from Rankine–Hugoniot relations and the Rayleigh line assuming the
medium to be a perfect gas. The same procedure is followed for CJ detonation by
additionally incorporating the term Q in the energy equation. Further, if it is assumed
that the pressure behind a detonation at the CJ point is much higher than the initial
value of pressure p_0, the velocity of a detonation can be related to the heat release Q
by the expression:

$$V_{CJ}^2 = 2(\gamma^2 - 1)Q \tag{6.11}$$

Here Q is the heat release in J/kg, and γ is the specific heat ratio which is taken to
be the same in the reactive medium and in the combustion products. The derivation
of the above equation is given in Annexure F.

In addition to the heat release term, the specific heat ratio γ is seen from Eq. 6.11
to influence the detonation velocity V_{CJ}. As the specific heat ratio γ decreases, the
velocity also decreases. Substances having lower values of γ have relatively more
complex molecular structure. For example, monatomic helium has a value of γ of
1.67, while the diatomic air or oxygen has γ of 1.4. As the molecular structure of the
gas becomes more complicated, a larger fraction of the heat released gets locked into
the molecules as its internal energy and the availability as kinetic energy for driving
the detonation wave decreases.

The density, pressure, and velocity behind a CJ detonation given by ρ_{CJ}, p_{CJ}, and u_{CJ} respectively are derived in Appendix F (Eqs. F.11 and F.14) and are obtained as:

$$\frac{\rho_{CJ}}{\rho_0} = \frac{\gamma + 1}{\gamma} \tag{6.12}$$

$$\frac{p_{CJ}}{p_0} = \frac{\gamma M_{CJ}^2}{\gamma + 1} \tag{6.13}$$

$$\frac{u_{CJ}}{V_{CJ}} = \frac{\gamma}{\gamma + 1} \tag{6.14}$$

Here M_{CJ} is the Mach number of the detonation wave and $M_{CJ} = V_{CJ}/a_0$, a_0 being the velocity of sound in the initially unreacted medium.

The density and pressure from Eqs. 6.12 and 6.13 are seen to be lower than for a shock having the same Mach number as the CJ detonation. This is due to the expansion of gases behind the shock wave as heat is released in the expanding gas. The CJ velocities and pressures in a detonation in stoichiometric mixture of some typical gaseous fuels with air at the ambient pressure of 1 atmosphere and 25 °C is given in Table 6.1.

For mixtures or compounds of fuels and oxidizers in solid or liquid phase, known as condensed explosives, the procedure for calculating the detonation velocity and pressure is the same as for the gaseous fuels. However, the equations of state for the solid and liquid explosives and for the high-pressure gases formed cannot be approximated as ideal gases, and we need to consider the tightly packed molecules. The ideal gas equation of state would not be valid, and complex equations of state are required to be used for determining the shock Hugoniot and the reaction Hugoniot. The velocities of detonation in the condensed explosives are 4 to 5 times higher than in gaseous fuel–air mixtures, while the pressures are some two to four orders of magnitude higher. We shall deal with detonations in condensed explosives in Chap. 8.

Table 6.1 Detonation velocities and pressures of stoichiometric mixture of gaseous fuels with air (1 atm., 25 °C)

S. No.	Gaseous fuel	V_{CJ} (km/s)	p_{CJ} (atm.)
1	Methane CH_4	1.8	16.05
2	Ethane C_2H_6	1.8	17.0
3	Propane C_3H_8	1.8	17.27
4	Butane C_4H_{10}	1.8	17.44
5	Acetylene C_2H_2	1.87	18.0
6	Hydrogen H_2	1.96	14.6

6.3.2 One Dimensional Structure of a Detonation

The processes involved in a detonation can be understood from Fig. 6.8. Here the reaction Hugoniot corresponding to heat release Q and the shock Hugoniot (in the absence of heat release, i.e., $Q = 0$), which passes through the initial state p_0 and ρ_0 at 'O' are shown. The Chapman–Jouguet detonation velocity (V_{CJ}) corresponds to the Rayleigh line OU. The line OU is tangent to the reaction Hugoniot at the point U.

An extension of the line OU, tangent to the reaction Hugoniot at U, meets the shock Hugoniot at S, as shown in Fig. 6.8. The values of p and $1/\rho$ immediately behind the shock wave traveling at velocity V_{CJ} must be on the shock Hugoniot before heat is released and therefore corresponds to point S. Values of pressure and density (p_S and ρ_S) for the shocked medium at S are seen to be higher than those corresponding to the CJ detonation given by point U (p_{CJ} and ρ_{CJ}) in Fig. 6.8. The gas expands from condition at S to the state at U as heat from the chemical reaction heats it up.

The structure of a Chapman–Jouguet detonation that emerges from the above figure is of a shock wave traveling at the Chapman–Jouguet velocity V_{CJ} and compressing the medium to a high pressure and density p_S and ρ_S, respectively. The high pressure and high temperature from the compression of the shock to p_S and T_S cause spontaneous chemical reactions after a certain value of induction time. The heat release from the chemical reactions further increases the temperature to a high value T_{CJ} corresponding to the temperature in a detonation. However, it would take some time for the chemical reactions to start releasing heat, and this time was referred to as the induction time. A schematic of the temperature increase from the initial ambient value T_0 to T_S after the shock and the subsequent increase to T_{CJ} after an induction time is shown in Fig. 6.9. The temperature remains constant at T_S during the induction period when there is hardly any heat release.

The evolution of pressure and density is shown in Figs. 6.10 and 6.11. The pressure rises to a high value p_S from the initial value p_0 at the shock and remains at this value during the induction period since this is the incubation time for chemical reactions

Fig. 6.8 Shock Hugoniot and reaction Hugoniot

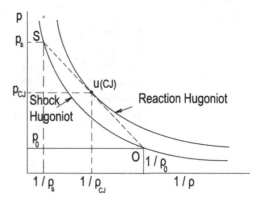

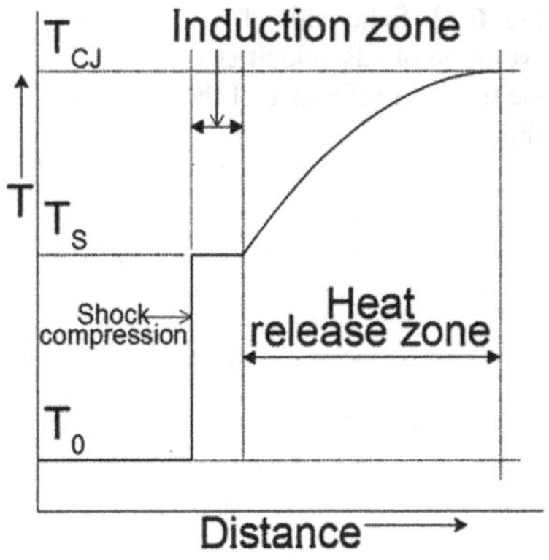

Fig. 6.9 Schematic of temperature evolution

Fig. 6.10 Schematic of evolution of pressure

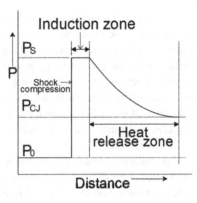

Fig. 6.11 Schematic of evolution of density

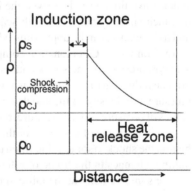

to get started. The expansion of gases due to the subsequent heat release causes the pressure to drop to the value p_{CJ}. Similarly, the density increases from the initial ambient value ρ_0 to ρ_S at the shock, remains at this value during the induction period, and thereafter decreases during the expansion process associated with heat release to ρ_{CJ}.

Fig. 6.12 Schematic of
evolution of gas velocities in
the frame of reference of the
shock

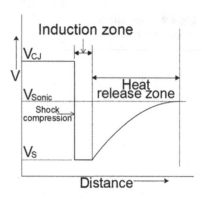

The velocity of the gases in the frame of reference of the shock wave is shown
in Fig. 6.12. The shock wave front travels at the detonation velocity. In the frame of
reference of the shock wave considered stationary, the unreacted gases at the ambient
conditions would therefore approach it at the Chapman–Jouguet velocity V_{CJ}. The
velocity decreases to V_S upon compression by the shock. The velocity V_{CJ} upstream
of the shock is supersonic, while V_S downstream of it is subsonic in the frame of
reference of the shock. During the induction period the velocity continues to be V_S.
It increases from the subsonic value of V_S as expansion takes place from heat release
to the sonic value V_{SONIC}.

The thickness of a detonation comprises of the shock thickness and the thickness
associated with the induction zone and the zone of heat release. The thickness of
a detonation would therefore be more than an order of magnitude greater than the
shock wave thickness. However, the increased thickness of a detonation would still
be much smaller than the length scales of interest in the detonation phenomenon,
and a detonation can be approximated as a thin front or wave as was assumed earlier.

In summary, a CJ detonation comprises of a shock front traveling at CJ velocity.
The compression at the shock causes heat release from chemical reactions to occur
after an induction time. The expansion process associated with heat release causes the
pressure and density of the shocked particles of gas to fall to the Chapman–Jouguet
values of pressure and temperature, respectively. The velocity of the shocked particles
of gas in the frame of reference of the shock wave increases from the subsonic value
behind the shock to sonic value when the heat release is over. A schematic of the Mach
number changes in the frame of reference of the detonation is shown in Fig. 6.13.

The sonic velocity at the tail-end of the heat release in the frame of reference of
the wave front does not allow any disturbance to catch up with it with the result that
the CJ detonation travels at constant velocity as a steady-state detonation.

In the case of an overdriven detonation (region I in Fig. 6.5), the velocity is
higher than for a CJ detonation and the larger compression from the higher shock
velocity causes the pressure p_S and pressure at the end of the heat release to be
more than corresponding to those in CJ detonation. The pressure being higher than

Fig. 6.13 Schematic of
Mach number variations in
the frame of reference of the
detonation

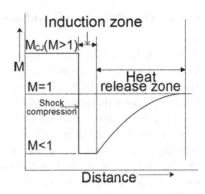

CJ value results in the end velocities to be lower than sonic value obtained in CJ detonation. The subsonic velocities allow the disturbances to catch up and the velocity of the overdriven detonation decays till it reaches the steady-state Chapman–Jouguet velocity.

6.3.3 ZND Structure of a Detonation

The changes that take place in the shock compression followed by heat release are incorporated in the Zeldovich–von Neumann–Doring (ZND) structure of a detonation. Instead of a simplified picture with shock wave causing spontaneous chemical reactions and heat release which drives a detonation, the ZND structure considers the process of shock heating and the heat release taking place after an induction zone. A schematic of a simplified detonation structure from heat release at the shock front is compared with the ZND structure of a detonation in Fig. 6.14.

Fig. 6.14 Schematic of
detonation

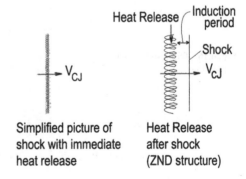

6.3.4 Detonation Cell and Multi-headed Detonation Front

The shock wave, which initiates the combustion process in a detonation, is not a single shock wave front traveling at a constant velocity. The front consists of multiple shocks. We had seen in the chapter on blast waves about the interaction of an incident blast wave with the ground and noted that at larger values of incident angle the reflected shock wave spurts off from the ground. A Mach stem shock connects the ground with the point of interaction between the incident and reflected shock waves forming a triple shock interaction. Figure 6.15 compares a regular reflection of an incident wave with a Mach reflection. The point O at which the three shocks—incident, reflected, and Mach stem shocks meet is known as a triple point.

The multiple shocks at the detonation front comprise of Mach interactions with incident, reflected and Mach stem shocks. The Mach stem shock is a strong shock while the incident and reflected shocks are weaker. The reflected shock moves in a transverse direction compared to the axial movement of the incident and Mach stem shocks along the detonation as shown in Fig. 6.16. The reflected shock is therefore also known as a transverse shock.

The lateral movement of the transverse shock and the axial movement of the incident and Mach stem shocks lead to the triple point moving at an angle to the axial direction. The trajectory of the triple point is shown by the dotted straight line in Fig. 6.17. Since the Mach stem shock is the strongest of the three shocks, the zone of combustion behind it will be nearer to it while for the relatively weaker incident and reflected shocks the zone of combustion is located further away. The detonation is therefore not a clean one-dimensional shock followed by a one dimensional heat release. It consists of a series of shocks with combustion distributed non-uniformly behind the different shocks as shown in Fig. 6.17.

Fig. 6.15 Comparison of a regular reflection with a Mach reflection

Fig. 6.16 Triple shock in a detonation

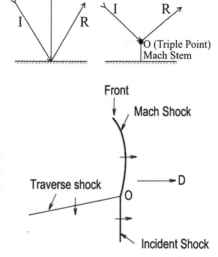

Fig. 6.17 Trajectory of triple point (dotted line) and heat release zones

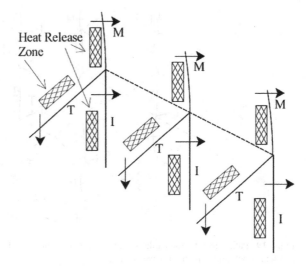

Fig. 6.18 Detonation cell

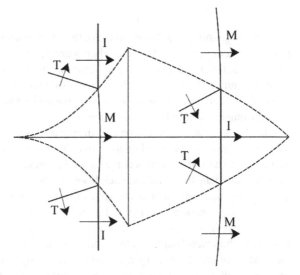

   The Mach number of the three shocks (Mach stem, incident, and transverse) is also not a constant but keeps fluctuating between some upper and lower limits. As a consequence the triple point does not move along a straight line but rather as an exponentially increasing or decreasing curve. The trajectory or path traced by the triple points resembles a diamond or a fish scale and is shown in Fig. 6.18. The Mach stem shock is represented by '*M*', while the incident and transverse shocks are shown as '*I*' and '*T*' respectively. The fish scale pattern formed by the path of the triple point is known as a detonation cell.

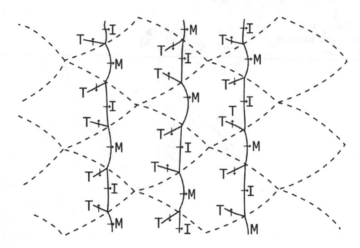

Fig. 6.19 Path traced by the triple point shown by broken lines: M denotes Mach stem shock, I is incident shock and T is the transverse shock

At the beginning of a detonation cell, the Mach number of the leading shock in it is a Mach stem shock. Its Mach number is greater than the Chapman–Jouguet value (M_{CJ}). It continually decays as it progresses within the cell and at the exit of the cell, its Mach number is about 0.6–0.8 times the value of M_{CJ}. In other words, at the start of a cell of detonation, the wave within it is overdriven while at it end the shock wave is not strong enough to form a detonation.

The shock, within a detonation cell, during its motion in the first half of the cell is a Mach stem shock corresponding to an overdriven detonation. This shock decays and in the second half of the cell becomes a weaker incident shock. The shock extending from this incident shock into the neighboring cell starts as a Mach stem shock. The transverse shocks intersect at the start of a detonation cell. A series of such cells formed in a detonation is shown in Fig. 6.19.

The multiple shocks at the detonation front give it a multi-headed shock structure. The Chapman–Jouguet detonation, though discussed as a steady-state wave in the context of the shock Hugoniot and reaction Hugoniot, is actually unsteady and propagates in spurts when observed in minute detail. At the point of intersection of the two adjacent transverse waves, viz. the start of the detonation cell, there is a spurt in the Mach number of the frontal shock. The sudden increase of the Mach number is due to surge in the chemical reactions and associated heat release from the higher temperatures due to the interaction of the adjacent transverse shocks. The Mach number of the frontal shock thereafter decays within the cell of detonation (Fig. 6.20).

The presence of the transverse waves in a detonation is responsible for the existence of the detonation cells and the transverse shocks periodically energize the frontal shocks. If by some mechanism such as by having a flexible confinement or a porous medium at the walls, the transverse shocks can be absorbed or removed, a

Fig. 6.20 Decay of Mach
number of the frontal shock
within the detonation cell

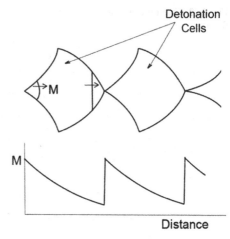

Fig. 6.21 Characteristic size
λ of a detonation cell

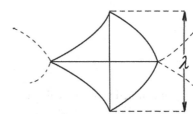

detonation can no longer propagate as a steady state wave. This procedure has been applied to arrest a detonation.

The size of the detonation cell is represented by its width λ (Fig. 6.21) and is an important parameter characterizing a detonation. It depends on the reactivity of the mixture and its physical and chemical properties. A minimum number of cells are required in a detonation to maintain the interaction of the transverse waves at the confluence of the triple points and thus provide for the propagation of a detonation.

6.4 Deflagration and Burning Velocities

While a detonation propagates at supersonic speeds from the shock–induced combustion, the deflagration or a flame, corresponding to the lower branch of the Hugoniot, propagates at low subsonic speeds. The mode of propagation is by diffusion of heat and active chain carriers generated in the chemical reaction into the region of unburned gases. Figure 6.22a illustrates a flame moving at a velocity S_u into the stationary medium of unburned gas (reactant). In Fig. 6.22b, the flame is shown in a frame of reference in which it is stationary. The reactants move toward the flame at speed S_u, and the burned gases move away from it at speed S_b.

Fig. 6.22 One-dimensional
flame

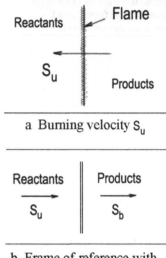

a Burning velocity S_u

b Frame of reference with
flame stationary

   A magnified view of the flame in the frame of reference with the flame stationary
is shown in Fig. 6.23. Heat, from the hot burned gases at temperature T_b, heats up
the unburned gases from the unburned value T_u.

   The heating of the unburned gases from the initial value of temperature T_u to an
intermediate hot gas temperature T_I causes the chemical reactions to start and some
small amount of heat release takes place. This corresponds to preheat discussed in
Chap. 5. Heat release occurs in the zone between the preheat temperature T_I and
the final temperature T_b. The heat release provides an outward convex shape for the
temperature distribution. A change in the shape of temperature distribution is seen
at T_I. Instead of a concave shape corresponding to receipt of heat, a convex shape
from generation of heat within it is obtained. T_I is called as inflection temperature
and separates the convex heat generation from the concave heat receiving portion. A
change in the sign of $d/dx(dT/dx)$ takes place at T_I, where x is the spatial distance
along the direction of propagation of the flame.

   The concentration of fuel C_f begins to drop rapidly after the preheat zone when
the rate of chemical reactions become significant. This is shown by the dotted line
in Fig. 6.23.

   The magnitude of heat conducted per unit surface area into the preheat zone (UI
in Fig. 6.23) is from the zone of heat release (IB) and is given by

$$k \frac{dT}{dx}\bigg|_I \tag{6.15}$$

Fig. 6.23 Temperature and concentration variations across a flame

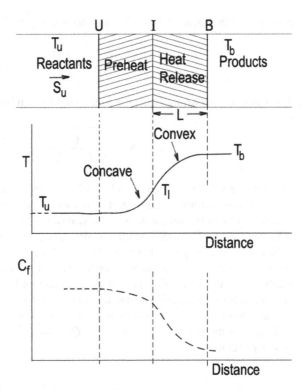

Here the thermal conductivity of the gas is denoted by k. The rate of heat transferred to the unburned gas given by Eq. 6.15 increases the temperature of the mass flow rate of the unburned gases $\rho_u S_u$ from the initial value of the unburned gas temperature T_u to the inflection temperature T_I. The enthalpy balance per unit area is

$$\rho_u S_u C_P (T_I - T_u) = k \frac{dT}{dx}\bigg|_I \tag{6.16}$$

Here C_P denoted the specific heat of the gas at constant pressure.

The temperature gradient is approximated as

$$\frac{dT}{dx}\bigg|_I = \frac{T_b - T_I}{L} \tag{6.17}$$

where L is the thickness (IB) of the one-dimensional reaction zone shown in Fig. 6.23.

If the heat release per unit volume is denoted by $\dot{Q}$ (as in Chap. 4), we can, for unit cross sectional area, write the total rate of heat release rate as $\dot{Q}L$ and

$$\rho_u S_u C_P (T_b - T_I) = \dot{Q}L \tag{6.18}$$

Substituting the value of $T_b - T_I$ from above in Eq. 6.17, we get

$$\frac{dT}{dx}\Big|_I = \frac{\dot{Q}}{\rho_u S_u C_P} \tag{6.19}$$

The value of the burning velocity is obtained by combining Eqs. 6.19 and 6.16 to give

$$S_u = \sqrt{\frac{\dot{Q}}{T_I - T_u} \frac{k}{\rho_u^2 C_P^2}} \tag{6.20}$$

The burning velocity is seen from Eq. 6.20 to vary as the square root of the rate of heat generated per unit volume $\dot{Q}$. It increases with decrease in the specific heat and decrease in the unburned gas density. Typical values of burning velocities are between 0.30 and 0.5 m/s for stoichiometric hydrocarbon–air mixture and about 2 m/s for stoichiometric hydrogen–air mixture. For acetylene–air mixture, the burning velocity is about 1.5 m/s. These values are seen to be about three to four orders of magnitude smaller than the detonation velocities discussed earlier. Hydrogen and acetylene are very reactive fuels and provide the maximum burning velocities. Ammonia–air mixtures have about the lowest value of burning velocities.

Denoting the rate of heat release as $\dot{Q} = AqC^n e^{-E_a/R_0 T}$, the value of burning velocity becomes

$$S_u = \frac{k^{1/2}}{\rho_u C_P (T_I - T_u)^{1/2}} (Aq)^{1/2} C^{n/2} e^{-\frac{E_a}{2R_0 T}} \tag{6.21}$$

The concentration assuming the medium to be an ideal gas is

$$C = \frac{p}{R_0 T_u}$$

where p is the gas pressure and R_0 is the universal gas constant. Since the gas density is

$$\rho_u = \frac{p}{R T_u},$$

R, being the specific gas constant, the variation of burning velocity with pressure is

$$S_u \propto p^{(n/2)-1} \tag{6.22}$$

For most fuel–air mixtures, the exponent n is about 2 indicating the burning velocity to be independent of pressure.

Changes in the initial temperature would, however, change burning velocity as per the term $1/(T_I - T_u)^{1/2}$ in Eq. 6.21. As initial temperature T_u increases, the burning velocity increases. Typical values of burning velocities for stoichiometric composition of common gaseous fuel–air mixtures are given in Table 6.2.

Table 6.2 Typical burning velocities of stoichiometric composition of fuel–air mixtures (0.1 MPa, 25 °C)

Fuel-air	Burning velocity S_u (m/s)
Methane–air	0.5
Propane–air	0.46
Butane–air	0.4
Acetylene–air	1.6
Hydrogen–air	3.0

6.4.1 Burning Velocity and Flame Speed

The burning velocity is to be distinguished from flame speed. The flame speed refers to velocity of propagation of the flame velocity as perceived by a stationary observer on the ground. The burning velocity is obtained when the medium ahead of it is stationary. The flame speed would be different from the burning velocity corresponding to the motion of the reactants ahead of the flame.

6.4.2 Thickness of Flame

The thickness of flame can be taken as the region over which its temperature increases from the initial value T_u to the burnt gas temperature T_b. This includes preheat and reaction zones and is schematically indicated by t_f in Fig. 6.24.

If the rate of heat release per unit mass of the reactants in the chemical reaction is denoted as Q joules/kg, the net heat release per unit surface area of the flame is $\rho_u S_u Q$ J/(m^{2}s), which must equal $k(T_b - T_u)/t_f$ to give

Fig. 6.24 Flame thickness

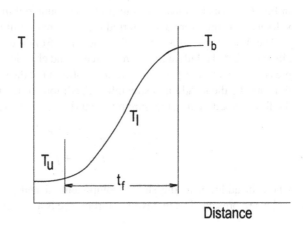

$$k\frac{T_b - T_u}{t_f} = \rho_u S_u Q t_f \tag{6.23}$$

The heat release per unit mass Q can be written as

$$Q = C_P(T_b - T_u) \tag{6.24}$$

where the specific heat C_P is in J/(kg K).

Substituting Eq. 6.24 in Eq. 6.23 and simplifying, we get

$$t_f^2 = \frac{k}{C_P \rho_u S_u} \tag{6.25}$$

As pressure increases, the density ρ_u increases, and the flame thickness reduces.

Since the thermal diffusivity $\alpha = k/\rho_u C_P$, the flame thickness

$$t_f = \frac{\alpha}{S_u} \tag{6.26}$$

Thus the flame thickness in a given medium is inversely proportional to the flame speed. It depends on the ratio of the rate at which heat diffuses to the velocity at which the flame travels. Typically the thermal diffusivity of a stoichiometric hydrocarbon air mixture is of the order of 2×10^{-5} m^2/s, while its burning velocity is about 0.2 m/s giving the flame thickness to be about a fraction of a millimeter. The flame thickness is sufficiently small for justifying the flame to be considered as a combustion wave.

6.4.3   Laminar and Turbulent Burning Velocities

The burning velocity S_u considered the molecular transport of heat and chemical constituents to provide for the profile of temperature and concentration illustrated in Fig. 6.23. Such molecular transport is encountered in laminar flow. The burning velocity S_u is therefore also referred to as laminar burning velocity.

The flow of reactants from spills and leaks from pressurized containers is generally turbulent. Turbulent flows are unsteady and chaotic with fluctuations in velocity, pressure, density, temperature, and concentration of the species in the chemical reactions unlike the steady and streamline conditions in laminar flows. As an example, the flow velocity u and pressure p for a turbulent flow are given by:

$$u = \bar{u} + u' \tag{6.27}$$

$$p = \bar{p} + p' \tag{6.28}$$

where in addition to the steady components $\bar{u}$ and $\bar{p}$, fluctuating components u' and p' are present. The intensity of turbulence is a reflection of the magnitude of the

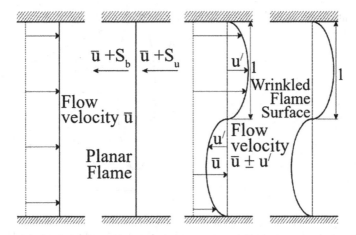

Fig. 6.25 Wrinkling of flame

fluctuating components and is the root mean square (RMS) of the fluctuating velocity to the mean velocity.

Turbulence is characterized by a length scale which is representative of the size of eddies or fluctuations in the flow. The size of eddies in the flow could vary over several orders of magnitude depending on the characteristic length scale of the problem being considered. When the length scale of turbulence is larger than the thickness of the flame, the fluctuations could wrinkle or distort a flame while if the length scale is smaller than the flame thickness, it would enhance the transport of heat and mass in the flame. Small values of length scales are known as Kolmogorov length scale of turbulence.

The wrinkling of a flame when the length scale of turbulence is greater than the flame thickness is shown in Fig. 6.25. The distance over which the eddy or variations in the fluctuating velocity persist is the length scale of turbulence. In regions wherein the flow velocity is greater than the mean value $\bar{u}$, the flame front distorts to a convex shaped flame while in regions wherein the velocity is less than $\bar{u}$, a concave shaped flame is formed. Cone-shaped flame fronts (Fig. 6.26) with base equal to the length scale of turbulence result due to fluctuations. The height of a cone-shaped front corresponds approximately to the time taken for the eddy of length l to be consumed by the flame (viz. a time of l/S_u) multiplied by the fluctuating velocity ($\bar{u}$). The height of the equivalent distorted conical flame is therefore $u'l/S_u$.

The distorted flame front gives a much larger flame surface area with the result that the rate of consumption of the unreacted gas goes up. This is equivalent to a much higher value of flame velocity which can be called as the turbulent flame velocity S_T. The ratio of the turbulent flame speed (for length scale greater than the flame thickness) compared to the laminar flame speed S_u is therefore:

$$\frac{S_T}{S_u} = \text{Area of distorted flame to the laminar planar flame}$$

Fig. 6.26 Idealized conical
distorted flame

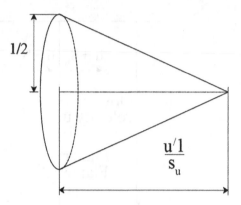

Assuming the distorted surface to be conical, the slant height of the cone (Fig. 6.26) is:

$$\sqrt{\left(\frac{l}{2}\right)^2 + \left(\frac{u'l}{S_u}\right)^2} = \frac{l}{2}\sqrt{1 + \left(\frac{2u'}{S_u}\right)^2} \tag{6.29}$$

The laminar flame of surface area corresponding to a circle of radius $l/2$ now has a conical surface of the above slant height giving the ratio S_T/S_u as:

$$\frac{S_T}{S_u} = \sqrt{1 + \left(\frac{2u'}{S_u}\right)^2} \tag{6.30}$$

As the intensity of turbulence increases, the fluctuation u' increases leading to a higher value of the turbulent burning velocity. For very large values of u', the turbulent burning velocity S_T becomes proportional to u' itself (Eq. 6.20).

The smaller Kolmogorov length scale of turbulence also leads to higher burning velocities by increasing the transport of heat through enhanced values of thermal conductivity in Eq. 6.20. The net effect of the small and large scales of turbulence and the intensity of turbulence in the turbulent flow is to enhance the burning velocity. The turbulent burning velocities could be one to two orders of magnitude greater than the laminar burning velocities.

6.4.4 Turbulent Flame Brush

The fluctuations in velocity u' in turbulent flow will lead to a spread of the flame thickness and therefore a thicker flame thickness. The flame thickness is given by α/S_u. Though the turbulent burning velocity is higher than the laminar flame velocity, the higher value of the turbulent thermal diffusivity causes the thickness of the turbulent flame to be greater. The higher thickness turbulent flame is often spoken of as a turbulent flame brush.

6.4.5 Pressure Changes Across a Flame

Consider a flame of thickness t_f across which the temperature increases from T_u to T_b (Fig. 6.27). In the frame of reference of the flame being stationary, the medium of density ρ_u moves toward it at speed S_u while the burned medium of density ρ_b moves away from it at speed S_b.

If we consider a small section dx at a distance x from the unburned side of the flame but within the flame thickness t_f and denote the density in the section dx to be ρ and the speed to vary from S at x to $S + dS$ at $x + dx$, the change in the pressure dp across the small distance dx is

$$dp = -\rho S dS \qquad (6.31)$$

Integrating the above equation over the thickness of the flame, i.e., from $x = 0$ for which $S = S_u$ to $x = t_f$ for which $S = S_b$, we get the pressure change across the flame. Denoting the pressure of the unburned gas as p_u and the burned gas pressure as p_b, we have

$$p_u - p_b = \rho S(S_b - S_u)$$
$$= \rho_u S_u^2 (S_b/S_u - 1) = \rho_u S_u^2 (\rho_u/\rho_b - 1) \qquad (6.32)$$

Fig. 6.27 Flame of thickness t_f

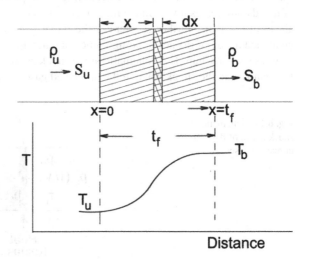

The unburned gas is at a much lower temperature than the burned gas. Its density ρ_u is therefore greater than the burned gas density ρ_b. Consequently, the pressure across a flame decreases.

The fall in pressure is, however, small. For a stoichiometric methane–air flame, the change in pressure is about 1.3 Pa. The pressure drop increases as the burning velocity increases as per Eq. 6.28. Thus for a turbulent flame, the pressure drop would be higher. Considering the very small pressure drops, the wave is referred to as constant pressure deflagration.

6.5 Fast Flame at Lower Chapman–Jouguet Point: Sub CJ or Quasi Detonation

Laminar and turbulent flames corresponding to the weak deflagration branch of the reaction Hugoniot have been discussed in the previous section. The question of whether a steady state deflagration could be formed corresponding to the lower CJ point (CJ Flame) like a steady Chapman–Jouguet detonation at the upper CJ point requires some clarification. Such a flame, if formed, will correspond to the maximum velocity since the slope of the Rayleigh line to the reaction Hugoniot is a maximum at the lower CJ point. Further, such a flame could be steady, i.e., travel at constant speed since the sonic plane at the end of combustion (CJ plane) would not allow the disturbances to catch up and decay it. Such flames at the lower CJ point with pressures lower than the initial value of pressures ahead of the flame have not been observed.

However, the laminar flame propagating at velocities less than a few meters per second becomes turbulent and its velocity increases as given in Sects. 6.4.3 and 6.4.4. The high velocity of this rapidly accelerating flame pushes the unburned gas ahead of it and compresses it very much like a piston would do. The motion of the medium ahead of the flame further accelerates the flame velocity. Compression waves are generated ahead of the accelerating flame, and they merge together to form a shock wave as seen in Chap. 1. A schematic of the shock compression process ahead of the flame at L' and the shock wave in the unreacted gases is shown in Fig. 6.28.

Fig. 6.28 Schematic of shock ahead of flame in unreacted gases

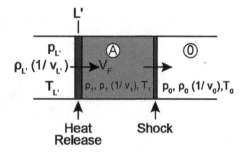

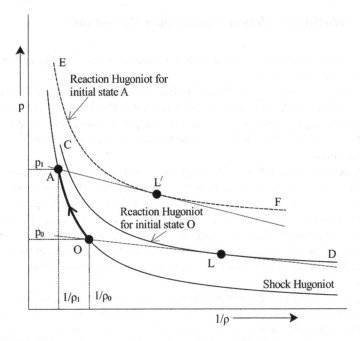

Fig. 6.29 Sub CJ detonation or quasi detonation at the lower CJ point

The shock-compressed gases are not heated to the high temperatures at which spontaneous chemical reactions can occur. The initial state of the unburned gases due to shock heating is still on the same shock Hugoniot; but it shifts to a higher pressure and higher density. The initial state of the gas before it reacts is now at point A, corresponding to pressure p_1 and density ρ_1 instead of the initial state of unburned gases at point O (p_0, ρ_0). This is shown in Fig. 6.29 and also in Fig. 6.28. The reaction Hugoniot corresponding to this initial state A is shown by the dotted line EF in Fig. 6.29 instead of CD corresponding to the initial state O in the absence of the heating. Heating by the shock of the unburned gas by the compression increases the energy release. The lower Chapman–Jouguet point is at L' at which the Rayleigh line from the state A is tangent to it.

A net increase in pressure is seen to take place for the final state L' when compared with the initial state O though for the process of the flame the pressure at L' is less than at A. This means with respect to the shock compressed state, the process is one of expansion as is to be expected in the flame.

Such compressed flames, when referred to the initial state at O and not the initial shock compressed state at A, are called as quasi-detonations. They travel at constant speed and give rise to a net compression with respect to the free stream condition. The solution procedure would require a simultaneous solution for the precursor shock and the reaction Hugoniot.

6.5.1 Velocity and Pressure in a Quasi Detonation

A simple procedure is given in this section to determine the velocity and pressure behind the quasi-detonation.

We note that the final pressure behind the flame $p_{L'} < p_1$ and $p_{L'} > p_0$. The velocity of the gas behind the flame in the frame of reference of the flame at L' (CJ condition) is sonic. Denoting this velocity as u_2 we have it equal to sound velocity at L' which is given by $a_2 = \sqrt{\gamma R T_{L'}}$. Here $T_{L'}$ is the temperature of products of combustion behind the flame.

We have seen in Sect. 6.4.5 that the pressure drop across a flame is very small. In fact for a second-order reaction, we found no change of pressure and said a deflagration is a constant pressure wave. Let us denote the small change in pressure across the flame by $\Delta p = p_1 - p_{L'}$. Let the flame travel at speed V_F. Hence in the frame of reference of the flame, we have the shocked unreacted gases at state (p_1, ρ_1) approach it at velocity V_F and leave it with velocity u_2. When the pressure drop across the flame becomes negligibly small, i.e., $\Delta p \to 0$, the momentum balance across the flame gives $V_F = u_2$.

But u_2 is the sonic speed corresponding to the CJ conditions behind the flame. We can therefore say that the speed of the flame V_F is sonic with respect to the conditions behind the flame. The speed of propagation of the quasi-detonation is therefore sonic defined for conditions behind the flame. A quasi-detonation is therefore a sonic flame behind the precursor shock wave.

The Mach number of this sonic flame with respect to free stream conditions ahead of the shock is

$$M = \frac{V_F}{\sqrt{\gamma R T_0}} = \sqrt{\left(\frac{T_F}{T_0}\right)}$$

The temperature of a stoichiometric mixture at 1 atm. and 25 °C of a hydrocarbon–air flame is typically about 2400 K. The ambient temperature is about 300 K. Hence the Mach number of the flame based on the sound speed in the ambient is $\sqrt{8} \approx 2.8$. The Mach number of CJ detonation in hydrocarbon air mixture is of about 5.5 (Table 6.1). The quasi-detonation velocity is therefore about half the value of the CJ detonation velocity.

The pressure in a quasi-detonation for the particular assumption of Δp across the flame to be zero is the pressure behind the shock viz., p_1. This value can be readily determined from the shock relations since the shock and flame travel together at the same velocity, and the Mach number of the flame was determined with reference to the sound speed in the free stream. Hence the shock Mach number is $M_S = M = 2.8$. The pressure ratio p_1/p_0 is

$$\frac{p_1}{p_0} = \frac{2\gamma}{\gamma + 1} M_S^2 - \frac{\gamma - 1}{\gamma + 1}$$

which for $\gamma = 1.4$ gives the value as 9. For a CJ detonation the value of p_{CJ} is 17 atm (Table 6.1). The pressure in a quasi-detonation is about half the CJ detonation pressure.

The quasi-detonation is therefore a sonic flame whose pressure rise from the ambient value is about half of the CJ value and travels at half the CJ speed. Quasi-detonations are observed in practice in both gaseous and condensed phase explosives.

6.6 Detonations and Flames: Destructive Influence

Two types of combustion waves comprising detonation involving compression across the combustion wave and deflagration or flame in which pressure slightly decreases across it are seen to be possible in reactive gas mixtures. While the detonation propagates at supersonic speeds, the deflagration or flame travels at low subsonic speeds. A shock wave is an integral part of the detonation and gives rise to the compression heating after which heat is generated by chemical reactions. A detonation is therefore an explosion controlled by the shock wave.

The transport of heat and reactants by diffusion results in the propagation of a flame. Turbulence increases the transport process in flames and leads to wrinkling of the flames. Flame speeds are increased by about an order of magnitude due to turbulence. Blockages and protrusions in confined geometries can augment turbulence and give rise to higher flame speeds.

The steady-state Chapman–Jouguet detonation corresponds to the upper point at which the Rayleigh line is tangent to the reaction Hugoniot. For states above the upper Chapman–Jouguet point, the detonation is no longer steady but decays to the steady Chapman–Jouguet value. Flames both laminar and turbulent, are in the expansion part of the reaction Hugoniot in the weak deflagration region. A steady state corresponding to the lower Chapman–Jouguet point is only formed with shock preceding the flame as in a quasi-detonation and results in an overall compression even though with reference to the conditions upstream of the flame there is a drop in pressure corresponding to an expansion. Quasi-detonations propagate at speeds about half of a CJ detonation.

The damage from an explosion involving detonations would be much more severe than a deflagration considering the large pressures and very high speeds. We shall consider the initiation of flames and detonations and the transition from flame to detonation in the next chapter.

A detonation and a quasi-detonation is inherently destructive and not suited for applications in power plants and propulsion. Part of the chemical energy released by the chemical reactions gets dissipated in the shock. The maximum useful work available in a constant velocity CJ detonation is the work done by the isentropic expansion of the high pressure and high temperature gases in the detonation reduced by the work taken up for the shock compression. Figure 6.30 illustrates an idealized one-dimensional CJ detonation in a pressure p versus specific volume v diagram. The shock Hugoniot and the reaction Hugoniot are shown by AB and CD respectively.

Fig. 6.30 Work from a CJ detonation

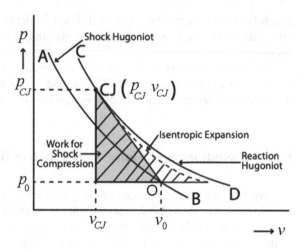

The initial state of the gas is at O (p_0, v_0) on the shock Hugoniot AB. The Chapman–Jouguet point is shown on the reaction Hugoniot by CJ wherein the pressure and specific volume are p_{CJ} and v_{CJ} respectively.

The work done in the shock compression of unit mass of gas is equal to

$$\frac{1}{2}(p_0 + p_{CJ})(v_0 - v_{CJ})$$

and is shown in gray in Fig. 6.29. The maximum work available from the expansion of unit mass of the high pressure gases at the detonation conditions p_{CJ} and v_{CJ} corresponds to its isentropic expansion to the ambient pressure and is given by $\int_{v_{CJ}}^{\infty} p\,dv$ and is shown by the hatched lines. The maximum useful work from a detonation is therefore

$$\int_{v_{CJ}}^{\infty} p\,dv - \frac{1}{2}(p_{CJ} + p_0)(v_0 - v_{CJ})$$

per unit mass. The detonation has a penalty when useful work is desired from it. It is therefore invariably used mainly for its destructive power.

Examples

6.1 *Von Neumann Pressure Spike and CJ Pressure*: Determine the von Neumann pressure spike and the Chapman–Jouguet pressure in a detonation of a hydrocarbon–air mixture at atmospheric pressure of 0.1 MPa given that the mean specific heat ratio for the mixture and combustion products are 1.39 and the heat release is 2.2 MJ/kg of the mixture. The sound speed in the unburned mixture at the ambient temperature is 320 m/s. The initial density of the mixture is 1.1 kg/m³.

Solution: We first calculate the Chapman–Jouguet detonation velocity corresponding to the upper tangential point on the reaction Hugoniot. It is obtained from Eq. 6.11 as

$$V_{CJ}^2 = 2(\gamma^2 - 1)Q$$

where Q is the heat release in J per kg of the mixture. Substituting the values of the specific heat ratio γ and Q, we get

$$V_{CJ}^2 = 2 \times 0.9321 \times 2.2 \times 10^6$$
$$V_{CJ} = 2025 \text{ m/s}$$

The Mach number of the CJ detonation $M_{CJ} = 2025/320 = 6.33$.
The CJ pressure of the detonation p_{CJ} is determined by Eq. 6.13 to be

$$\frac{p_{CJ}}{p_0} = \frac{\gamma M_{CJ}^2}{\gamma + 1}$$

Substituting the values, we get

$$p_{CJ} = 0.1 \times 1.39 \times 6.33^2/2.39 = 2.33 \text{ MPa}$$

The von Neumann pressure spike will correspond to the shocked gases before combustion takes place. Assuming the same mean value of the specific heat ratio γ for the unburned gases ahead of the detonation and the shocked gases, we have for the pressure rise in a shock wave:

$$\frac{p}{\rho_0 \dot{R}_S^2} = \frac{2}{\gamma + 1} - \frac{\gamma - 1}{\gamma(\gamma + 1)} \frac{1}{M_S^2}$$

Here M_S is the Mach number of the shock front which is the same as the Mach number M_{CJ} of the detonation, and $\dot{R}_S$ is the same as V_{CJ}. Substituting the values of ρ_0, $V_{CJ}(= \dot{R}_S)$ and $M_S(= M_{CJ})$ in the above expression gives the pressure p behind the shock as

$$p = 0.8338 \times 1.1 \times 2025^2 2 = 3.76 \text{ MPa}$$

The von Neumann pressure spike behind the detonation is seen to be 3.76 MPa which is very much higher than the value of the Chapman–Jouguet detonation pressure of 2.33 MPa.

6.2 *Heat Release and Maximum Work Availability in a Detonation*: Determine: (i) Heat release in a CJ detonation and (ii) the maximum work available from a CJ detonation. Methane–air of stoichiometric proportions at a temperature of 25°C and 100 kPa pressure could be considered.

Solution:

(i) *Heat Release in the CJ Detonation:* Table 6.1 gives the values CJ detonation velocity and CJ pressures in a stoichiometric mixture of methane and air as 1.8 km/s and 16.05 atm. We can find the value of energy release Q in the detonation from Eq. 6.11

$$V_{CJ}^2 = 2(\gamma^2 - 1)Q$$

We need the value of the specific heat ratio γ in the above equation to get the value of Q. We therefore write the equation for the reaction between methane and air at stoichiometric proportions as

$$CH_4 + 2(O_2 + 3.76\,N_2) = CO_2 + 2H_2O + 2 \times 3.76\,N_2$$

We need the values of specific heat C_P at constant pressure and at constant volume C_V to determine γ. The specific heat C_P of methane and air at 1 atm. and 25°C is given in the Table of thermophysical properties as 1.039 J/(g K) and 4.4 J/(g K). Converting it in units of J/(mole K), we get $\hat{C}_{P,AIR} = 1.039 \times 28.8 = 29.92$ J/(mole K) while $\hat{C}_{P,CH4} = 4.4 \times 16 = 70.4$ J/(mole K). Here the molecular mass of air and methane of 28.8 g/mole and 16 g/mole are used.

The values of molar-specific heats $\hat{C}_{V,AIR}$ and $\hat{C}_{V,CH4}$ are obtained from the thermodynamic relation $\hat{C}_P - \hat{C}_V = R_0$ where R_0 is the universal gas constant equal to 8.314 J/(mole k). We get the values of $\hat{C}_{V,AIR} = 21.61$ J/(mole K) and $\hat{C}_{V,CH4} = 62.1$ J/(mole K).

We have 1 mole of methane and $2 \times (1 + 3.76) = 9.52$ moles of air. Hence the mean specific heat of the stoichiometric mixture of methane air is

$$\hat{C}_{P,MIX} = \frac{70.4 + 9.52 \times 29.82}{1 + 9.52} = 33.77 \text{ J/(mole K)}.$$

Similarly, the value of $\hat{C}_{V,MIX} = 25.46$ J/(mole K). The value of the specific heat ratio γ is

$$\frac{\hat{C}_{P,MIX}}{\hat{C}_{V,MIX}} = \frac{33.77}{25.96} = 1.326.$$

Assuming the same value of the specific heat ratio γ for the products of the detonation, we have heat release from the detonation Q J/kg as $V_{CJ}^2/2(\gamma^2 - 1)$. Substituting the value of $V_{CJ} = 1800$ m/s and $\gamma = 1.326$, we get $Q = 2.137 \times 10^6$ J/kg.

It is instructive to compare the above value of heat release with the heat release predicted using the standard heats of formation in the equation

$$CH_4 + 2(O_2 + 3.76\,N_2) = CO_2 + 2H_2O + 7.52N_2$$

Using the standard heats of formation given in Chap. 4, the heat release in the reaction is $-[-393.5 + (-286.7) - (-74.9)] = 605.3$ kJ. The mass of the reactants is $16 + 2(32 + 3.76 \times 28) = 290.56$ g. Hence the heat release per unit mass is 2.083 kJ/g $= 2.083 \times 10^6$ J/kg. This value is a little (2.5%) lower than that determined from the equation relating V_{CJ} with Q. The difference is mainly due to the assumption of γ remaining the same for the reactants as well as the detonation products. The values, however, are about the same.

(ii) *Maximum Work Available from CJ Detonation:* The maximum work corresponds to the reversible expansion work from the CJ state to the ambient. The CJ pressure for the methane air detonation is 16.05 atm. Since the work done between two states for all reversible processes is the same and corresponds to the maximum work, we consider the case of isentropic expansion from the high-pressure CJ state to the ambient pressure. The expansion work is

$$W_{max} = \int_{p_{CJ}}^{p_\infty} p \, dv$$

where p_{CJ} is the CJ detonation pressure and p_∞ is the ambient pressure. For the isentropic expansion $pv^\gamma = $ constant. Hence the pressure p at any point is the expansion is given by $p_{CJ} v_{CJ}^\gamma = pv^\gamma$ where v_{CJ} is the specific volume corresponding to CJ state and v is the specific volume at pressure p. Substituting the value of p in the integral by $p_{CJ}(v_{CJ}^\gamma/v^\gamma)$, the maximum work becomes

$$W_{max} = p_{CJ} v_{CJ}^\gamma \int_{CJ}^{v_{exp}} \left(\frac{dv}{v^\gamma}\right)$$

Integrating, we get

$$W_{max} = p_{CJ} v_{CJ}^\gamma \left[\frac{v_{exp}^{1-\gamma} - v_{CJ}^{1-\gamma}}{1 - \gamma}\right] = \frac{p_{CJ} v_{CJ} - p_\infty v_{exp}}{\gamma - 1}$$

The value of $p_{CJ} = 16.05$ atm while $p_\infty = 1$ atm $= 1.013 \times 10^5$ Pa. The value of $\gamma = 1.326$. We need to determine the values of v_∞ and v_{CJ}.

The molecular mass of the unreacted mixture corresponds to $CH_4 + 2$ $(O_2 + 3.76 N_2)$ and is $[16 + 2(32 + 3.76 \times 28)]/[1 + 2(1 + 3.76)] = 27.61$ g/mole. The specific volume of the reactant at 1.013×10^5 Pa and 25°C is

$$v_\infty = \frac{1.013 \times 10^5}{(8.314/0.02671) \times 298} = 1.092 \text{ m}^3/\text{kg}$$

The value of the specific volume of detonation products v_{CJ} can be found from the density ratio across a CJ detonation which has been found to be

$$\frac{\rho_{CJ}}{\rho_\infty} = \frac{\gamma + 1}{\gamma}$$

Since $\gamma = 1.326$, we get $\rho_{CJ}/\rho_\infty = 1.754$. But $\rho_{CJ}/\rho_\infty = v_\infty/v_{CJ}$ and we get the value of

$$v_{CJ} = \frac{v_\infty}{1.326} = \frac{1.092}{1.326} = 0.824 \ \text{m}^3/\text{kg}.$$

The velocity of the CJ detonation is given as 1800 m/s, and the sound speed in the unreacted medium is

$$\sqrt{\gamma RT} = \sqrt{1.326 \times (8.314/0.02761) \times 298} = 345 \ \text{m/s}.$$

Hence $M_{CJ} = 1800/345 = 5.217$.

Substituting value of M_{CJ} and γ in the expression for pressure ratio across a CJ detonation, we have

$$\frac{p_{CJ}}{p_\infty} = \frac{\gamma M_{CJ}^2}{\gamma + 1} = 15.5$$

The gases at CJ state expand isentropically to the ambient pressure. The value of the specific volume of the expanded gas corresponding to the expansion to ambient pressure is given as

$$v_{\text{exp}} = \left(\frac{p_{CJ}}{p_\infty}\right)^{1/\gamma} v_{CJ} = (15.5^{1/1.326}) \times 0.824 = 6.54 \ \text{m}^3/\text{kg}$$

Substituting the values of $p_{CJ}, p_\infty, v_{CJ}, v_{\text{exp}}$ and γ in the expression for work, we get

$$W_{Max} = \frac{1.013 \times 10^5 (15.5 \times 0.824 - 6.54)}{1.326 - 1} = 1.94 \times 10^6 \ \text{J/kg}$$

This expansion work of 1940 kJ/kg compares with the heat release in the detonation of 2137 kJ/kg. Part of the expansion work gets to form the shock. The work required for the shock was seen from the shock and reaction Hugoniot as $(1/2)(p_{CJ} + p_\infty)(v_\infty - v_{CJ})$. On substituting the values we get the compression work required as

$$\frac{1}{2} \times 1.013 \times 10^5 \times (15.5 + 1) \times (1.092 - 0.824) = 224 \ \text{kJ/kg}.$$

Therefore 224 kJ out of the expansion work of 1940 kJ (i.e., about 11%) is required for the shock compression. The net maximum work output is $1940 - 224 = 1716$ kJ/kg of the mixture. Not only is the availability of maximum work much lower than the energy of the detonation (1940 kJ/kg compared to 2137 kJ/kg, i.e., 90%), but further penalty of about 11% is required for the shock compression. The detonation is therefore not a good contender for generating power or for operating propulsion systems.

It may be noted in this example that the gases are expanded to a specific volume of 6.54 m^3/kg, whereas the specific volume of the ambient at the same ambient pressure is 1.092 m^3/kg. The larger specific volume at the same ambient pressure is due to the higher temperature of the expanded gases.

6.3 *CJ and Quasi Detonation Velocities and their Pressures:* A given explosive gas mixture at 100 kPa pressure and 27 °C has a specific heat ratio of 1.35. The energy released per unit mass is 2500 kJ/kg, and the temperature of the products is 2500°C. Assuming the specific heat ratio of the products remains the same as in the reactants, determine the steady propagation velocity of: (i) CJ detonation and (ii) quasi-detonation in the explosive gas mixture. The molecular mass of the reactants is 25 g/mole while the molecular mass of the products is 30 g/mole. Determine the pressure of the CJ detonation and the quasi-detonation.

Solution: The CJ velocity is given by

$$V_{CJ}^2 = 2(\gamma^2 - 1)Q$$

where Q is the heat released in J/kg and γ is the specific heat ratio. Substituting the values, we get the CJ velocity as 2028 m/s.
The sonic velocity in the products is $\sqrt{\gamma RT}$, where R is the specific gas constant and T is the temperature of products. The value of $R = 8.314/0.03 = 277$ J/kg K. Since the temperature is given as $2500 + 273 = 2773$ K, the sound speed in the products is 1018 m/s. The speed of the reaction front is therefore 1018 m/s, and this is the velocity of the quasi-detonation.
The CJ velocity is 2028 m/s while quasi-detonation velocity is 1018 m/s. The pressure in the CJ detonation is given by

$$\frac{p_{CJ}}{p_0} = \frac{\gamma M_{CJ}^2}{\gamma + 1}$$

The sound speed in the unreacted gases is $\sqrt{\gamma RT_0} = 367$ m/s. Therefore $M_{CJ} = 2028/367 = 5.53$. The CJ pressure is given by

$$\frac{p_{CJ}}{p_0} = \frac{\gamma M_{CJ}^2}{\gamma + 1}$$

With p_0 given as 100 kPa, we get $p_{CJ} = 17.6 \times 100$ kPa = 1.76 MPa.
The quasi-detonation pressure is the pressure of the shocked gases ahead of it. The shock Mach number is $1018/367 = 2.77$, and the pressure of the shocked

gas is

$$\frac{p}{p_0} = \frac{2\gamma M_S^2}{\gamma + 1} - \frac{\gamma - 1}{\gamma + 1}$$

Substituting $M_S = 2.77$ in the above gives $p/p_0 = 8.67 \times 10^5$ Pa $= 0.867$ MPa.

Hence the CJ pressure is 1.76 MPa, while the quasi-detonation pressure is 0.867 MPa.

6.4 *Mach Number of CJ Detonation and Velocity of Quasi Detonation:* In the CJ detonation of a stagnant reactive gas mixture the sonic velocity of the products of combustion is 1200 m/s. The sonic velocity in the reactants is 400 m/s. The products follow the detonation front with a velocity of 800 m/s. Determine the CJ velocity, and the Mach number of the CJ detonation.

What is the velocity of the quasi-detonation in the mixture?

Solution: Let the detonation velocity be given as D m/s. In the frame of reference of the detonation wave, the unreacted gases approach the detonation at a velocity D m/s while the products move away from it at a velocity $(D - u)$ m/s. In the frame of reference of the detonation, velocity behind it for the CJ detonation is the sonic velocity in the products.

In the given problem, $D - u = 1200$ or $D = 1200 + 800 = 2000$ m/s.

The CJ detonation velocity is 2000 m/s.

The Mach number of the CJ detonation is $D/a_0 = 2000/400 = 5$.

The velocity of the quasi-detonation is the sonic velocity of the products and is 1200 m/s.

6.5 *Comparison of the explosion of ammonium nitrate at Beirut port in August 2020 with the explosion of ammonium nitrate at Tianjin port in August 2015:* Ammonium nitrate was involved in the 2020 explosion at Beirut and in the 2015 explosion at Tianjin. While the quantity of ammonium nitrate was 2750t at the Beirut explosion, only a third of the quantity of about 800 t was involved in the Tianjin explosion. However, the damage in both these explosions was about the same with a crater of diameter somewhat greater than about 100 m being formed at the site of the warehouses where ammonium nitrate was kept. Severe damage was also observed over a distance of about 2 km for both the explosions. What could be the reason for the effects of mass of explosive ammonium nitrate not correlating directly with the damage caused in the explosion?

Solution: The diameter of the crater and the damage at a given distance from the source of energy release in an explosion was discussed in Chaps. 2 and 3. Both the crater diameter and the overpressure at a given distance for a strong blast were seen to scale as the explosion length. Hence the diameter of crater formed and the distance to which severe damage should have occurred in the Beirut explosion would be $(2750/800)^{1/3}$ times, i.e., about 1.5 times that in the explosion at Tianjin.

Both the explosions took place in two steps with a low-intensity explosion being followed by a very high-intensity explosion. In Tianjin explosion, ammonium nitrate was kept in bags in a number of containers. Some chemicals were also present in the warehouse and the ammonium nitrate ignited from the high-intensity fire from nitrocellulose. The high intensity flame from nitrocellulose could have caused a rapid fire in the ammonium nitrate, which in the confine-ment of the containers and the warehouse created strong shocks and became a CJ detonation in the subsequent and severe second explosion.

In the Beirut explosion, the bags of ammonium nitrate were stacked together in bags and the environment may have degraded it as it was stored in the warehouse for about six years. The source of ignition, whether by the heat accumulating gradually within the bulk of ammonium nitrate or by incendiaries from fire crack-ers, was not as strong as from nitrocellulose in the Tianjin explosion. In fact the first explosion at Beirut was much weaker with the formation of a whitish smoke suggesting a weak burning is initiated in a closed environment. The interaction of the flame with the walls of the confinement and the ammonium nitrate might have produced much weaker shocks and a quasi-detonation in the subsequent second explosion instead of the CJ detonation perhaps formed in the Tianjin explosion. The effects of a larger mass of ammonium nitrate in the Beirut explo-sion seem most likely to have been compromised with a quasi-detonation whose effectiveness is about half that of the CJ detonation. Hence the net destructive potential was about the same in both the explosions.

Nomenclature

C	Concentration (moles/m^3)
C_P	Specific heat at constant pressure [kJ/(kg K)]
$\hat{C}_P$	Specific heat at constant pressure [kJ/(kmole K)]
h	Specific enthalpy (kJ/kg)
h_0	Specific enthalpy of unburned medium (kJ/kg)
h_1	Specific enthalpy behind combustion wave (kJ/kg)
k	Thermal conductivity of unburned gas [kJ/(m s K)]
L	Extent of chemical heat release zone (m)
M	Mach number of gases behind combustion wave in the frame of reference of wave
M_{CJ}	Mach number of the Chapman–Jouguet detonation wave
M	Molecular mass (g/mole)
p	Pressure behind combustion wave (Pa)
p_0	Pressure of the unburned medium ahead of the combustion wave (Pa)
Q	Heat release per unit mass of mixture (kJ/kg)
q	Heat release per unit mole (kJ/mole)
$\dot{Q}$	Rate of heat release per unit volume of the mixture [kJ/(m^3 s)
S	Speed/Velocity (m/s)

S_u Burning velocity (m/s)
S_b Velocity of burned gases in frame of reference of flame (m/s)
T Temperature (K)
t_f Flame thickness (m)
V_{CJ} Velocity of Chapman–Jouguet detonation (m/s)
v_0 Velocity of combustion wave; velocity of unburned gases in frame of reference
          of combustion wave (m/s); specific volume (m^3/kg)
v_1 Velocity of unburned gases in frame of reference of combustion wave (m/s)
α Thermal diffusivity of unburned gas (m^2/s)
γ Specific heat ratio
λ Transverse spacing of detonation cells (m)
ρ Density behind combustion zone (kg/m^3)
ρ_0 Density of unburned gas (kg/m^3)
ω Rate of chemical reaction [moles/(cc s)]

Subscripts

b Burned
CJ Chapman–Jouguet conditions
f Flame
I Inflection point
o Initial condition
S Shock
u Unburned gas

Further Reading

1. Courant, R. and Freidricks, K. O., *Supersonic Flow and Shock Waves*, Inter-science, New York, 1948.
2. Kuo, K. K., *Principles of Combustion*, Wiley, New York, 1968.
3. Landau, L. D. and Lifshitz, E. M., *Fluid Mechanics*, Pergamon Press, London, 1966.
4. Lee, J. H., *The Detonation Phenomenon*, Cambridge University Press, New York, 2008.
5. Liepman H. W., and Roshko, A., *Elements of Gas Dynamics*, Wiley, New York, 1957.
6. Lewis, B. and von Elbe, G., *Combustion, Flames and Explosion of Gases*, Academic Press, New York, 1961.
7. Penner, S.S., *Chemistry Problems in Jet Propulsion*, Pergamon Press, London, 1957.
8. Quintiere, J.G., *Fundamentals of Fire Phenomenon*, Wiley, Sussex, 2006.
9. Shapiro, A.H., *The Dynamics and Thermodynamics of Compressible Flow*, Vol. 1, The Ronald Press, New York, 1953.
10. Shchelkin K. I. and Troshin, Y.K., *Gas Dynamics of Combustion*, Mono Book Corp., Baltimore, MD, 1965.

11. Sokolik, A.S., *Self Ignition, Flame and Detonation in Gases*, Akad. Nauk. SSR, English Translation, Jerusalem, 1963.
12. Strehlow, R. A., *Fundamentals of Combustion*, 2nded., McGraw-Hill, New York, 1984.
13. Strehlow, R. A., The nature of transverse waves in detonation, *Combustion and Flame*, 9, 109–119, 1969.
14. Stull, D. R., *Fundamentals of Fire and Explosion*, AIChE Monograph Series, Vol. 73, No. 10, 1977.
15. Turns, S. R., *An Introduction to Combustion: Concepts and Applications*, 2nd ed., McGraw Hill, Boston, MA, 2006.

Exercise

6.1 (a) Determine the thickness of a laminar flame propagating in a stoichiometric mixture of propane and air. The pressure and temperature of the propane–air mixture are 1 bar and 300K respectively. The laminar burning velocity is 0.45 m/s and the thermal diffusivity is 12×10^{-4} cm^2/s.

(b) Calculate the change in pressure across the above flame.

(c) Determine the rate of energy release from this flame (in kW) when travelling in a pipe of diameter 25 mm.

6.2 (a) Calculate the velocity of a Chapman–Jouguet detonation in a stoichiometric methane–air mixture given the following:

(i) Methane–air mixture:

Pressure = 1 bar

Temperature = 300 K

(ii) Combustion products:

Temperature: 2400 K

Pressure: 16.05 bar

Mean molecular weight: 24 kg/kmol

Specific heat ratio: 1.24.

(b) Determine the velocity of the combustion products following the detonation wave.

6.3 A peculiar phenomenon known as popping is reported in the combustion of fuel hydrazine (N_2H_4) with oxidizer N_2O_4 in liquid propellant rockets. In this phenomenon, stratified pockets of hydrazine vapor in the mixture of hydrazine and N_2O_4 vapor detonate and lead to spikes in the pressure. Assuming the products of decomposition of hydrazine form ammonia and hydrogen at a temperature of 1000 K with a molecular mass of 20 g/mole and specific heat ratio of 1.3, determine the magnitude of the pressure spike assuming that the initial pressure of the pockets of hydrazine is 800 kPa. Consider both the cases of Chapman–Jouguet and quasi-detonations.

The specific heat ratio of hydrazine vapor could also be assumed as 1.3. The heat released in the decomposition of hydrazine is 3000 J/g.

Chapter 7
Formation of Flames and Detonations in Gaseous Explosives

In Chap. 5, the conditions for thermal explosion of a reactive substance were determined based on the heat generated and the heat loss from it. Terms like auto-ignition and ignition temperatures were derived using a lumped mass assumption. It is rarely possible for the entire bulk of the substance to get ignited; rather, a flame or a detonation, as described in the last chapter, would get initiated from an ignition or initiation source and it would thereafter propagate. We consider the formation of flames and detonations and the conditions of the gaseous explosive for which the formation and propagation is possible in this chapter.

7.1 Initiation of Flame

Consider energy E_0 to be deposited in a small volume v_0 of a reactive medium. The energy raises the temperature within the volume v_0 of the medium and initiates chemical reactions in the small volume and releases heat. Some of the heat and reactive species formed in the reaction are also transferred out of this volume into adjacent regions of the unburned medium. This leads to conditions conducive for chemical reactions to occur in these adjacent regions, and they begin to burn. This process continues, and the flame propagates over the entire volume of the substance.

The transport of heat and activated radicals from the initial volume into the neighboring regions results in a decrease of its temperature and in a depletion in its reactive species. The energy deposition in the initial volume of the medium and the continuing chemical reactions in it offset the effect of the losses by generating heat and active radicals. If, however, the rate of loss of heat and reactive species is greater than the rate at which these are produced by chemical reactions, then the temperature and the concentration of active radicals decrease and the initial volume can no longer sustain the flame and its spread.

© The Author(s), under exclusive license to Springer Nature Switzerland AG 2021
K. Ramamurthi, *Modeling Explosions and Blast Waves*,
https://doi.org/10.1007/978-3-030-74338-3_7

For a self-sustaining flame to be produced by an ignition source, it is therefore necessary that the net heat generated and active species formed from chemical reactions should be greater than the losses. This implies that the energy deposited by the ignition source must be capable of forming a volume of chemical reactions greater than a threshold value so that the generation is greater than the dissipation. The unreacted medium adjacent to the initiated volume of the hot gases should not quench it. The minimum volume required to be formed for ignition is called as flame kernel or ignition kernel.

7.1.1 Divergence and Loss

Consider a spherical shell of unburned gases of width δr around a spherical volume of radius r. The spherical volume contains the burned gases as shown in Fig. 7.1. When the spherical shell burns, its thickness increases because the density decreases from the unburned value ρ_u to the burned value ρ_b. The radius of the burned shell becomes:

$$r + \delta r \frac{\rho_u}{\rho_b}$$

The ratio of the surface area of the shell after the burning to the value before it burns is

$$\left(\frac{(r + \delta r (\rho_u/\rho_b))^2}{(r + \delta r)^2} \right) \tag{7.1}$$

If we consider $\delta r \ll r$, the above equation becomes

$$\left(1 + 2 \frac{\delta r}{r} \frac{\rho_u}{\rho_b} \right) \left(1 - 2 \frac{\delta r}{r} \right) \tag{7.2}$$

The divergence or increase in the area due to burning is proportional to $2\delta r/r$. The value of δr must be of the order of flame thickness given in Chap. 6. The ignition

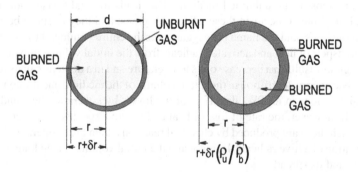

Fig. 7.1 Divergence from burning

source must be such that the losses from the divergence and other causes should not cause the flame to be quenched. We therefore address the quenching of flame for determining the minimum size of the ignition kernel and ignition energy.

7.1.2 Quenching of Flame

Quenching is related to losses, and the quenching of flame can be visualized by considering failure of a flame to propagate in a pipe of diameter d (Fig. 7.2). If the thickness of the flame is t_f, as shown, the heat generated in the flame is

$$\left(\frac{\pi d^2 t_f}{4}\right) \dot{Q} \tag{7.3}$$

where $\dot{Q}$ is the rate of heat generated per unit volume by the chemical reactions.

The rate of heat loss from the flame to the wall is

$$k(\pi d t_f)\frac{dT}{dr} \tag{7.4}$$

Here k is the mean thermal conductivity of the combustion products in the width t_f. Taking the temperature gradient dT/dr at the wall as

$$\frac{dT}{dr} = \frac{T_b - T_w}{d/2} \tag{7.5}$$

where T_b is the hot burned gas temperature and T_w is the wall temperature (ambient), we get the heat loss rate $\dot{Q}_L$ as

$$\dot{Q}_L = k\pi d t_f \frac{T_b - T_w}{d/2} \tag{7.6}$$

Quenching of the flame takes place when the heat generation rate within it is less than heat loss rate. Denoting the threshold diameter for the critical conditions of quenching as $d = d_q$, we get

Fig. 7.2 Quenching of a flame from heat loss to the walls

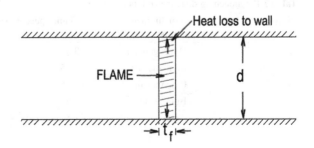

$$\frac{\pi}{4}d_q^2 t_f \dot{Q} = k\pi d_q t_f \frac{T_b - T_w}{d_q/2} \tag{7.7}$$

where d_q is the quenching diameter or quenching distance.

The heat generation rate per unit volume is given by

$$\dot{Q} = \frac{\rho_u S_u C_P (T_b - T_u)}{t_f} \tag{7.8}$$

where ρ_u is the unburned gas density, S_u is the flame speed, C_P is the specific heat at constant pressure and T_u is the unburned gas temperature. Substituting the above in Eq. 7.7 and simplifying, we get for the threshold condition of quenching:

$$\frac{d_q^2}{t_f} = 8\left(\frac{k}{\rho_u S_u C_P}\right) \tag{7.9}$$

Here we have taken the wall temperature T_w equal to the unburned gas temperature T_u. The thermal diffusivity of the unburned gas α is $k/\rho_u C_p$, and α/S_u denotes the flame thickness t_f (Chap. 6). From Eq. 7.9, we therefore have $d_q^2/t_f^2 = 8$ giving the quenching thickness to be more than the flame thickness. Denoting the ratio of the quenching diameter d_q and flame thickness t_f as C, we get from Eq. 7.9

$$\frac{\rho_u S_u C_P d_q}{k} = \frac{1}{C} \tag{7.10}$$

The expression $\rho_u C_P S_u d_q/k$ denotes the ratio of convection to conduction heating and is known as Peclet number (Pe).

Among stoichiometric mixtures of fuel gases in air, hydrogen–air mixture has the smallest value of the quenching diameter of about 0.5 mm. The values of quenching distance for stoichiometric composition of a few fuel–air mixtures are given in Table 7.1. The flame speeds are also given in Table 7.1.

Knowledge of the quenching distance is useful, not only for estimating energy for ignition, but also for design of flame isolators. In Davy's safety lamp, which is used in an atmosphere of explosive gases, the screen size is kept less than the quenching diameter so that the flame will not ignite the ambient explosive gas.

Table 7.1 Quenching distance and flame speed

S. No.	Stoichiometric composition	Flame speed S_u (m/s)	Quenching distance d_q (mm)
1	Hydrogen–air	3	0.6
2	Methane–air	0.45	2.5
3	Propane–air	0.46	2.0
4	Butane–air	0.4	3.4
5	Acetylene–air	1.5	0.8

7.1.3 Minimum Ignition Energy

The minimum ignition energy (MIE) required to initiate a flame must be able to form a threshold or critical volume of hot reacting gases, called as ignition kernel, such that the heat loss from the volume to the unburned medium adjacent to it will not quench the ignition kernel. The heat and radicals leaving the ignition kernel, as seen in Sect. 7.1.1, propagate the flame in the medium.

Consider the ignition kernel to be spherical and of diameter d as shown in Fig. 7.1. The diameter of the kernel should be greater than the quenching distance d_q in order to be not quenched by diffusion of heat away from it. The energy E propagating out of the ignition kernel required for flame to propagate in an adjacent layer of thickness δr

$$E = \pi d_q^2 \delta r \times q$$

where $\pi d_q^2 \delta r$ is the volume of the annular spherical flame surrounding the ignition kernel and q is the energy released per unit volume of the combustible gas medium. δr is the thickness of the flame $t_f (= k/\rho_u C_P S_u)$. This is shown in Fig. 7.3.

Considering the heat released per unit volume of the medium as

$$q = \rho_u C_P (T_b - T_u) \tag{7.11}$$

the energy E is given by

$$E = \pi d_q^2 \frac{k}{\rho_u C_P S_u} \rho_u C_P (T_b - T_u) = \frac{k}{S_u} (T_b - T_u) \pi d^2 q \tag{7.12}$$

A minimum energy of $k(T_b - T_u)\pi d_q^2 / S_u$ from the flame kernel of diameter d_q should be available for ignition of the surrounding layer. MIE is therefore given as

$$\mathrm{MIE} = \frac{k}{S_u} (T_b - T_u) \pi d^2 q \tag{7.13}$$

Fig. 7.3 Energy from ignition kernel causing chemical reactions in annular element

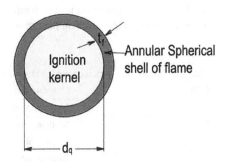

Substituting the value of k/S_u from Eq. 7.10, we get the minimum ignition energy as

$$\text{MIE} = \pi d_q^3 \rho_u C_P T_b C \left(1 - \frac{T_u}{T_b} \right) \tag{7.14}$$

Since $T_u/T_b \ll 1$, and ρ_u, C_P and T_u do not vary significantly for the different gaseous fuels–air mixtures, we get

$$\text{MIE} \propto d_q^3 \tag{7.15}$$

The above trend of MIE varying as cube of quenching distance is observed for the fast burning gas mixtures having high temperature combustion products.

An increase of the initial temperature of the medium reduces the ignition energy based on Eq. 7.14. If the energy is delivered very slowly, most of the energy is lost by diffusion from the annular volume of thickness equal to the flame thickness surrounding the kernel and the energy required for the ignition increases.

The minimum ignition energy of a few typical stoichiometric gas mixtures is given in Table 7.2. The stoichiometric mixtures of hydrogen and acetylene with air have the lowest MIE of about 0.02 mJ. The hydrocarbon–air mixtures (except acetylene) have MIE of about an order of magnitude higher. The higher reactivity of hydrogen and acetylene with air is responsible for their very low MIE.

7.1.4 Limits of Flammability

When the concentration of the fuel in a fuel–air mixture is reduced such that the quantity of the fuel in the mixture is very lean, the energy released per unit mass of the mixture drastically decreases. The temperature of the products of combustion accordingly gets reduced. When the energy released and the associated burned gas temperature reduces below a threshold value, chemical reactions cannot be sustained. A steady flame cannot therefore be supported in the fuel-lean mixture.

Similarly, if the concentration of fuel in the fuel–air mixture is very high such that the oxygen is not sufficient to form adequate amounts of oxidized products, the energy released in the reaction becomes negligibly small. The chemical reactions are

Table 7.2 Minimum ignition energy of stoichiometric mixtures with air

S. No.	Gaseous mixture	MIE (mJ)
1	Hydrogen–air	0.018
2	Methane–air	0.28
3	Ethane–air	0.24
4	Butane–air	0.25
5	Acetylene–air	0.017

then unable to maintain a flame. Lean fuel concentrations and rich fuel concentrations in fuel–air mixtures, which cannot form a flame, are said to be outside the limits of flammability. This is shown in the plot of heat release as a function of equivalence ratio in Fig. 7.4.

The fuel-lean concentration below which a flame cannot exist in the mixture is known as the lean or lower flammability limit. Fuel concentration above which a flame is not possible is called as rich or upper flammability limit. The lower and upper flammability limits are expressed as percentage volume of the fuel gas in the volume of the mixture (the volume of fuel in a given volume of the fuel–air mixture expressed as a percentage). A schematic of the lower and upper flammability limits due to the rapid decrease of the energy release or equivalently the temperature of the combustion products is illustrated in Fig. 7.5 by L and U respectively. S denotes the stoichiometric composition.

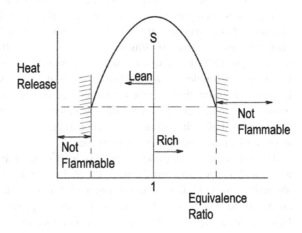

Fig. 7.4 Small values of heat release for a very fuel-lean and a very fuel-rich mixtures

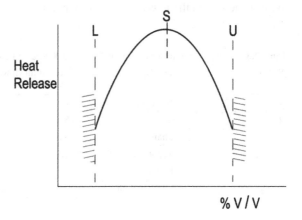

Fig. 7.5 Illustration of limits of flammability

The concentration of reactants was expressed in Chap. 4 in moles/cc. This could be expressed in units of volume of fuel/volume of mixture by using the ideal gas equation and converting moles/cc into an equivalent partial pressure of the gaseous fuel in the mixture. If the temperature and pressure of the gas mixture are T (K) and p (Pa) respectively, and the partial pressure of the fuel is p_f (Pa) and the concentration is C_f moles/m^3, we have

$$p_f = C_f R_0 T \qquad (7.16)$$

where R_0 is the universal gas constant in J/(mole-K). Noting that the volume fraction is proportional to the partial pressure fraction at the given value of temperature, we get

$$\% \frac{\text{volume fuel}}{\text{mixture volume}} = \% \frac{p_f}{p} \qquad (7.17)$$

The existence of lean and rich limits of flammability is essentially due to the chemical energy release being inadequate to maintain a flame. In the case of hydrocarbon–air mixtures, it is observed that the lean flammability limit corresponds to a threshold energy release of about 50 kJ/mole of the mixture and a threshold burnt gas temperature of about 1500 °C. In the case of hydrogen–air mixture, a much lower heat release of about 20 kJ/mole and a limit temperature of 800 K is seen to correspond to the lower flammability limit. The lean (L) and rich (U) limits of flammability at ambient pressure of 1 atm and 25 °C for different gas mixtures in air are given in Table 7.3. The lean limit of hydrocarbon gases is observed from Table 7.3 to generally decrease as the molecular mass of the gas increases.

It is also seen from the above table that hydrogen–air mixtures are flammable over a very wide range (between 4% and 75% v/v hydrogen). The lean limits are, however, about the same or higher for the hydrogen–air mixtures than for the hydrocarbon gas–air mixtures. It is the lean fuel–air mixture that is generally formed in an accidental leak. Since the lean limit for hydrogen is about the same or higher than for the hydrocarbon gases, hydrogen–air mixtures do not pose to be as hazardous compared to other fuels from the lower flammability point of view.

Table 7.3 Lean and rich limits of flammability at 100 kPa and 25 °C

S. No.	Gas mixture	Lean limit (lower) L (% v/v)	Rich limit (upper) U (% v/v)
1	Hydrogen–air	4	75
2	Methane–air	5	14
3	Ethane–air	3.5	15
3	Propane–air	2.1	9.5
4	Butane–air	1.8	8.4
5	Acetylene–air	2.6	12.8

7.1.5 MIE and Flammability Limits

The minimum ignition energies were given for stoichiometric mixtures of the fuel–
air mixtures in Sect. 7.1.3. As the mixtures depart from stoichiometric condition
and progressively become more and more fuel-lean or fuel-rich, their reactivity
decreases. The flame thickness and quenching thickness increase. The minimum
ignition energy therefore increases. At the lean and rich flammability limits, the
mixture is not ignitable. This implies that MIE would be infinity. A plot of MIE as a
function of volume of fuel in the fuel–air mixture would therefore be U-shaped with
minimum value of MIE for stoichiometric composition and infinitely large values
at the fuel-lean and fuel-rich limits. Figure 7.6 gives variation of ignition energy for
mixtures of hydrogen, methane, and propane in air as the amount of the fuel in the
mixture is varied.

It is seen from Fig. 7.6 that values MIE for mixtures near the flammability limits are
much higher than the stoichiometric values. In fact, the MIE for fuel concentrations
near the lean limit is about the same for the hydrocarbon gases such as methane
and propane with air as for the hydrogen–air mixture though under stoichiometric
conditions the MIE for hydrogen–air mixture is less than one-tenth of the MIE
for methane/propane–air mixtures. This indicates that hydrocarbon or hydrogen air
mixtures formed as lean fuel mixtures near the lean limits of flammability are equally
vulnerable to get ignited. The very wide range of flammability limits for hydrogen–air
mixture compared to methane and propane air mixtures is also seen in Fig. 7.6.

7.1.6 Low-Pressure Flammability Limits

When the pressure of a fuel–air mixture falls below a threshold value for which the
heat release is insufficient to support a flame, a flame can no longer be initiated nor

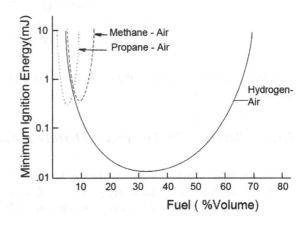

Fig. 7.6 Variation of
ignition energy with
concentration

propagate in the mixture. The pressure below which flammability conditions do not
exist is termed as low-pressure limits of flammability.

7.1.7 Influence of Initial Temperature on Flammability Limits

An increase of the initial temperature of a gaseous mixture will enhance its heat of
combustion and the temperature of the combustion products. If the lower flammability
limit, determined at the standard temperature of 25 °C, is denoted as L_{25} (%v/v), then
considering the increased enthalpy of combustion from the increased sensible heat
of the mixture at temperature T(°C), the energy balance gives

$$L_T \Delta H_C = L_{25} \Delta H_C - C(T - 25) \tag{7.18}$$

Here the term $C(T - 25)$ denotes the sensible heat over and above 25 °C, C being
the specific heat of the gas mixture. L_T is the lower flammability limit at T (°C) and
ΔH_C is the heat of combustion at the standard condition of 25 °C and 1 atm pressure.
The lower flammability limit (L_T) is reduced due to an increase of temperature.
Equation 7.18 is simplified to give

$$\frac{L_T}{L_{25}} = 1 - \frac{C}{L_{25} \Delta H_C}(T - 25) \tag{7.19}$$

Similar to the decrease of the lower flammability limit as the initial temperature
increases, an increase occurs for the upper flammability limit with increase of temperature of the gas mixture. Approximate values of the lower and upper flammability
limits (L_T and U_T) at any temperature T (°C) are given as a function of the flammability limits (L_{25} and U_{25}) at 25 °C by

$$\frac{L_T}{L_{25}} = 1 - 0.00072(T - 25) \tag{7.20}$$

$$\frac{U_T}{U_{25}} = 1 - 0.00072(T - 25) \tag{7.21}$$

A wider range of flammable mixtures is formed at higher values of its initial temperature.

7.1.8 Flammability Limit for a Mixture of Gases

The flammability limit of a mixture of gases (L) comprising of constituents 1, 2, 3,
…, n having fractional volume $f_1, f_2, f_3, \ldots, f_n$ such that $f_1 + f_2 + f_3 + \cdots + f_n = 1$, is given by Le Chatelier's rule, which states that

$$L = \frac{1}{\frac{f_1}{L_1}, \frac{f_2}{L_2} + \cdots + \frac{f_n}{L_n}} \tag{7.22}$$

Here $L_1, L_2, \ldots, L_n$ denote the lower flammability limits of constituents $1, 2, \ldots, n$. The above harmonic relation given by Eq. 7.22 is valid for mixtures having constituents of similar chemical structure. The relation is used for both lower (L) and upper (U) flammability limits.

7.1.9 Upward and Downward Flammability Limits

The standard set-up for measuring the flammability limits consists of a tube of about 10 cm diameter and a length of about 150 cm in which the gas mixture is enclosed and ignited. The tube is usually held vertically, and the gas mixture in the tube is ignited at the bottom. The condition of the mixture for which the flame is unable to travel the length of the tube is taken to be outside the flammability limits.

If the gaseous mixture is ignited at the top of the tube, instead of igniting at the bottom, the flame would travel downward. The limits of flammability determined from experiments on downward propagation of a flame are narrower than those obtained for upward propagation of a flame. This is because the buoyant motion of the hot gases in the case of upward propagation transports the heat and preheats the gas mixture. The lean limits are therefore lowered, and the upper limits increased in accordance with Eqs. 7.20 and 7.21. Measurements of flammability limits in horizontal tubes give values of limits to be between those measured for vertically upward and vertically downward propagating flames.

7.2 Minimum Oxygen Concentration: Maximum Safe Oxygen Concentration

Flammability limits are expressed in units of volume of the combustible (fuel) per unit volume of the mixture of fuel and air. If we presume that the gaseous fuel and air to be ideal gases, the flammability limits normally obtained at the ambient pressure and temperature mixture can be written in terms of molar concentration, i.e., moles of fuel per unit mole of the mixture.

Generally, the lower flammability limit L for the hydrocarbon–air mixture is between 0.012 and 0.05 volume of fuel/volume of mixture or 1.2–5 %V/V. The lower flammability limit is $L =$ Volume of fuel/ (volume of fuel + volume of air) which equals moles of fuel/(moles of fuel + moles of air).

As an example, for propane, $L_{C_3H_8} =$ moles of C_3H_8/(moles of C_3H_8+ moles of air). Since the volume of fuel is small at the lean flammability limit, we can approximate the above as $L =$ moles of fuel/(moles of air).

Completed combustion products are formed due to the excess air under conditions of the lower flammability limit. The reaction for the specific case of propane air at the lean limit is

$$LC_3H_8 + (O_2 + 3.76N_2) = 3LCO_2 + 4LH_2O + 3.76N_2 + (1 - 5L)O_2$$

The heat generated corresponds to the formation of 3L moles of CO_2 and 4L moles of H_2O in the products. If we had a stoichiometric reaction in which fully oxidized products of combustion are formed, the reaction is

$$C_3H_8 + 5(O_2 + 3.76N_2) = 3CO_2 + 4H_2O + 5 \times 3.76N_2$$

This implies that in a reaction of the combustible mixture at limit L, the energy generated is much smaller as only a fraction 3L moles of CO_2 and 4L moles of H_2O are generated leading to the formation of a limit mixture corresponding to the lower flammability limits.

The same completed combustion products, i.e., 3L of CO_2 and 4L of H_2O would have resulted if instead of 5 moles of O_2 in the stoichiometric reaction, 5L moles of O_2 were used. Thus $5 \times L$ is the minimum molar fraction or concentration of oxygen required to form a flame. The molar fraction of oxygen is also the same as the volume of oxygen per unit volume of the mixture. This value of $5 \times L$ is the minimum oxygen concentration (MOC) required to support the flame in the mixture.

In general, for a hydrocarbon C_nH_m, the chemical reaction for complete combustion is

$$C_nH_m + vO_2 = nCO_2 + \left(\frac{m}{2}\right)H_2O$$

where v is the stoichiometric coefficient. The product $v \times L$ is the minimum oxygen concentration (MOC).

The minimum oxygen concentration (MOC) is therefore obtained as the oxygen required for stoichiometric combustion corresponding to the lean limit L. Thus, for example, for a propane–air mixture, the lower limit L is 2.1% volume of propane in the propane–air mixture. For stoichiometric combustion, however, the reaction is given by

$$C_3H_8 + 5O_2 \rightarrow 3CO_2 + 4H_2O.$$

The MOC for propane–air is

$$\frac{2.1 \times 5}{100} = 10.5\% \text{ volume oxygen/volume of the mixture.}$$

MOC is a useful term and gives a measure of inert gas such as N_2 or CO_2 or water vapor that is to be added to a flammable mixture to dilute the oxygen in it and make its energy release less than the value obtained at the lower flammability limits. Substances having higher values of specific heats are more effective as they also contribute to a decrease of the temperature of the products. MOC is sometimes

Table 7.4 Lean and rich limits of flammability at 100 kPa and 25 °C

Hydrocarbon-air	L (%V/V)	Stoichiometric coefficient v	MOC (%V/V)
Methane–air	5	2	10
Ethane–air	3.5	3.5	12.5
Propane–air	2.1	5	10.5
Butane–air	1.8	6.5	11.7
Do-decane–air	0.6	18.5	11.1

referred to as maximum safe oxygen concentration (MSOC). The values of MOC for some of the hydrocarbon fuels are given in Table 7.4.

An MOC value of 10 is generally used for hydrocarbon-air mixtures.

7.3 Flammability Limits of Vapors From Volatile Liquids

7.3.1 Formation of Flammable Vapor–Air Mixture from Volatile Liquids

The vapor pressure of a volatile liquid increases when heated. At the normal boiling temperature, the vapor pressure is the same as ambient pressure. The boiling temperature changes with changes of the ambient pressure. The vapor pressure at the saturation temperature, which is same as the boiling temperature, is known as saturation pressure. The value of saturation pressure p_s increases as the saturation temperature T_s increases as shown in Fig. 7.7.

Beyond a critical value of temperature and pressure, the liquid and vapor phases are equally dense, and there is no distinction between a vapor and a liquid. This is indicated by the point C in Fig. 7.7.

Fig. 7.7 Equilibrium between a volatile liquid and its vapor

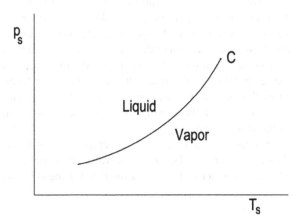

Fig. 7.8 Flammability of kerosene fuel at ambient pressure of 0.1 MPa

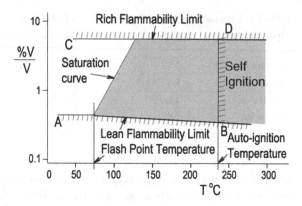

On the heating of a volatile liquid fuel, the vapor issuing from its surface mixes with the ambient air. The concentration of the fuel vapor in the mixture of vapor and air, if greater than the lean limit of flammability, can result in the formation of a flammable mixture. However, if the concentration of vapor exceeds the upper flammability limit, no flame can be formed. The variation of the flammability limits with changes in temperature of the vapor is shown in Fig. 7.8. While the volumetric concentration of the lean flammability limit decreases with temperature, the concentration of the upper flammability limit increases with increase of temperature as per discussion of the variation limits with temperature in the last section. The shaded region between the lean (AB) and rich limits of flammability (CD) in Fig. 7.8 shows the region of flammable vapor–air mixture formed from a volatile kerosene fuel. The left side of the flammable region is bound by the saturation curve. The lower and upper sides are bound by the lean flammability limit and the rich flammability limit, respectively.

7.3.2   Flash and Fire Point Temperatures

The temperature of the liquid fuel at which a vapor–air mixture corresponding to the lean flammability limit is formed due to its vapor mixing with the ambient air is known as the flash point temperature. This is shown by in Fig. 7.8 and corresponds to the point of intersection of the saturation curve with the lean flammability curve AB. It is an important property quantifying the fire hazard of volatile fuels. It represents the temperature of the liquid at which the vapor formed with air can get ignited. Since the mixture with air corresponds to the lean flammability limit, the fuel vapor gets consumed in the burning and the flame cannot be sustained. A momentary flash of flame is formed.

In order to sustain the flame, the temperature of the volatile fuel needs to be higher than the flash point. The threshold value of temperature at which a copious amount of fuel gets generated to feed a sustained flame is known as fire point temperature.

The saturation temperature decreases with a reduction of the ambient pressure. The flash and fire points would therefore be lower at reduced ambient pressures such as at higher altitudes.

When the temperature is increased beyond the auto-ignition temperature (Fig. 7.8), spontaneous burning takes place in the flammable mixture. The fire point temperature is normally below the auto-ignition temperature.

The flash and fire point temperatures of volatile fuels such as kerosene, used for jet engines and liquid propellant rockets, are increased to make it safer for use by incorporating small mounts of additives in the fuel. The modified kerosene is known as jet fuel (Jet-A, Jet A-1, Jet B) for jet engines and rocket propellant (RP) for rockets. Small amounts of additives are also added to reduce the gumming tendency of kerosene and its tendency to get charged by static electricity.

Droplets of fuel or mist or fog, formed by super-cooling of the vapor, can catch fire even though the temperature is below the flash point temperature. In this case, the vaporization of the droplets of the volatile fuel forms adequate concentration of the vapor to sustain a flame.

7.4 Initiation of Detonation: Detonation Kernel

7.4.1 Requirement of Strong Shock Wave

A detonation is seen in Chap. 6 to comprise of a shock wave driven by the energy released by the chemical reactions taking place behind it. A shock of sufficient strength is required to heat the gases to sufficiently high temperatures corresponding to its auto-ignition temperature and thereby spontaneously release the chemical energy required to drive the shock wave. Unlike the formation of a flame, which was seen to require the transport of heat from the ignition kernel to its adjacent layers, the initiation of a detonation requires the formation of a sufficiently strong shock wave by the ignition source. The energy required for initiating a detonation is therefore expected to be higher than initiating a flame.

The Mach number of a shock wave M_s, formed by rapid release of energy, was seen in Chap. 2 to decay as it progressed over a distance R_s according to the relation given below:

$$\left(\frac{R_s}{R_0}\right)^3 = \frac{1}{4\pi I \gamma} \frac{1}{M_S^2} \tag{7.23}$$

where R_0 is the explosion length and I is the energy integral fraction. The variation of the shock Mach number M_S with distance R_s from the source for a given value of R_0 is shown in Fig. 7.9.

The temperature generated in a shock wave of different Mach numbers can be determined from Eq. 2.44 in Chap. 2. For an initial temperature of 300 K and a

Fig. 7.9 Decay of a shock
wave with distance R_s from
the energy source

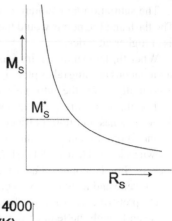

Fig. 7.10 Hot gas
temperature in a shock

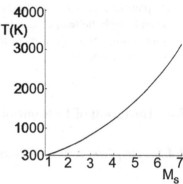

specific heat ratio γ of 1.4, the temperature in a shock wave for Mach numbers up to
7 is given in Table 7.5 and plotted in Fig. 7.10.

Typical auto-ignition temperatures of fuel–air mixtures are about 800 K. The auto-
ignition temperature of a stoichiometric mixture of hydrocarbon gases and hydrogen
in air at an initial temperature and pressure of 300 K and 0.1 MPa is given in Table 7.6.
A minimum threshold value of shock Mach number would therefore be required for
the auto-ignition to occur. The threshold value of shock Mach number is denoted by
M_S^* and is shown in Fig. 7.9. For spontaneous ignition the shock Mach number must
be greater than M_S^*.

Table 7.5 Temperature in a shock (ambient temperature = 300 K; $\gamma = 1.4$)

M_S	T (K)
7	3140
6	2385
5	1740
4	1210
3	805
2	500

Table 7.6 Auto-ignition temperatures of stoichiometric mixture of gases initially at 300 K and 0.1 MPa

Fuel	Temperature (K)
Methane	815
Ethane	790
Propane	725
Acetylene	578
Hydrogen	675

7.4.2 Requirement of a Minimum Kernel for Detonation

The energy release from chemical reactions, if spontaneous auto-ignition of the shocked gases takes place in a shock wave, could contribute to maintain its strength and thereby form a detonation. The energy release by the chemical reactions of the shocked particles of the gas takes place after a distance Δ behind the shock front corresponding to the induction time. The value of this induction distance Δ would depend on the Mach number of the shock and hence on the distance R_S that the shock has traveled from the source of energy release. A schematic of the zone of chemical energy release, when the shock has traveled a distance R_S, is the shaded zone in Fig. 7.11. Denoting the density in the combustion zone, which is a function of the radius as $\rho(r)$ and the energy release per unit mass of the burnt gas as Q, the chemical energy release is given as:

$$E_{\text{Chem}} = \int_0^{R_S - \Delta(R_S)} Q\rho(r)4\pi r^2 dr \tag{7.24}$$

If the energy release by the ignition source is E_S, and the energy release from chemical reaction is

$$\int_0^{R_S - \Delta(R_S)} Q\rho(r)4\pi r^2 dr$$

Fig. 7.11 Zone of energy release after an induction distance

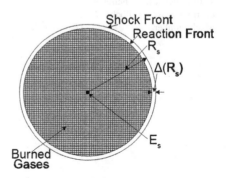

Shock Front
Reaction Front
R_s
$\Delta(R_s)$
E_s
Burned Gases

the increase of the kinetic and internal energies of the shocked mass of gas could be written as:

$$E_S + \int_0^{R_S-\Delta} Q\rho 4\pi r^2 dr = \int_0^{R_S} \left[\frac{u^2}{2} + (e - e_0)\right] \rho 4\pi r^2 dr \qquad (7.25)$$

Here u denotes the particle velocity, and e is the internal energy per unit mass. The initial value of internal energy of the medium is denoted as e_0. While the Mach number of the shock due to the source E_S decreases as the shock propagates as was also seen in Eq. 7.23 and Fig. 7.9, the Mach number of the shock would monotonically increase from the chemical energy release term

$$\int_0^{R_S-\Delta} 4\pi \rho Q r^2 dr$$

The decaying influence is therefore arrested by the heat release as long as spontaneous chemical reactions with small values of induction distance Δ are possible.

The variation of Mach number due to the combined influence of the energy from the source and chemical reactions is illustrated in Fig. 7.12. The influence of the source and chemical energy release is also shown in the figure as blast wave decay and chemical heat release. At large distances, the chemical energy drives the detonation as a constant velocity Chapman–Jouguet detonation at M_{CJ}.

The energy requirements for initiating a detonation is to form a shock wave of sufficient strength such that the energy release by the chemical reactions of the gas processed by the shock wave overcomes the decaying influence of the source as it progresses. In this way, the decaying influence of the energy release by the source is arrested. Since the spontaneous energy release by the chemical reactions is possible only for values Mach numbers greater than a threshold value of M_S^*, the energy from the ignition source should satisfy the following two conditions:

Fig. 7.12 Decaying and energizing influence of the source energy and chemical energy release on shock Mach number

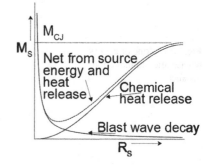

1. The shock wave formed from the igniter energy release should not have decayed down to values below M_S^* by the time the chemical energy release dominates and overcomes the decaying trend of the shock Mach number.
2. The igniter must form a shock wave whose Mach number is greater than M_S^* when it has traveled a certain minimum distance away from the source of energy release. By this time the chemical energy starts getting liberated.

7.4.3 Detonation Kernel in Analogy to Flame Kernel: Energy Required

The role of the igniter in the initiation of detonations is to therefore to form an adequate volume of strongly shocked gases so that the chemical energy release from this volume can arrest the decaying shock wave generated by the igniter energy alone. This threshold volume is known a detonation kernel in analogy to the ignition kernel associated in the formation of a flame. The critical size of the detonation kernel which corresponds to the minimum energy for initiating a detonation should be able to progress the shock Mach number as it travels further to reach the steady Chapman Jouguet value.

Experiments conducted with different explosive gaseous mixtures show the blast wave to directly decay to a Chapman Jouguet detonation when the energy release is very much higher than the minimum threshold energy required for initiating a detonation. When the energy release is less than the threshold value, the blast wave decays and no detonation is formed. At the critical or minimum threshold value, the blast wave initially decays to a Mach number below M_{CJ} to M_S^* and thereafter progresses to a detonation. These three regimes are referred to as supercritical, subcritical, and critical and is illustrated in Fig. 7.13.

In summary, the energy release has to be impulsive so that a shock is formed. The energy need not necessarily be of thermal origin, and any stimulus which can form a sufficiently strong shock wave (such as a strong blow) can result in a detonation in a combustible medium. However, the strength of the shock wave and the volume

Fig. 7.13 Supercritical, subcritical, and critical regimes of initiation

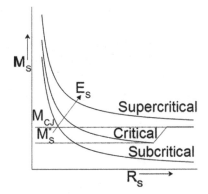

of the shocked kernel of gases should exceed a threshold value so that the chemical energy from the auto-ignition prevents the decay of the shock wave.

The energy requirements for initiating a detonation are several orders of magnitude higher than the minimum ignition energy for forming a flame. For stoichiometric hydrogen–air mixture, the energy required is about 400 J for a initiating detonation compared to 0.02 mJ for a flame. The initiation of detonation in propane–air mixture requires about 100 kJ of energy compared to a minimum ignition energy of 0.5 mJ.

7.4.4 Limits of Detonation

It is not possible to generate and propagate a detonation in all reactive mixtures. Similar to the lower and upper flammability limits, there exist certain values of concentration, pressure, and dilution with inert substances beyond which a detonation cannot be formed. The chemical reactions, if at all initiated by the shock waves, are not vigorous and adequate for these conditions of concentration, pressure, and dilution to sustain the shock. The threshold conditions for which a detonation is no longer possible are known as limits of detonation in analogy to limits of flammability. The limits of detonation are narrower than the limits of flammability considering the higher rates of energy release to maintain the shock wave. As an example, while the concentration limits of flammability for hydrogen–air mixtures are between 4 and 75% volume of hydrogen per volume of the mixture, the limits of detonation of the hydrogen air mixtures are between 18 and 60%. This is shown in Fig. 7.11. Though the lean limit of detonation is generally specified as 18%, large-scale detonations have been obtained for 11% volume of hydrogen in the hydrogen–air mixture (Fig. 7.14).

Fig. 7.14 Limits of detonation compared with limits of flammability

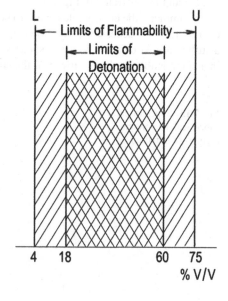

The lean limits of detonation of hydrocarbon–air mixtures are in the region of 2 to 3% by volume. These are substantially lower than the value of 18% for the hydrogen–air mixture.

7.5 Transition of Flame to Detonation

The energy required to be impulsively deposited by the ignition source to initiate a detonation in an explosive gas mixture is observed to be very much larger than the energy required for forming a flame in it. The impulsive release of energy requires that the power density of the energy release by the source must be large such that a blast wave is generated by the source. Such energy sources may not be readily encountered in practice. It appears therefore that energy release sources have to be specifically designed to produce large energy release at sufficiently high power densities in order to initiate a detonation in a given explosive.

However, it is always possible to form a flame with very insignificant levels of energy release, and the flame can transit to a detonation in a confined geometry. The flame moves away from the ignition source and while doing so compresses the unburned gas ahead by compression waves in a way similar to a piston in a cylinder. The burning velocity would be larger in the compressed unburned gases. The unburned gases are also pushed ahead, and the flame traveling in the velocity-induced unburned gases will have a flame velocity that is higher than the burning velocity. This results in the acceleration of both the flame and the unburned gases ahead of the flame. The interaction of the compression waves among themselves and with the confinement generates an interacting system of shock waves. Slipstreams are formed at the point of interaction of the shocks. The resulting flow of the unburned gases becomes very turbulent. The flame, which was originally laminar, becomes turbulent with large wrinkles in it. Depending on the confinement and the shock wave ahead of the flame a quasi-steady detonation could be formed, and this subsequently changes to a detonation.

The large wrinkles, formed at the flame front, could also engulf a considerable amount of the unburned mixture and when this gets heated up and explodes, the spontaneous energy release from it could be considerable to form shock waves. The shock waves could be further accelerated by energy release rates behind them to form a strong shocks and hence detonations. Obstacles placed in the path of a flame help in accelerating the flame and shock waves and often result in the formation of detonations.

Detonations are more readily formed in confined geometries by the above process of flame to detonation transition. The distance traveled by a flame before it becomes a detonation is known as "Run up distance for a detonation" and the process as deflagration to detonation transition (DDT).

Examples

7.1 *Lower Flammability Limit, MOC and Inerting*:

(a) The lower flammability limit of kerosene vapor in a mixture of kerosene vapor and air is 0.6% vol/vol. Assuming the chemical formula for kerosene as $C_{12}H_{26}$, express the lower flammability limit in mg (milligram) of kerosene vapor per liter (mg/L).

(b) Determine the minimum oxygen concentration (also known as maximum safe oxygen concentration) for kerosene vapor in air using the data given in (a) above.

(c) If an explosive cloud of kerosene–air mixture is formed at atmospheric conditions of 0.1 MPa and 25 °C in which the % volume of kerosene vapor is 1.2, determine the volume of inert CO_2 that must be added to the cloud to prevent it from catching fire.

Solution:

(a) *Concentration at the lower flammability limit in mg/liter*: Let the lower flammability limit of kerosene vapor–air mixture be denoted by L % volume of kerosene vapor in the volume of the mixture. L is given as 0.6%, i.e., in $1l$ of the mixture, the volume of kerosene is 0.006 l:

$$L = \frac{0.006 l_{kerosene}}{1 l_{mixture}}$$

Kerosene has the chemical formula $C_{12}H_{26}$.
The molecular mass of kerosene $M = 144 + 26 = 170$ g/mole
1 mole of an ideal gas at the standard condition of 0.1 MPa and 25 °C has a volume of

$$\frac{R_0 T}{p} = \frac{8.314 \times 298}{10^5} = 0.0248 \text{ m}^3 = 24.8 \text{ liters.}$$

Considering the kerosene vapor to be an ideal gas, we note that 24.8 liters of vapor will have a mass equal to the molecular mass = 170 g.
Therefore, 0.006l has a mass of $170 \times 0.006/24.8 = 0.0411$ g.
This corresponds to $1l$ of the mixture of kerosene vapor and air.
Hence $1l$ of the mixture of kerosene vapor and air has a mass of kerosene 41.1 mg/liter.
The lean limit L has a concentration of 41.1 mg/liter.

(b) *Minimum oxygen concentration*: The minimum oxygen concentration (MOC) or the maximum safe oxygen concentration (MSOC) corresponds to the lean limit of flammability (L). It gives the oxygen concentration corresponding to the stoichiometric reaction of the lean limit fuel. The stoichiometric reaction of kerosene–air is

$$C_{12}H_{26} + 18.5(O_2 + 3.76N_2) \rightarrow 12CO_2 + 13H_2O + 69.56N_2$$

At limit L, $C_{12}H_{26}$ is 0.006 vol./vol. of the mixture.

Since moles are proportional to volume, the minimum oxygen concentration is $0.006 \times 18.5 = 0.111$ volume of oxygen/vol. of mixture.

MOC $= 11.1\%$ by volume of oxygen. Oxygen concentration less than the above value of 11.1% will lead to heat release less than at the limits and a flammable mixture would not be possible.

(c) *Inert carbon dioxide to be added to the cloud to prevent its ignition*: The existing cloud has kerosene volume fraction:

$$\frac{V_{ker}}{V_{mix}} = 0.012$$

which is higher than at the limits (0.006). We need to add CO_2 such that

$$\frac{V_{ker}}{V_{CO_2} + V_{mix}} = 0.006$$

We therefore get

$$\frac{V_{CO_2} + V_{mix}}{V_{mix}} = \frac{0.012}{0.006} = 2$$

This gives $V_{CO_2}/V_{mix} = 1$, therefore

$$V_{CO_2} = V_{mix}$$

The cloud must be diluted with a volume of CO_2 equal to the volume of the mixture of kerosene vapor and air.

A lower quantity of CO_2 would suffice in practice considering the higher values of specific heat of CO_2 which contributes to a lowering of the temperature of the combustion products.

7.2 *Flammability Limit from Flash Point Temperature*: The flash point of liquid toluene is 4.4 °C. At this temperature, the value of its saturated vapor pressure is 1.24 kPa. Determine the lower flammability limit of a toluene vapor–air mixture at standard atmospheric conditions in % volume toluene vapor in the total volume of the mixture.

Solution: The lower limit of flammability (L) corresponds to the concentration of the vapor formed at the flash point temperature. Since the vapor pressure at the flash temperature is 1.24 kPa and the net pressure of the vapor and air is the ambient pressure of 100 kPa, the volume fraction is $p_v/p_a = 1.24/100 = 0.0124$. The lean limit of flammability is 1.24% vol/vol.

7.3 *Energy to Initiate CJ Detonation*: Determine the energy required to directly initiate a Chapman–Jouguet detonation in a gaseous fuel air mixture at 100 kPa and 300 K. The sound speed in the fuel air mixture is 330 m/s while the specific heat ratio $\gamma = 1.4$. The Mach number of the CJ detonation is 5.

Solution: The dependence of induction time τ on temperature T is given by the correlation $\tau = 3.4 \times 10^{-6} \exp(1100/T)$ s, where T is the temperature in K. The Mach number M_S of a strong blast wave generated by an energy source of explosion length R_0 at a distance from the energy source R_S is obtained (Chap. 2) as

$$M_S^2 = \frac{0.134}{(R_S/R_0)^3}$$

A detonation kernel should be formed to supply energy to the shock front and overcome the decaying influence of the energy source as the shock front progresses. The energy, released in the spherical detonation kernel of radius R_S, should be such that (i) the shock Mach number M_S formed at R_S can auto-ignite the mixture and (ii) the size R_S reached by the shock gives sufficient time for the energy to be released in the volume enclosed by the shock at R_S.

We require a CJ detonation to be directly formed; hence the Mach number of the shock M_S that is generated by the energy source equals M_{CJ}. The velocity of the medium behind the shock in the frame of reference of the shock is u and with respect to the shock velocity,

$$\frac{u}{\dot{R}_S} = \frac{\gamma - 1 + 2/M_S^2}{\gamma + 1}$$

With $\dot{R}_S = 5 \times 330 = 1650$ m/s and $M_S = 5$, $u = 565.7$ m/s.
The temperature of the medium behind the shock is

$$\frac{T}{T_0} = \frac{p}{p_0} \times \frac{\rho_0}{\rho} = \left(\frac{2\gamma}{\gamma + 1} M_s^2 - \frac{\gamma - 1}{\gamma + 1} \right) \times \left(\frac{\gamma - 1 + 2/M_s^2}{\gamma + 1} \right)$$

The ratio works out as 9.918 giving $T = 2975$ K.
The induction time τ in seconds is

$$\tau = 3.4 \times 10^{-6} \exp \left(\frac{1100}{T} \right)$$

$$= 3.4 \times 10^{-6} \exp \left(\frac{11,000}{2975} \right)$$

$$= 1.37 \times 10^{-4} \text{ s}$$

The induction distance must therefore be $u \times \tau = 565.7 \times 1.37 \times 10^{-4} = 0.078$ m. This must be the distance R_S required for the detonation kernel.
From the blast decay equation

$$M_S^2 = \frac{0.134}{(R_S/R_0)^3}$$

we get the value of R_S/R_0 at $M_S = 5$ as

$$\frac{R_S}{R_0} = \left(\frac{0.134}{25}\right)^{1/3} = 0.175$$

But the value of $R_S = 0.078$ m. Hence

$$R_0 = \left(\frac{0.078}{0.175}\right) = 0.445$$

The explosion length is defined as

$$R_0 = \left(\frac{E_0}{p_0}\right)^{1/3}$$

Hence the energy $E_0 = 10^5 \times 0.445^3 = 0.088 \times 10^5 = 8.8 \times 10^3 = 8.8$ kJ. The energy required to initiate the CJ detonation is 8.8 kJ.

7.4 *Explosion of a Partially Ventilated Aircraft Fuel Tank During Aircraft Take-off*: The (following explosion problem is formulated in the context of the Boeing 747 aircraft that exploded on flight TWA 800 eleven minutes after take off from New York airport on July 17, 1996.

A ventilated fuel tank of an aircraft of volume $1/2$ m^3 contains small amounts of volatile liquid kerosene fuel. The temperature of the kerosene is 35 °C and does not change with time immediately after take-off considering the large thermal capacity of the tank compared to the very small quantity of kerosene contained in it. When the aircraft takes off from the ground and gains altitude, the reduced pressure of the ambient air, which is communicated to the tank, promotes the formation of a combustible kerosene vapor–air mixture. If the lean limit of flammability of the kerosene–air mixture by volume is 1.8% and does not vary significantly with temperature, determine the altitude of the aircraft at which a flammable kerosene vapor–air mixture is formed in the tank. You can assume the following:

(i) The temperature of the air in the tank continues to be 35 °C.
(ii) The saturated vapor pressure p_v of kerosene at 35 °C is 598 Pa.

(iii) The variation of ambient pressure with altitude (height) is given below:

Height (m)	Pressure (MPa)
0	0.100
1000	0.0899
2000	0.0795
3000	0.0701
4000	0.0617
5000	0.0540
6000	0.0472
7000	0.0411
8000	0.0356
9000	0.0308

Solution: The temperature of liquid kerosene is specified as 35 °C. Vapor pressure of kerosene at this temperature is given as 598 Pa.

In the given volume of the tank at pressure p_a (same as the ambient pressure since the tank is ventilated), the fractional volume of kerosene vapor f is

$$f = \frac{p_v}{p_a}$$

At sea level, the volume fraction of kerosene vapor in the tank is $598/10^5 = 0.00598$, which is very much lower than the lean flammability limit L of 0.018.

For the given value of vapor pressure of kerosene p_v, the value of pressure p_a in the tank corresponding to the lean limit $L = 0.018$ is given by

$$L = \frac{p_v}{p_a}$$

$$p_a = \frac{p_v}{L} = \frac{598}{0.018} = 33,222 \text{ Pa} = 0.0332 \text{ MPa}.$$

From the table on pressure variations with altitude, the above pressure of 0.0332 MPa is obtained for an altitude of 8480 m. A flammable mixture is formed in the tank when the aircraft is at an altitude of 8.48 km and could explode.

Nomenclature

C	Constants used in Eqs. 7.10, 7.18 and 7.19
C_P	Specific heat at constant pressure [kJ/(kg K)]
d	Diameter (m)
d_q	Quenching diameter or quenching thickness (m)
E	Combustion energy (kJ)
E_0	Energy deposited for ignition/detonation in a reactive medium (kJ)
f	Volumetric fraction of a particular gas in a mixture
I	Energy integral used in Eq. 7.23
k	Thermal conductivity of medium [kJ/(ms °C)]

L   Fuel lean concentration limits of flammability (% volume/volume)
MIE   Minimum ignition energy (J)
M_S Mach number of shock wave
M_S^* Threshold Mach number of shock wave to form a detonation
p Pressure (Pa)
Pe Peclet number denoting ratio of convection to conduction heat transfer
$\dot{Q}_L$ Heat loss by conduction from flame (J)
R_0 Universal gas constant [J/(mole K)]
r Radius (m)
S_u Flame propagation speed in unburned medium (m/s)
T Temperature
t_f Flame thickness (m)
U Fuel rich concentration limits of flammability (% volume/volume)
α Thermal diffusivity (m^2/s)
δ Small increment
ΔH_C Energy liberated during combustion (J)
γ Specific heat ratio

Subscripts

b Burned gases
f Fuel
S Saturation
w Wall

Further Reading

1. Borisov, A. A. and Loban, S. A., Detonation limits of hydrocarbon–air mixtures in tubes, *Combustion, Explosions and Shock Waves*, 13, 618–621, 1978.
2. Deonation Database: www.galcit.caltech.edu/detn_db/html/
3. Kuo, K. K., *Principles of Combustion*, John Wiley and Sons, New York, 1968.
4. Lee, J. H., *The Detonation Phenomenon*, Cambridge University Press, New York, 2008.
5. Lee. J.H. and Ramamurthi, K., On the concept of the critical size of a detonation kernel, *Combustion and Flame*, 27, 331–340, 1976.
6. Lee, J. H., Soloukhin, R., and Oppenheim, A. K., Current views on gaseous detonations, *Astronautica Acta*, 14, 565–584, 1969.
7. Lewis, B. and von Elbe, G., *Combustion, Flames and Explosion of Gases*, Academic Press, New York, 1961.
8. Liepman, H.W., and Roshko, A., *Elements of Gas Dynamics*, Wiley, New York, 1957.
9. Lovachev, L. A., Babkin, V.S., Bunev, V.A., V'Yun, A.V., Krivulin, V.N., and Baratov, A.N., Flammability limits: an invited review, *Combustion and Flame*, 20, 259–289, 1973.
10. Murthy Kanury, A., Limiting case of fire arising from fuel tank/ pipeline ruptures, *Fire Safety Journal*, 3, 215–226, 1980.

11. Oppenheim, A.K., *Dynamics of Combustion*, Springer, Berlin, 2006.
12. Strehlow, R.A., *Fundamentals of Combustion*, 2nd ed., McGraw-Hill, New York, 1984.
13. Stull, D. R., *Fundamentals of Fire and Explosion*, AIChE Monograph Series, vol. 73, no. 10, 1977.
14. Turns, S. R., *An Introduction to Combustion: Concepts and Applications*, 2nd ed., McGraw Hill, Boston, MA, 2006.
15. Urtiev, P. A. and Oppenheim, A.K., Experimental observations of transition to detonation in an explosive gas, *Proceedings Royal Society London* A, 295, 13–28, 1966.

Exercise

7.1 Determine the lower limit of flammability of a cloud containing 0.6% hexane, 2.0% methane, and 0.7% ethylene by volume and the balance air. The lean limits of flammability of hexane, methane, and ethylene are 1.1, 5.0, and 2.7 volume % fuel in air. Comment on the flammability limit of the mixture with respect to the flammability limits of the constituent gases.

7.2 (a) Determine the equivalence ratio at the lower limit of flammability of a cloud of propane (C_3H_8) and air. The lower flammability limit of propane in the propane–air mixture is 2.1% v/v.

   (b) For the above problem, express the lower flammability limit in mg (milligram) of propane per liter of the cloud (mg/L).

7.3 A fireman needs to enter a house containing a stagnant explosive gas mixture of stoichiometric butane and air in order to evacuate the inhabitants. He has to carry a powerful light source. A safety metal screen is to be provided around the light source such that the heat released from the light source will not ignite the explosive gas in the ambient explosive gas medium. What must be the maximum distance between the metal wires provided in the screen? The thermal diffusivity of the explosive mixture is 0.5×10^{-4} m^2/s, and the flame speed in it is 0.5 m/s. State any assumptions made.

7.4 Determine the quenching diameter (d_c) and the ratio of the quenching diameter to flame thickness (d_c/t_f) of a stoichiometric mixture of a combustible gas mixture given the following data:

   Thermal diffusivity of the gas mixture (α): 0.5×10^{-4} m^2/s
   Flame speed (S_u) = 0.5 m/s
   Volumetric heat release rate of the stoichiometric mixture = 250×10^3 kJ/(m^3 s)
   Temperature of the hot burned gases = 1500 °C
   Ambient temperature = 30 °C
   Mean thermal conductivity of the hot gases = 0.2 J/(m s °C)

7.5 If the quenching diameter of the hydrogen–air mixture near the limits of flammability at the temperature of 30 °C is 2 mm and the ignition temperature of the mixture is 600 °C, estimate the energy release from an accidental spark which would be sufficient to ignite a mixture of hydrogen and air formed near the flammability limits.

The specific heat at constant pressure and density of the relevant hydrogen–air mixture are 1.3 kJ/(kg K) and 0.95 kg/m^3.

7.6 (a) A hydrogen car is by mistake parked in a small non-ventilated garage of volume 50 m^3. If the ambient temperature and pressure in the garage is 100 kPa and 30 °C, determine the maximum mass of hydrogen, which can leak from the fuel tank of the car without being a fire or an explosion hazard? The lower flammability limit of hydrogen in air is 5% by volume of hydrogen in the hydrogen–air mixture at 100 kPa and 30 °C. You can assume that hydrogen disperses uniformly in the garage and also neglect the volume occupied by the car.

(b) If the temperature in the garage were to drop to 0 °C, would a larger or smaller leakage of (mass of) hydrogen from the car than given in "a" above be permitted without compromising on fire and explosion safety and why?

7.7 Consider a turbulent flame propagating at a steady velocity of 2.5 m/s in a fuel–air in a pipe 50 mm in diameter. The density of the mixture is 1.3 kg/m^3, and the heat release per unit mass is 3000 kJ/kg. Determine the rate of heat release in KW

7.8 A stoichiometric mixture of butane gas–air mixture is formed within a bund surrounding a chemical reactor. It is desirable to ensure that the mixture formed at 100 kPa and 27 °C should be suitably inerted with nitrogen gas such that it will not catch fire. The fire handbook gives the minimum oxygen concentration (MOC) for butane–air to be 12%. In order to ensure explosion safety, it is proposed to reduce the oxygen concentration to 10% and thus have a factor of safety. Determine the volume of nitrogen gas at 100 kPa and 27 °C that needs to be added to the stoichiometric mixture per unit volume of the mixture.

7.9 Toluene is methyl benzene $CH_3C_6H_5$. It is a volatile liquid, and its vapor pressure in kPa is given by $p_v = \exp(689/T)$, where T is the temperature of liquid toluene in Kelvin. If the flash point temperature of the volatile liquid toluene is 4 °C, determine the minimum oxygen concentration for the toluene vapor–air mixture. The ambient pressure is 100 kPa.

Chapter 8
Condensed Phase Explosions

Condensed phase explosives contain the fuel and the oxidizer integrated together as a solid or a liquid. The fuel and oxidizer could be combined in the form of a chemical compound (i.e., fuel and oxidizer exist as a single molecule) or could be intimately mixed together as a heterogeneous substance. In the latter, the fuel and oxidizer are mixed well to be qualified as a single substance. We shall, in this chapter, consider the explosion behavior of the different types of condensed explosives.

Detonations in reactive substances were discussed in Chaps. 6 and 7, and it was noted that process of detonation in condensed phase explosives would be similar to that in gaseous fuel–air mixtures. The unreacted Hugoniot for the condensed phase materials and the reacted Hugoniot, corresponding to the very high pressure of the combustion products, cannot be determined by using ideal gas equation as done for the gas phase detonations. The use of more complicated equations of state becomes necessary, and the mathematical formulation becomes more involved. We shall first look at the constitution of the condensed explosives and thereafter address their explosion behavior.

8.1 Hydrocarbon Fuels Constituting Condensed Phase Explosives

8.1.1 Single, Double, and Triple Bonds

A fundamental understanding of the nature of fuel used for the condensed explosives is a prerequisite to determine the properties of the explosives. Most fuels are of organic origin and consist of hydrogen and carbon and are called as hydrocarbons. The atomic structure of these hydrocarbons consists of electrons being shared between the carbon and hydrogen atoms.

© The Author(s), under exclusive license to Springer Nature Switzerland AG 2021
K. Ramamurthi, *Modeling Explosions and Blast Waves*,
https://doi.org/10.1007/978-3-030-74338-3_8

When one pair of electrons is shared between the carbon atoms, the bond is said to be a single bond. For two shared pairs of electrons between the carbon atoms, the bond is labeled as a double bond, while with three shared paired of electrons, the bond is said to be a triple bond. The nature of bond between the carbon atoms characterizes the hydrocarbon fuel.

8.1.2   Alkanes, Alkenes, Alkynes, and Alkadienes

When carbon atoms in the hydrocarbon are attached to each other by single bonds, the hydrocarbon is said to be saturated and is known as alkanes. If there are one or more double and triple bonds between the carbon atoms, the hydrocarbon is said to be unsaturated. Unsaturated hydrocarbons with one double bond are called alkenes, while those with a single triple bond between the carbon atoms are called as alkynes. In the presence of two double bonds between carbon atoms, the hydrocarbon is known as alkadiene. Examples of alkanes, alkenes, alkynes, and alkadienes are given below:

(a) Alkanes:

Methane   Ethane   Propane

The general formula for alkane is $C_n H_{2n+2}$. Alkanes, which are saturated, are also known as paraffins. The family of paraffins comprises gases, namely methane, ethane, propane, butane, and includes volatile liquids, namely octane ($C_8 H_{18}$) and dodecane ($C_{12} H_{26}$) [kerosene]. Heavy oils such as lubricating oil ($C_{50} H_{102}$) also belong to the paraffin family. As the number of carbon atoms increases, the molecules become more complex and dense, and their vapor pressure decreases. The alkanes have small negative values of heat of formation considering their simple single bond structure.

(b) Alkenes:

Pentene

These have a single double bond with the general formula C_nH_{2n}. The structure of a typical alkene, pentene, is given above.

(c) Alkynes:

Alkynes have a single triple bond and have the general formula C_nH_n. Acetylene is an alkyne, and its chemical structure being

$$H - C \equiv C - H$$

Acetylene

The triple bond provides higher reactivity to acetylene compared to hydrocarbons having single and double bonds. The alkenes and alkynes are unsaturated and are also known as olefins and di-olefins (dienes), respectively.

(d) Alkadienes:

These have two double bonds with the double and single bonds occurring alternatively between the adjacent carbon atoms. A typical alkadiene is butadiene with four carbon atoms shown below:

$$H_2C = CH - CH = CH_2$$

Butadiene

(e) Ring compounds:

Instead of the straight single chain hydrocarbons, the bonds when bent to form a ring structure are known as cyclo-compounds. Propane $(H - CH_2 - CH_2 - CH_2 - H)$ having the CH_2 in the ring structure

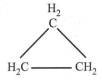

Ring structure

is known as cyclo-propane. Similarly, cyclo-hexane has the CH_2 of hexane $(H - CH_2 - CH_2 - CH_2 - CH_2 - H)$

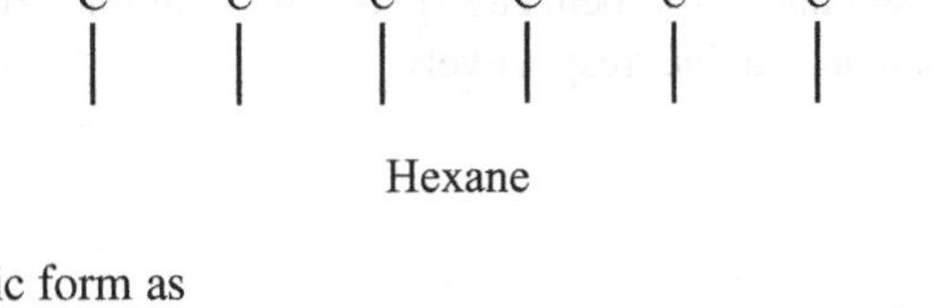

$$- C - C - C - C - C - C -$$

Hexane

arranged in a cyclic form as

$$H_2C$$

Ring structure

The alkenes, alkynes, and alkadienes could also have a ring structure to give cyclo-alkenes, cyclo-alkynes, and cyclo-alkadienes. The straight chain paraffins and cyclo-paraffins (alkanes), olefins (alkenes), di-olefins (alkynes), and alkadienes are known as aliphatic compounds.

8.1.3   Aromatic Structure: Benzene

A particular ring compound having six carbon atoms with three double bonds and three single bonds, shown below, is known as benzene.

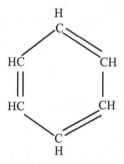

Benzene

The hydrocarbons having the benzene ring structure are known as aromatic compounds since they possess strong aroma or smell. It is also possible to have hydrocarbons with two and three benzene rings, and examples of these substances are naphthalene and anthracine, respectively.

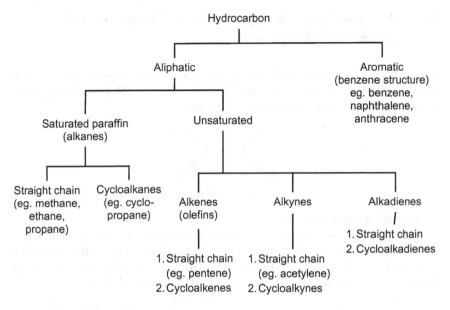

Fig. 8.1 Classification of hydrocarbons

8.1.4 General Classification of Hydrocarbons

Based on the above discussions of aliphatic and aromatic hydrocarbons, the overall classification of fuel hydrocarbons is given in Fig. 8.1.

We shall use the above structure of hydrocarbons to understand the different condensed explosives.

8.2 Explosives from Hydrocarbons

An explosive, it may be recalled, is a substance having inbuilt fuel and oxygen. The addition of oxidizers to hydrocarbons through nitrate (ONO_2) and nitro (NO_2) radicals provides them with oxygen. While nitrate (ONO_2) and nitro (NO_2) radicals are used with straight chain aliphatic compounds, nitro (NO_2) radicals are generally combined with cyclo-aliphatic and aromatic compounds. Typical condensed phase explosives and their structure are given in the following.

8.2.1 Nitromethane, Nitroglycerine, and Nitroglycol from Aliphatic Hydrocarbons

Nitromethane is the simplest nitrocompound based on the aliphatic paraffin, viz. methane CH_4. It has the chemical formula CH_3NO_2, the nitro (NO_2) radical replacing the hydrogen atom of methane to give the structure:

$$H\diagdown \atop H - C - N {\diagup O \atop \diagdown O} \atop H \diagup$$

Nitromethane

Nitromethane is a liquid explosive.

Nitroglycerine, also a liquid explosive, is obtained by nitration of glycerine. Glycerine is formed from straight chain propane C_3H_8 by substituting the three hydrogen atoms by hydroxyl radicals OH to give $C_3H_5(OH)_3$. Replacing OH of glycerine by ONO_2 results in nitroglycerine $C_3H_5(ONO_2)_3$. This is shown below:

Propane	Glycerine (Propane Triol)	Nitroglycerine $C_3H_5(ONO_2)_3$
$-C-C-C-$	$H_2C - OH$ $H C - OH$ $H_2C - OH$	$H_2C - ONO_2$ $H C - ONO_2$ $H_2C - ONO_2$

Nitroglycol, is similarly had from ethylene glycol $C_2H_4(OH)_2$ to give nitroglycol:

$$ONO_2 - CH_2 - CH_2 - ONO_2$$

Nitroglycol

Nitromethane, nitroglycol, and nitroglycerine are based on saturated hydrocarbons methane, ethane, and propane and are liquid explosives.

8.2.2 Nitrocellulose from Cellulose

Cellulose contains carbon, hydrogen, and oxygen and has the molecular formula $[C_6H_7O_2(OH)_3]_n$ where n is the number of repeating chemical units. Part of the hydroxyl radical in cellulose is replaced by nitrate ONO_2 radical to give nitrocellulose $[C_6H_1 0 - xO_{5-x}(ONO_2)_x]_n$. A general formula for nitrocellulose is $[C_6H_7O_2(OH)_y(ONO_2)_x]_n$ where $x + y = 3$.

Nitrocellulose is also known as gun cotton. The amount of nitration is defined as the mass of nitrogen in nitrocellulose. It is generally about 12.6% though a maximum nitration of about 13% is possible.

8.2.3 Penta Erythritol Tetra Nitrate (PETN) from Straight Chain Aliphatic Compound

This is a very powerful solid explosive popularly known as PETN. It is obtained by nitration of the straight chain aliphatic compound penta erythritol to give penta erythritol tetra nitrate. Erythritols are sweet alcohols like glycol with penta erythritol having a molecular formula $C_5H_8(OH)_4$. PETN is obtained by replacing OH radical by ONO_2 and has the molecular formula $C_5H_8(ONO_2)_4$.

8.2.4 RDX and HMX from Cyclo-Aliphatic Hydrocarbons

RDX, denoting R&D explosive, is a compound consisting of a cyclo-ring structure cyclo-trimethylene trinitramine. Amine radical NH_2 is derived from ammonia, and the formula for RDX is $(CNH_2)_3(NO_2)_3$. It has the chemical formula $C_3H_6N_6O_6$. The chemical structure of RDX is

RDX

Her Majesty's Explosive (HMX) is cyclo tetramethylene tetranitramine. It has the chemical formula $(CNH_2)_4(NO_2)_4$. The chemical structure of HMX is

HMX

RDX and HMX are powerful solid explosives.

8.2.5 Trinitrotoluene (TNT) from Aromatic Benzene Ring

Toluene is methyl benzene. Three NO_2 radicals (trinitro) are substituted for three hydrogen atoms in toluene to give trinitrotoluene.

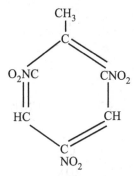

Toluene ($CH_3C_6H_5$)
(Methyl benzene)

TNT – $CH_3C_6H_2(NO_2)_3$

TNT is a solid. It melts at 80 °C, which is very much lower than the temperature at which it begins to chemically react and therefore can be used in the liquid state to mix with other ingredients. It is relatively stable compared to other explosives. Dynamite is made by using an absorbent substance such as saw dust or Fuller's earth to form solid sticks.

TNT is used as a standard for assessing the potential of an explosive to generate a blast. We shall deal with it in Chap. 12.

Since TNT melts at a low temperature of 80 °C, the molten TNT can flow readily into crevices and zones where heat and friction could cause it to explode. Adiabatic heating due to compression of voids and porous regions of it can result in a detonation. It needs to be handled with care.

8.2.6 Picric Acid (PA) from Phenyl

Removal of one hydrogen from the aromatic benzene ring results in the phenyl radical. The addition of OH group in place of the removed H gives phenol, also known as carbolic acid. The nitration of phenol by replacing 3 H by 3 NO_2 gives trinitrophenol, which is known as picric acid. The molecular formula of picric acid is $C_6H_3N_3O_7$. The molecular structure is given below:

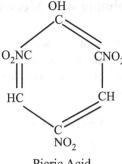

Picric Acid

8.2.7 Tetryl

Addition of N and NO_2 to TNT gives tetryl. The molecular formula of tetryl is $C_7H_5N_5O_8$.

8.2.8 TATB

Amines were used in RDX and HMX with the cyclo-aliphatic compounds. Amines could also replace the hydrogen atom of the aromatic group. When three amine groups and three nitrate groups are attached to the carbon atoms in the benzene ring, the explosive is known as triamino trinitro benzene (TATB – $C_6H_6N_6O_6$). The structure is shown below:

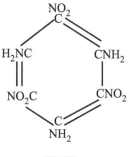

TATB

TATB is stable at a temperature up to 260 °C unlike TNT which melts at 80 °C. It is pressed and then machined to the required shape. It is safer to handle, store, and transport when compared to TNT. It is therefore used more widely as an insensitive high energy explosive.

8.3 Explosives with Radicals of Azide, Fulminate, Acetylide, and Stephnate with Metals

(a) **Azides**: The azide radical N_3 is of the form $N^- = N^+ = N^-$ and forms metal azide salts such as sodium azide, lead azide, and silver azide. The metal azide salts do not contain oxygen; they, however, decompose violently to give off nitrogen gas as indicated below:

$$2NaN_3 \rightarrow 2Na + 3N_2$$
$$2AgN_3 \rightarrow 2Ag + 3N_2$$

While the sodium azide decomposes when heated to 275 °C, lead azide and other heavy metal azides readily detonate when heated or shaken. This is due to the inherent instability of the N_3 radical and the rapid decomposition. Organic azides are also formed and are very reactive.

(b) **Fulminates**: Fulminate ion CNO with structure ($C\equiv N-O$) also forms salts with metals like the azides and these readily detonate. The salts are known as fulminates, e.g., mercury fulminate $Hg(CNO)_2$ and silver fulminate $Ag(CNO)_2$. These are very sensitive to friction.

(c) **Acetylides**: Acetylide radical from acetylene ($C\equiv C$) with the triple bond between the carbon atoms forms salts with metals such as calcium acetylide CaC_2 or silver acetylide AgC_2. These are being unstable readily detonate. Acetylene gas C_2H_2 was seen to be very reactive.

(d) **Stephnate**: In addition to the above three unstable radicals comprising azides, fulminates, and acetylides, other radicals derived from aromatic benzene ring or aliphatic hydrocarbons suitably nitrated can form salts with metals, which are explosive. A typical radical from the aromatic hydrocarbon is stephnate ($C_6HN_3O_8$) shown below:

Stephnate

Lead styphnate $Pb(C_6HN_3O_8)$ readily reacts when subject to heating or to static electricity.

8.4 Inorganic Explosives: Black Powder

Nitrates and perchlorates of metals and nonmetals are reactive. Typical examples are potassium and ammonium nitrate (KNO_3, NH_4NO_3), potassium perchlorate ($KClO_4$), and ammonium perchlorate (NH_4ClO_4). All these substances are solids in the form of crystals. They decompose into gases when heated.

Potassium nitrate is mixed with carbon and sulfur to form a composition known as black powder, which burns readily. It is also known as gun powder. The standard composition of black powder is 75% potassium nitrate, 15% charcoal, and 10% sulfur by mass. This gives the molar composition of potassium nitrate, carbon, and sulfur to be in the proportion 19 : 32 : 8 and the chemical formula for black powder as

$$19KNO_3 + 32C + 8S$$

The molar composition of black powder is therefore

$$KNO_3 + 1.69\,C + 0.42\,S$$

Inorganic ammonium perchlorate [$AP - NH_4ClO_4$] is used with polymers to make solid propellants for rockets. Ammonium dinitramide [$ADN - NH_4N(NO_2)_2$], another inorganic substance, is not as sensitive as AP and is safer to use. It also provides higher energy.

8.5 Characteristics of Explosive Compositions

The condensed phase explosives are seen to be derived from (a) nitration of hydro-carbons, (b) combination of the readily reactive radicals with metals, and (c) inorganic reactive substances. The standard heats of formation of these explosives and whether they are fuel-rich, oxidizer-rich, or are stoichiometric are given in Table 8.1. A composition was seen to be stoichiometric (see Chap. 4) when completely oxidized products of combustion are formed, and this condition provides the maximum value of energy in the reaction. Nitroglycerine, for example, having chemical formula $C_3H_5(ONO_2)_3$ decomposes to form

$$C_3H_5(ONO_2)_3 \rightarrow 2^1/_2H_2O + 3CO_2 + {}^1/_2O_2 + 1^1/_2N_2$$

It is slightly oxygen-rich since a small amount of oxygen is formed in the products. HMX, on the other hand, has the molecular formula $(CH_2)_4(NNO_2)_4$, and its reaction is given by

$$(CH_2)_4(NNO_2)_4 \rightarrow 4H_2O + 4CO + 4N_2$$

Table 8.1 Properties of explosives derived from hydrocarbons

S. No.	Condensed phase explosive	Heat of formation ΔH_f^0(kJ/mole)	Stoichiometry fuel-rich oxygen-rich	Oxygen fraction ξ	Heat of combustion (kJ/kg)
I	*Aliphatic–straight chain*				
I.1	Nitromethane (NM)	−113	Fuel-rich	0.57	6100
I.2.	Nitroglycerine (NG)	−380	Oxygen-rich	1.06	7350
I.3	Nitroglycol	−259	Stoichiometric	1	6350
I.4.	Nitrocellulose (NC)	−261	Fuel-rich	0.63	3030
I.5.	PETN	−538	Fuel-rich	0.86	5600
II	*Aliphatic-cyclo*				
II.1	RDX	+62	Fuel-rich	0.66	5600
II.2	HMX	+75	Fuel-rich	0.66	5630
III	*Aromatic*				
III.1	Trinitrotoluene (TNT)	−26	Fuel-rich	0.36	4620
III.2	Picric acid (PA)	−224	Fuel-rich	0.52	3585
III.3	Tetryl	+20	Fuel-rich	0.42	4570
III.4	TATB	−154	Fuel-rich	0.33	4031

Since incompletely oxidized product CO is formed, HMX is fuel-rich. The amount of oxygen available is less than required for stoichiometric combustion.

If we denote the fraction of O_2 available in the explosive to that required for stoichiometric combustion as ξ, $\xi < 1$ implies fuel-rich composition, while $\xi > 1$ denotes oxygen-rich composition. The values of ξ for the different explosives are also given in Table 8.1. The majority of the condensed phase explosives are seen to be fuel-rich.

The heat of combustion per unit mass of explosive is determined by the procedure given in Chap. 4 and is given in the last column of Table 8.2. Explosives derived from aliphatic hydrocarbons release higher amounts of energy during combustion compared to those derived from aromatic hydrocarbons. For these explosives (derived from aliphatic hydrocarbons), the energy release during the combustion is seen to generally increase with increase of ξ. This is shown in Fig. 8.2. In the case of compositions based on aromatic hydrocarbons, the increase in the quantity of energy as the fraction of oxygen ξ increases is much smaller than for the aliphatic-based hydrocarbons (Fig. 8.3).

Fig. 8.2 Variation of energy release with oxygen content for aliphatic-based explosives

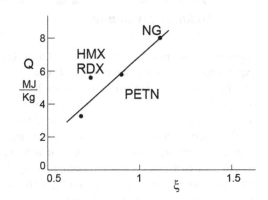

Fig. 8.3 Variation of energy release with oxygen content for aromatic-based explosives

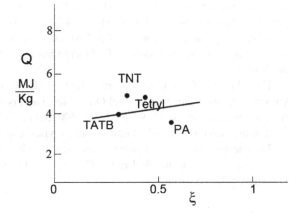

Table 8.2 Heat of formation and energy release from inorganic explosives

S. No.	Explosive	Heat of formation ΔH_f^0 (kJ/mole)	Energy release (kJ/kg)
1	Lead azide	+469	1650
2	Mercury fulminate	+386	1755
3	Lead styphnate	−855	1850

The energy release for inorganic compositions can be determined if their chemical reaction is specified and the heats of formation of the explosive and the products are known. The energy release for typical inorganic explosives is given in Table 8.2. The energy release is much smaller than for the hydrocarbon-based compositions.

8.5.1 Enhancing Oxygen Content by Addition of Oxygen-Rich Compounds: AN-NM Slurry, ANFO, Gelatine Dynamite

Ammonium nitrate (AN) is widely used as a fertilizer. There have been accidental explosions involving detonation of AN such as the Texas City Disaster in April 1947 and the Beirut Port Explosion in August 2020 and are discussed in Chap. 1. AN is oxygen-rich with $\xi = 1.5$. It has a heat of formation of -183 kJ/mole, and the energy release during its decomposition is 4860 kJ/kg.

The energy release of an explosive was seen to be generally higher when the oxygen content in it was more. In order to achieve higher energy release, the oxygen content in a fuel-rich explosive is enhanced by the addition of the oxygen-rich AN. This is because more completely oxidized products of combustion get formed when the oxygen content of a fuel-rich propellant is increased. As an example, the energy content of the very fuel-rich Nitro-Methane (NM) is increased by adding AN to it and making it into a strong explosive (slurry of solid suspension of AN in liquid NM). The slurry explosive is known as AN-NM and is widely used for blasting rocks. It was also used in the explosion of the Murrah building at Oklahoma City, discussed in Chap. 1.

The addition of AN to hydrocarbon fuel oil also forms an explosive slurry. It is known as Ammonium Nitrate Fuel Oil (ANFO) and is used for blasting rocks. The typical composition used is 94% AN and 6% fuel oil. It has also been unfortunately used in the making of bombs known as improvised explosive devices (IED).

The energy release of dynamite is also increased by adding AN to it. The addition of oxygen-rich NG to the fuel-rich NC increases its energy content. It forms a gel known as gelatine dynamite.

8.5.2 Reduction in Oxygen Content: Plastic Explosives

The energy released from an explosive can be brought down by reducing its oxygen content such as by introducing a polymer in it. Teflon and thermoplastics are also used. The polymer or plastic-bonded explosives (PBX) are less energetic and safer to use, and explosives such as HMX are used in certain applications to give HMX-based PBX. Crystals of HMX are also coated with thermoplastic fuels, and this makes it safer to use. More energetic polymers such as glycidyl azide polymer (GAP) are promising polymers for PBX.

8.6 Volume of Gas Generated from Condensed Explosives: Explosion Severity, Pyrotechnic Compositions, Thermites

The rate of volume of gas generated by explosives will determine the pressure and hence its destructive power when exploded. We would therefore like to determine the volume of gas generated per unit mass of the explosive.

Consider the oxidizer-rich explosive nitroglycerine $C_6H_5(ONO_2)_3$ for which the oxygen fraction $\xi = 1.06$. Its chemical reaction can be expressed in a simple manner without considering the dissociation at the high temperature of the products as

$$C_3H_5(ONO_2)_3 \rightarrow 3CO_2 + 2^1\!/_2H_2O + 1^1\!/_2N_2 + 0.25O_2$$

Completed products of combustion, viz. CO_2 and H_2O and 0.25 mol of excess O_2, are left in the products for each mole of nitroglycerine. All these products including nitrogen would be in the gas phase. The total number of moles of the gaseous products would be $3 + 2^1\!/_2 + 1^1\!/_2 + ^1\!/_4 = 7.25$ moles. The total volume of the gaseous products at the standard conditions of 100 kPa pressure and 298 K, assuming the H_2O to be in gaseous phase and the products to conform to an ideal gas, is

$$v = \frac{R_0 T}{p} \tag{8.1}$$

where v is the specific volume of the products per mole, R_0 is the universal gas constant, T is the temperature in Kelvin, and p is the pressure in Pa. Substituting the values of pressure and temperature in Eq. 8.1, we get

$$v = \frac{8.314 \times 298}{10^5} = 0.0248\,\text{m}^3/\text{mole}$$

Since 7.25 mol of gaseous products are formed from one mole of nitroglycerine, the volume of the gaseous products formed per mole of nitroglycerine at the standard condition is $7.25 \times 0.0248 = 0.18$ m^3. One mole of nitroglycerine has a molecular mass of $36 + 5 + 3 \times (16 + 14 + 32) = 227$ g/mole. Hence the gas formed per gram of nitroglycerine at the standard condition is $0.18/227 = 7.9 \times 10^{-4}$ m^3/g $= 0.79$ liter/g.

The energy liberated in the reaction per mole of the nitroglycerin is $-[-3 \times 393.5 - 2.5 \times 286.7 - (-333.7)] = 1563$ kJ where the standard heats of formation of the products and nitroglycerin are used. The energy liberated per unit mass of nitroglycerin is $1563/227 = 6.885$ kJ/g. The heat release causes the temperature of the products to go up, and the volume of the gases will change depending on the final pressure.

We therefore observe that if the nitroglycerine is kept in a confined space, the large volume of gas generated in the products will lead to high pressures; however, if it is kept in an unconfined space, the volume of gas would escape into the open, and no pressure buildup may be possible.

The calculations for the gas volume at standard conditions and energy release were based on the simple method of balancing the reaction given in Chap. 4. The heat release in J/g and volume of gas generated at standard conditions in liters/g for a few explosives are given in Table 8.3.

If we consider black powder for which the molar composition given in Sect. 8.4 is $KNO_3 + 1.69C + 0.42S$, the products formed can be written for the fuel-rich mixture following the method in Chapter 4 as

$$KNO_3 + 1.69C + 0.42S \rightarrow$$
$$0.29K_2CO_3 + 0.21K_2S + 0.21S + 0.5N_2 + 0.73CO_2 + 0.67CO$$

Here solid phase constituents K_2CO_3, K_2S and sulfur are assumed to be formed in the products. K_2CO_3 has a molecular mass of 138.2 g/mole, while K_2S and sulfur have molecular mass of 110.2 g/mole and 32 g/mole, respectively. The fraction of solids in the products that do not contribute to pressure is therefore $[(0.29 \times 138.2 + 0.21 f \times 110.2 + 0.21 \times 32)/(0.29 \times 138.2 + 0.21 \times 110.2 + 0.21 \times 32 + 0.5 \times 28 + 0.73 \times 44 + 0.67 \times 28)] = 0.52$. Only less than half the products are seen to contribute to generate pressure by the black powder composition, and it will therefore not be a high performing explosive.

Table 8.3 Volume of gas generated and energy released for explosives

S. No.	Explosive	Gas generated (liters/g)	Energy liberated (J/g)
1	Nitroglycerine	0.74	6700
2	Nitroglycol	0.74	6730
3	PETN	0.79	5940
4	TNT	0.76	4520
5	Picric acid	0.79	3745

Explosive mixtures that form significant amounts of condensed phase products are used for producing flashes and flares, sound, and heat rather than strong explosions. The solid constituents retain heat for a longer time than a gas and hence glow over a prolonged period. They also transfer heat more effectively when in contact with a solid explosive and are therefore useful for ignition devices. These explosive mixtures are known as pyrotechnics and thermites. While pyrotechnics comprise compositions involving oxidizers barium nitrate and potassium chlorate with metals aluminum, iron, strontium, etc., the thermites consist of metals reacting with metal oxides. The reaction of barium nitrate with aluminum and potassium chlorate with iron is given by

$$4Al + 3BaNO_3 \rightarrow 2Al_2O_3 + 3BaO + 1.5N_2$$
$$8Fe + 3KClO_4 \rightarrow 4Fe_2O_3 + 3KCl$$

The reaction of aluminum with $BaNO_3$ is seen to produce very small amounts of gaseous products, whereas the reaction of iron with $KClO_4$ has no component of gas in its products. Metal to metal oxide reactions, known as thermite reactions, take place in condensed phase without any constituents in gas phase. As examples, the reaction between metal aluminum and iron oxide is given by

$$2Al + Fe_2O_3 \rightarrow Al_2O_3 + 2Fe$$

and generates 15,600 kJ of heat per kg of aluminum, while the reaction of titanium with iron oxide

$$3Ti + 2Fe_2O_3 \rightarrow 3TiO_2 + 4Fe$$

releases 8100 kJ of heat per kg of titanium.

8.7 Deflagration and Detonation of Condensed Explosives

8.7.1 Deflagration in Confined and Unconfined Spaces

The condensed explosive burns at its exposed surface. The burn rate, as in the case of gaseous explosives, depends on the heat transferred to it from the hot combustion products. If the pressure of the hot combustion products increases, more heat is transferred to the unburned surface, and the burn rate therefore increases.

A condensed explosive, if burnt in a confinement, does not allow the product gases formed in huge quantities to escape with the result that the ambient pressure at the surface of the explosive increases and the burn rate increases. However, in an unconfined space, the ambient pressure being constant, the explosive burns at a constant rate. It often smolders like a piece of coal when burnt in an open space.

8.7.2   Detonation in Confined and Unconfined Spaces

When the surface of the condensed explosive is subjected to a strong shock wave, it detonates with the heat release energizing the shock. The energy transfer is by the shock compression and not by the heat transfer. The detonation features are similar to those derived from the shock Hugoniot and reaction Hugoniot for gaseous explosives. Constant velocity Chapman–Jouguet (CJ) detonations or constant low velocity quasi-detonations are formed behind a precursor shock wave. The CJ detonation velocities are in the range of 6000 to 8000 m/s, while the lower velocity quasi-detonations have velocities between 2000 to 3000 m/s. For PETN the CJ detonation velocity is about 5900 m/s, while the velocity of the quasi-detonation is about 2500 m/s. The lower velocities are from the deflagration at the lower CJ state on the reaction Hugoniot of the compressed condensed explosive as in the case of the gaseous detonations. While the low velocity detonation can transit to CJ detonation by reinforcement of the shock front by the confinement, it is not possible for the CJ detonation to transit into a constant velocity sub CJ detonation.

The very rapid rate of energy release in some explosives like those based on azides, acetylides, styphnates causes shock waves to form before any appreciable expansion occurs and the explosive detonates spontaneously even in the absence of a strong precursor shock being imposed at it surface.

8.7.2.1   Influence of Nature of Confinement

Figure 8.4 shows a circular strand of solid explosive enclosed on its cylindrical surfaces by different types of confinements. In one a rigid metal casing is used (Fig. 8.4a), while in the second a flexible confinement comprising layers of paper or cardboard is used (Fig. 8.4b). The strand could also be in an unconfined environment (Fig. 8.4c).

When the explosive is rigidly confined and detonated, there is no yielding at the surface, and no radial expansion of the high-pressure detonation gases is possible. However, in the absence of the confinement (Fig. 8.4c), the radial expansion of the gases at the boundary causes a reduction in detonation velocity, and the detonation front curves outward as shown. With a flexible confinement, the degree of expansion would depend on the yielding of the material used for confinement. At distances away from the confining boundary, the radial expansion is not felt.

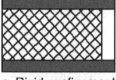 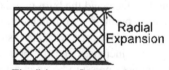 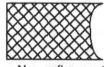

a Rigid confinement   b Flexible confinement   c No confinement

Fig. 8.4 Different confinements

A small diameter strand of explosive would therefore be more adversely influenced by the radial expansion of the detonation products. With charge diameters less than certain threshold values of diameter, the radial expansion would not permit constant velocity CJ detonation to propagate in it. This threshold value of diameter is spoken of as critical charge diameter. The critical diameter would depend on the explosive and the nature of confinement. If the charge is constrained by a fairly rigid metal casing, the value of the critical charge diameter would be very much lower than when confined by an yielding confinement or no confinement at all. We need to also remember that a detonation, as discussed for gaseous detonations, is not one-dimensional but consists of a muti-headed front. The reaction zone thickness is influenced by the unsteady flow behind the multi-headed shock front. The thickness is much larger than for a constant value of temperature behind the shock front. It is known as the hydrodynamic thickness of a detonation since the thickness of a detonation is influenced by the dynamics of the flow in the detonation. The hydrodynamic reaction zone gets quenched by the radial expansion in the case of the critical charge diameter.

8.7.2.2 Quantification of Confinement Through Relaxation Times

The role of acoustic impedance of materials on shock reflection and transmission was dealt with in Chap. 3. The magnitude of the impedance of the casing material relative to the value of the impedance of the solid explosive charge would decide the nature of reflection at the interface. Accordingly, a relaxation time for the confinement (τ_R) could be worked out and compared with the characteristic relaxation time of detonation (τ_D) in the explosive. If $\tau_R \ll \tau_D$, the confinement would adversely influence the propagating detonation.

The relaxation time of a confinement of a given mass M per unit surface area in contact with the explosive divided by the acoustic impedance of the explosive, viz.

$$\tau_R = \frac{M}{Z_E}$$

expresses the relaxation time in seconds since the units of M are kg/m^2, while the units of Z_E are N-s/m^3. The relaxation time of a detonation is its hydrodynamic thickness (δ_H) divided by the sound velocity (a) in it, viz.

$$\tau_D = \frac{\delta_H}{a}$$

8.7.3   Detonation and Heterogeneity of the Explosive

The interfaces corresponding to the grain boundaries in homogeneous explosives, ingested air bubbles in liquid explosives, and heterogeneities in solid and slurry explosives are regions or interfaces where the acoustic impedance changes. Depending on the change of acoustic impedance at the interface, different types of shock wave interaction are possible, and this has been discussed in Chap. 3. The formation of strong reflected shocks at the interfaces could easily trigger the onset of detonations in these regions.

The compression energy if unevenly distributed leads to local zones of high temperatures. Powdered explosives are therefore more susceptible to detonation. However, micronization of crystals of explosives will remove the defects in the grain boundaries associated with the larger crystals and will make it less sensitive to initiate detonations in it.

With large volumes of explosives, the bulk of the explosive acts as the confinement. For deflagration to detonation transition, confinement is essential for strengthening the shocks required for the initiation of the detonation.

8.8   Parameters of Explosive Influencing Detonation; Classification in Four Categories

The Chapman–Jouguet detonation velocities in the condensed explosives are given in Table 8.4. The density of the explosive and the pressure in the detonation are also given. The detonation velocity and pressure of the aliphatic cyclo paraffin-based explosives such as PETN, RDX, and HMX are seen to be higher than for the aromatic-based explosives such as TNT, PA, Tetryl, and TATB and the straight chain-based explosives NM and NG. The inorganic-based explosives have reduced detonation velocities and detonation pressures. This is understandable since the energy release in inorganic explosives is much smaller compared to the organic-based explosives.

From Table 8.4, it is seen possible to group the explosives according to the magnitude of detonation velocity and detonation pressure into the following four categories:

- Straight chain (liquid) explosives
- cyclo-paraffin-based explosives
- aromatic-based explosives and
- explosives with inorganic base.

It is also seen that as the density of the explosive increases, the detonation velocity and pressure increase in each of the above category of the explosives.

The high values of several km/s for CJ velocities and the intense pressures of several hundreds of kbar cause the detonation in condensed explosives to have a shattering effect on the surroundings. It produces a huge hammer blow on a solid surface in contact with the explosive and is known as brisance.

Table 8.4 Detonation velocity and detonation pressure

S. No.	Explosive	Density (kg/m³)	Chapman–Jouguet detonation Velocity (m/s)	Detonation pressure (kbar)
I	*Aliphatic: straight chain-based paraffin*			
	Nitromethane (NM) liquid	1128	6290	
	Nitroglycerine (NG) liquid	1590	7600	141
II	*Paraffin-based (solid)*			
	PETN	1670	7900	300
	Cyclo-Paraffin-based (solid)			
	RDX	1800	8750	347
	HMX	1900	9160	393
III	*Aromatic (solid)*			
	TNT	1640	6940	190
	PA	–	7300	–
	Tetryl	1700	7510	–
	TATB	1890	7860	250
IV	*Inorganic (solid)*			
	Lead azide	4000	5100	230
	Mercury fulminate	–	5400	–
	Lead stephanate	–	7050	–
	Ammonium nitrate	1050	4500	–
V	*Slurry*			
	ANFO (6%FO, 94%AN)	880	5500	74

The pressure behind the detonation can be readily obtained from the momentum equation across a detonation in the frame of reference of the detonation (Fig. 8.5). The medium of condensed explosive of density ρ_0 moves toward the detonation at velocity V_{CJ}, while the high-pressure gases move away at velocity $V_{CJ} - u$. Here u is the velocity of the high-pressure products that follow the detonation. The mass of the unreacted solid phase medium moving toward the detonation for unit area is $\dot{m} = \rho_0 V_{CJ}$. The pressure p in the detonation is due to the high-pressure and high-temperature gases.

It is equal to the rate of change of momentum and is

$$p = \dot{m}[V_{CJ} - (V_{CJ} - u)] = \rho_0 V_{CJ} u$$

Fig. 8.5 Detonation
propagating in solid phase
explosive in frame of
reference of the detonation

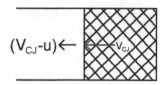

The pressure therefore increases with the density of the condensed explosive and the Chapman–Jouguet velocity of the detonation. Since the gas velocity following the detonation would be proportional to the CJ velocity, the pressure in a condensed phase detonation p is proportional to $\rho_0 V_{CJ}^2$.

8.9   High Values of Activation Energies

A sudden spurt of energy release was observed in Chap. 4 for chemical reactions having high values of activation energies. During the induction or preheat phase, there was no perceptible energy release when the activation energy was large; however, once the induction period was over, there was a spurt in the energy release. This aspect is illustrated again in Fig. 8.6. When the activation energy is small, the reaction starts at the low temperature itself and the heat release gradually increases as the temperature increases. A small rate of energy release cannot support a detonation.

The above aspect is well illustrated by considering the example of a pyrophoric substance such as phosphorous. Phosphorous immediately reacts (activation energy being very small) on meeting with the oxygen of air to form HPO and P_2O_2, both of which emit light. However, the small rate of heat release is such that it cannot drive a shock wave required for a detonation. Most of the condensed explosives have high values of activation energies. (TNT: 173 kJ/mole, NM: 224 kJ/mole).

Fig. 8.6 Rate of energy
release for small and large
values of activation energy

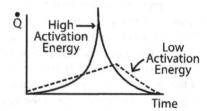

8.10 Ease of Formation of Detonation in Condensed Explosives

A stimulus is necessary to form a detonation. The stimulus could be heat, friction, impact, or pressure. The stimulus should lead to the formation of a shock, which causes chemical reactions and heat release and thereby drives the shock. The shock driven by heat release constitutes the detonation.

As seen in the discussions on gas phase detonations, a flame or a deflagration in a confined medium or otherwise could lead to its acceleration and its transition to a detonation. The hot spots in condensed explosives from the interaction of the burning with the heterogeneities and voids in it could likewise lead to a detonation. As an example, the liquid explosive nitromethane has been traditionally used as a cleansing solvent. It was not considered to be an explosive. However, when air bubbles are present, it detonates. In a major explosion at Pulashi, Illinois, on June 1, 1958, air bubbles got formed in 40 m^3 of nitromethane while being transported in a rail car and the nitromethane detonated.

Similarly ammonium nitrate (AN) was not known to form a detonation. It is widely used as a primary ingredient for the black powder, which is a burning composition. Large quantities of fertilizer grade AN (FGAN) [7700 tons] stored in a hull of a ship docked in the Texas City port detonated, causing serious damage and fatalities on April 16, 1947. In North Korea, the blast from a wagon containing AN is reported to have killed about 3000 people on April 22, 2004. A similar incident happened at Beirut port on August 4, 2020. Self-heating of the substance due to small values of heat gained, especially in unventilated surroundings, could lead to a detonation. Substances that spontaneously generate energy can detonate provided that shocks are formed and the shocks are supported by the energy release. However, if the energy release is very mild, a detonation cannot be formed even though burning can take place.

8.11 Low Explosives, Primary Explosives, and Secondary Explosives

Explosives, which liberate energy by burning but do not detonate, are known as 'low explosives'. Pyrotechnic mixtures, thermites, and rocket propellants belong to the class of low explosives.

Primary explosives detonate from stimulus of shocks, friction, pressure, or heat. The inorganic solid phase explosives comprising unstable azides, fulminates, acetylides, and stephnates are very sensitive to small stimulus of energy and are primary explosives.

Secondary explosives detonate when subject to strong shocks such as formed by primary explosives or otherwise. The aliphatic, aromatic, and slurry-based explosives, considered in this chapter, are secondary explosives. They generate much more energy than the primary explosives. They are therefore also known as high explosives. The classification of explosives into low, primary and high categories is shown below.

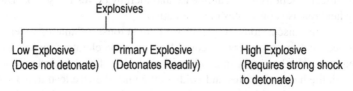

It may be noted that all condensed explosives can be made to burn and not generate a detonation if the shock formation in it could be prevented. In order to form a detonation, it is necessary to have an initiating device that forms strong shocks in it.

8.12 Overall Classification of Condensed Explosives

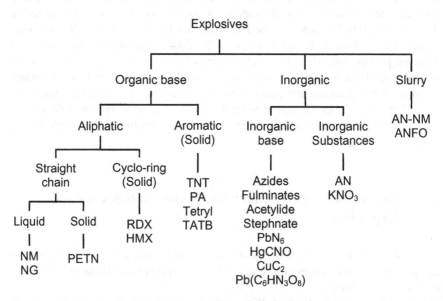

Condensed phase explosives are formed from different hydrocarbons combined with nitrate, nitro and amine radicals, metals with reactive radicals and from the reactive organic and inorganic substances. The detonation characteristics of condensed explosives are observed to depend on the nature of hydrocarbons from which it is obtained, the different reactive radicals, the heterogeneity, and the nature of the confinement housing the explosive. The classification of the condensed explosives based on the organic and inorganic base and the reactive radicals is given in the above chart.

Examples

8.1 *Rate of Energy Release in TNT*: Determine the energy release rate when a cylindrical strand of TNT of diameter 1 cm is enclosed in a housing on its cylindrical surface and is initiated at the end to form a C J detonation.
Energy of TNT = 4520 kJ/kg, Detonation velocity of TNT = 6940 m/s, density of TNT = 1640 kg/m^3.

Solution: Cross-sectional area of detonation

$$A = \frac{\pi}{4} \times 0.01^2 = 78.5 \times 10^{-6} \text{ m}^2$$

Rate of consumption of mass of

$$\text{TNT} = \text{density} \times \text{ area } \times \text{ detonation velocity}$$
$$= 1640 \times 78.5 \times 10^{-6} \times 6940 = 893.5 \text{ kg/s}$$

Rate of energy release $= 893.5 \times 4520 = 4.04 \times 10^6$ kW.

8.2 *Thermite Reaction: Heat generated and Heat Required to Initiate Reaction*: A thermite reaction between a metal and a metal oxide cannot be spontaneous like a condensed explosive. Determine the heat generated and heat to be supplied for the thermite reaction between aluminum and iron oxide. The thermo-physical data is given below:
Aluminum:

Molecular mass = 29.68 g/mole
Melting point temperature = 660 °C
Latent heat of fusion = 397 kJ/kg = 10.7 kg/mole
Specific heat = 0.921 kJ/(kg °C) = 0.025 kJ/(mole °C)

Iron:

Molecular mass = 55.85 g/mole
Melting point temperature = 1538 °C
Latent heat of fusion = 209 kJ/kg = 11.67 kJ/mole
Specific heat = 0.46 kJ/(kg °C) = 0.026 kJ/(mole °C)
Specific heat of iron oxide = 0.103 kJ/(mole K)

Standard Heats of Formation:

$$\Delta H^0_{f,Fe_2O_3} = -822.2 \text{kJ/mole}$$
$$\Delta H^0_{f,Al_2O_3} = -1669.8 \text{kJ/mole}$$

Solution: The thermite reaction is given by

$$2Al + Fe_2O_3 = 2Fe + Al_2O_3$$

The heat generated in the reaction is

$$-[\Delta H^0_{f,Al_2O_3} - \Delta H^0_{f,Fe_2O_3}]$$

since Al and Fe are elements with zero heats of formation. The heat of reaction is $-[-1669.8 - (-822.2)] = 847.6$ kJ. Hence the energy liberated per kg of the thermites is

$$\frac{847.6}{2 \times 0.02698 + (2 \times 0.05585 + 3 \times 0.048)} = 3967 \text{ kJ/kg of thermite.}$$

However, to start the reaction, aluminum, which has a lower melting temperature, has got to melt so that molten aluminum can react with the solid iron oxide at its surface. It would be better to have the thermite mixture as fine powder to increase the surface area of contact. The heat required to melt the 2 mol of aluminum in the reaction is

$$2 \times [C(T_b - T_i) + L]$$

where C is the specific heat of aluminum, T_b is its melting point temperature, and and T_i is the initial temperature. Taking the initial temperature as 25°C, we get the heat required as $2 \times [0.025(660\text{-}25) + 10.7] = 53.15$ kJ. The 1 mol of Fe_2O_3 is also heated to the melting point temperature of aluminum, viz. 660°C, and the heat required is $1 \times [0.013 \times (660 - 25)] = 65.4$ kJ. The net heat therefore supplied to start the reaction is $53.15 + 65.4 = 118.56$ kJ. The net heat from the reaction is $847.6 - 118.56 = 729.04$ kJ. The net energy from 1 kg of the thermite mixture is therefore

$$\frac{729.04}{2 \times 0.02698 + (2 \times 0.05585 + 3 \times 0.048)} = 3412 \text{ kJ/kg}$$

We still have substantial energy release. This heat raises the temperature of iron and aluminum oxide in the products. The thermite burns without a flame as the reactions proceed in the condensed phase. If a flame is desirable, barium nitrate, sulfur, or a polymer is added to the thermite composition. The addition reduces the energy requirement for ignition, and a flame is produced in the reaction.

Nomenclature

Q	Energy released during combustion (MJ/kg)
$\dot{Q}$	Rate of energy release (kJ/s)
ξ	Ratio of oxygen available in explosive to oxygen required to provide stoichiometric combustion
ΔH^0_f	Standard heat of formation
ΔH_C	Energy release in the reaction of the explosive

Further Reading

1. Akhavan, J., *The Chemistry of Explosives*, 2nd ed., The Royal Society of Chemistry, London, 2004.
2. Barrow, G. M., *The Structure of Molecules*, W. A. Benjamin Inc., New York, 1974.
3. Bridgeman, P. D., The effect of high mechanical stresses on certain solid explosives, *J. Chemical Physics*, 15(5), 311–313, 1947.
4. Cook, G.B., The initiation of explosion in solid secondary explosives, *Proceedings of the Royal Society London*, 246, 154, 1958.
5. Cook, M.A., *The Science of High Explosives*, Reinhold, New York, 1958.
6. Cooper, P.W. and Kurowski, S.R., Introduction to Technology of Explosives, Wiley-VCH, 1996
7. Davis, W. C., Craig, B. G., and Ramsay, J. B., Failure of Chapman Jouguet theory for liquid and solid explosives, *Physics of Fluids*, 8, 2169, 1965.
8. Johansson, C. H. and Persson, P. A., *Detonics of High Explosives*, Academic Press, London, 1970.
9. Kubota, N., *Propellants and Explosives: Thermochemical Aspects of Combustion*, Wiley-VCH, Weinheim, 2002.
10. Mader, C. L., *Numerical Modeling of Explosives and Propellants*, 3rd ed., CRC Press, Boca Raton, FL, 2008.
11. Stull, D. R., *Fundamentals of Fire and Explosion*, AIChE Monograph Series, Vol. 73, No. 10, 1977.

Chapter 9
Unconfined and Confined Gas Phase Explosions

Gaseous fuels and volatile liquid fuels are used for household and industrial purposes and for propulsion. They are transported by road, rail, and sea in tankers of different sizes. An inadvertent spill or leak of the gaseous or volatile liquid fuel in the open atmosphere or in a confined space or a partially confined space can form a fuel–air cloud that could explode. In this chapter, we deal with explosion of flammable gases and vapors in open atmosphere and in confined and partially confined spaces.

9.1 Unconfined Explosions

Let us consider a cloud of fuel gas/vapor mixture to be formed within the limits of flammability in open surroundings. Upon being ignited by a suitable ignition source, a flame is formed which traverses through the cloud and consumes the reactive gas mixture. We have seen that the burning velocities in most hydrocarbon–air mixtures are less than about one meter per second, and the rate of energy release would therefore be small. The formation of fire would be expected to cause damage of thermal origin comprising heat conduction, convection, and radiation.

The chemical reactions in the combustion of hydrocarbon–air and hydrogen–air mixtures are associated with large values of activation energies. This provides for a long induction time followed by a short period of energy release as discussed in Chap. 5. The spurt in the rate of energy release from the reaction could therefore result in the formation of strong pressure waves or blast waves and hence an explosion.

The three basic requirements for combustion or burning are (i) fuel, (ii) air, and (iii) ignition source. In the absence of any one of these three, a fire or an explosion cannot occur. A triangle representing the three requirements of fuel, air, and ignition source on its three sides is known as a fire or hazard triangle and is shown in Fig. 9.1.

In the case of a fire, the ambient turbulence and wind would make the flame turbulent. The wrinkling of the flame is further enhanced due to the instability

© The Author(s), under exclusive license to Springer Nature Switzerland AG 2021
K. Ramamurthi, *Modeling Explosions and Blast Waves*,
https://doi.org/10.1007/978-3-030-74338-3_9

Fig. 9.1 Fire/Hazard
triangle

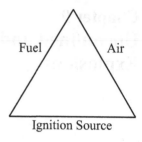

associated with the movement of a heavy medium into a lighter medium in the frame of reference of the flame. Turbulent burning velocities are much higher than laminar burning velocities, which we calculated in Chap. 7. The high values of turbulent burning velocities result in a rapid consumption of the explosive gas mixtures and result in rapid rate of increase in pressure when the explosive is confined in a given volume.

Detonation of unconfined clouds is possible when the concentration of fuel in such clouds lies between the lean and rich limits of detonation and a strong source of energy is present to form shock waves. Weak ignition sources such as electrostatic sparks or hot spots, in general, cannot form a detonation. Transition of a flame to detonation is also unlikely in the unconfined geometry as strong shocks cannot be formed. Obstructions or blockages in the passage of the flame or confinement are required as these promote shocks to be formed ahead of the flame by the compression wave interactions.

9.2 Confined Explosions

Here the fuel–air mixture is restricted to an enclosure of given volume. The temperature of the combustion products and the rate of propagation of the flame in the mixture depend on the fuel and its concentration in the fuel–air mixture.

9.2.1 Maximum Explosion Pressure

The maximum explosion pressure is determined assuming constant volume combustion in the given enclosure. The energy release (ΔQ) in the enclosure of volume V is determined from the standard heats of formation and the moles of the reactants and combustion products at the particular equivalence ratio. Adiabatic conditions are assumed. The energy release, so determined increases the internal energy of the gases to give

$$\Delta Q = \Delta U = mC_V(T_f - T_i) \tag{9.1}$$

where T_f is the temperature of the combustion products, T_i the initial (ambient) temperature, m the mass of reactants which equals the mass of the products, and C_V the mean specific heat of the products in kJ/(kg K). Writing the value of C_V as $R/(\gamma - 1)$ where R is the specific gas constant in kJ/(kg K), we get

$$\Delta Q = m \frac{R}{\gamma - 1}(T_f - T_i)$$

Using the ideal gas equation $pV = mRT$, the maximum pressure p_m in the explosion is determined as

$$\frac{\Delta Q}{V} = \frac{p_m - p_i}{\gamma - 1} \tag{9.2}$$

where V, in the above expression, denotes the volume of the enclosure and p_i the initial pressure. The energy release per unit volume of the mixture $\Delta Q/V$ (kJ/m^3) was seen in Chap. 4 to depend on the concentration and the properties of the fuel. The value of the maximum value explosion pressure will therefore depend on the particular flammable mixture used. For most of the stoichiometric mixtures of hydrocarbon gases with air, the maximum pressure p_m is between 0.7 and 0.8 MPa. In the case of a stoichiometric mixture of acetylene–air, the maximum explosion pressure is about 1 MPa. The maximum pressures decrease for fuel-rich and fuel-lean mixtures.

If a detonation is formed in the enclosure, the pressure rise will correspond to the detonation pressure, the values of which are between 1.4 and 1.8 MPa for the stoichiometric hydrocarbon mixtures. Reflection of the shock front of the detonation at the walls would increase the maximum value of pressure.

9.2.2 Violence or Rate of Pressure Rise

The flame, formed from the ignition source located at one end of the chamber, traverses the unburned gas mixture in the chamber and in the process raises the pressure and temperature in the enclosure. As the flame progresses, its speed increases in view of the motion of the gas ahead of it, the flame becoming turbulent. The heating up of unburned gases from the pressure wave propagating from the flame also increases the frame speed. The rate of energy release increases resulting in a progressive enhancement of the rate of pressure rise. However, the rate of pressurization decreases when the volume fraction of the gases being burnt decreases in the enclosure toward the end of the flame travel. The rate of pressure rise therefore reaches a maximum before it begins to decrease again. A typical rate of pressure rise in the enclosure is shown in Fig. 9.2.

The maximum rate of pressure rise $(dp/dt)|_m$ corresponds to the inflection point in the pressure time trace as shown in Fig. 9.2. Maximum values of dp/dt are obtained when the gas mixture is ignited at the center of the enclosure.

Fig. 9.2 Typical evolution
of pressure in confined
explosions

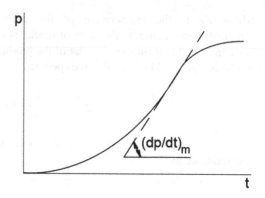

Fig. 9.3 Influence of
volume of enclosure on rate
of pressure rise
($V_3 > V_2 > V_1$)

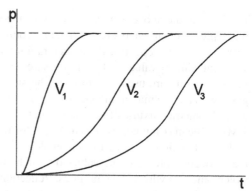

As the volume of the enclosure is increased, the maximum rate of pressure rise decreases and is illustrated in Fig. 9.3. This is due to larger rate of energy required to increase the rate of pressure rise in an enhanced volume of the reactive gas. The maximum pressure reached in the enclosure is, however, independent of the volume.

The maximum rate of pressure rise would correlate inversely with the volume of the enclosure. Considering that the flame has to traverse along the chamber and the faster it traverses, higher would be the rate of increase of pressure, a length scale associated with the volume should correlate with the maximum rate of pressure rise. The characteristic length scale of the chamber is $V^{1/3}$, and the maximum rate of pressure rise $dp/dt|_m$ would therefore be inversely proportional to $V^{1/3}$. The product

$$\left.\frac{dp}{dt}\right|_m \times V^{1/3}$$

would have a constant value for a particular gas mixture in an enclosure of a given configuration. The constant is denoted by K_G since the constant refers to a gas mixture. The unit of K_G is expressed in bar-m/s.

$$K_G = \left.\frac{dp}{dt}\right|_m \times V^{1/3} \tag{9.3}$$

The flame speed depends on the level of turbulence in the gas mixture and the motion of gas ahead of it. The constant K_G would therefore be influenced by the turbulence in the gas mixture and the shape of the enclosure. Similarly, the ignition energy transports heat into the unburned gas medium and influences the flame speed and thus affects the value of K_G. When there is no significant turbulence and the energy supplied for ignition is not too high (about 10J), the values of K_G for stoichiometric mixtures of methane–air, propane–air and hydrogen–air mixtures are typically about 55, 75, and 550 bar-m/s. Most of the vapors from volatile hydrocarbon in a stoichiometric mixture with air have values of K_G between 40 and 70 bar-m/s.

In enclosures having large aspect ratios (long chambers), the accelerating flame could significantly enhance the value of $dp/dt|_m$ and hence K_G. When obstructions are present either at the boundaries or otherwise, severe turbulence is created and the speed of the flame could approach the sonic velocity. This leads to much larger values of $dp/dt|_m$ and K_G. The maximum pressure p_m and the maximum rate of pressure rise $dp/dt|_m$ increase significantly when the flame transits to a detonation.

9.3 Methods of Decreasing Maximum Pressure and Maximum Rate of Pressure Rise

9.3.1 Relief Venting

Devices such as burst disks, explosion doors, or spring-loaded relief valves open up and provide vent area to decrease the pressure build-up in an enclosure. The mass flow rate through the vent relieves the high pressures. Unacceptable levels of pressure are avoided by providing the relief.

Figure 9.4 shows the decrease of pressure when the relief is designed to give way when the pressure in the enclosure reaches a value p_R. The relief devices are particularly useful when the enclosure cannot withstand the maximum explosion pressure. However, the vent area must be such that the mass flow rate through it is sufficient to arrest the pressure rise and bring down the pressure. Explosion doors and burst discs have much larger vent areas than relief valves and are therefore very effective.

If a sensor is used to detect the rate of pressure rise in the enclosure, it can be used to operate a vent device and protect the equipment and personnel.

9.3.2 Halons; Suppression of Rate of Pressure Rise

Halogenated compounds were seen in Chap. exrefchap5 to decrease the rate of energy release by consuming the active chain carriers. Injection of chloro-bromo-methane

Fig. 9.4   Relieving pressure
during an explosion

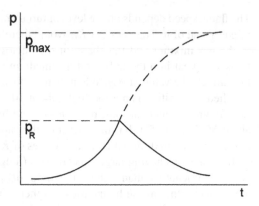

Fig. 9.4 Relieving pressure during an explosion

in the enclosure at the beginning of the pressure rise suppresses the rate of chemical reactions, the rate of energy release and hence the maximum rates of pressure rise.

9.4   Maximum Experimental Safety Gap

Equipment operating in an explosive environment are usually designed to withstand the explosion pressure and the maximum rate of pressure rise. However, it is also important that the equipment does not transmit the explosion to the explosive environment outside it. The explosion within the equipment such as from heated surfaces, hot bearings, or sparks (electrostatic or friction), etc. occurring in the equipment, should not be transmitted to the ambient explosive medium.

The dimensions of openings in the housing, containing the equipment, if smaller than the quenching distances (see Chap. 7), will not allow the flame to propagate outside the housing. The threshold size of the opening in the housing below which the flame does not propagate into the ambient explosive medium is also referred to as the maximum experimental safety gap (MESG). Davy's safety lamp and flame traps used in explosive environments have screens with openings of size less than the threshold value such that the flame cannot be transmitted to the explosive medium. Fine mesh screens are often used for hydrocarbon–air mixtures to ensure non-ignition of the mixture. Hydrogen–air flame has a very small quenching distance and is therefore more difficult to quench. Sintered bronze metals have been employed successfully for quenching flames in hydrogen–air mixtures.

While ensuring safety by preventing the transmission of flame to the ambient through provision of a small gap by wire screens, perforated sheets or porous materials in the flame arrestor, it is relevant to note that quenching is not possible over prolonged duration after a flame or explosion is initiated within the equipment. This is because the quenching surfaces, when heated, can themselves bring about an explosion in the ambient medium. If significant pressure difference exists across the

Fig. 9.5 Transmission of
explosion by pipeline
connecting two vessels

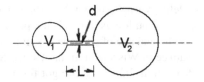

quenching gap, the squirting of the hot gases into the ambient explosive medium can
initiate an explosion from the hot jets.

9.4.1 Enclosures Joined by Pipes

In chemical process industries, two or more enclosures or reactors are very often
joined by a pipe conveying the explosive medium. Explosion in one vessel should
not be transmitted to the other vessel for which the diameter of the pipeline and its
length are to be suitably chosen.

Consider enclosures of volumes V_1 and V_2 connected by a pipeline of diameter d
and length L (Fig. 9.5). If both the enclosures contain the same gaseous mixture, the
possibilities of transmitting an explosion from one to the other are addressed in the
following.

Based on the discussions of MESG in the last section, explosions in volume V_1
will not be transferred to the volume V_2 when the diameter of the pipeline is less
than MESG. The hot gases from V_1 forces into V_2 as a jet through the pipeline of
diameter d and mixes with the gases therein. If the jet cools down during its travel in
the pipeline or during the mixing with the gases in volume V_2, no ignition would take
place. However, if the jet retains sufficient energy and initiates combustion in V_2, a
near-constant volume explosion is possible in V_2 considering the turbulent mixing
provided by the jet and the diffusion of the hot gases.

A unique value of MESG is therefore difficult to be specified. The characteristic
times of mixing and the associated heat loss times and the chemical reaction times, as
discussed in Chap. 5 on thermal explosions, govern the transmission of the explosion.
The energy content in the jet (jet ignition), the nature of the explosive gas, and the
volumes V_1 and V_2 influence the phenomenon.

9.5 Partial Confinement

If a gas spill occurs in an unconfined geometry in the presence of blockages, the
blockages often act as partial confinement to the spread of the flame and accelerate
it. Blockages could lead to the transition of the flame to a detonation and therefore
result in very rapid rates of pressure rise. A typical incident concerns the leak of
hydrogen gas from high-pressure hydrogen cylinders in downtown Stockholm from

a delivery truck on March 3, 1983. The obstructions led to detonation causing glass panes of buildings to break and people to be knocked down.

The leakage of propane gas in a manufacturing unit of Falk Corporation at Milwaukee in Wisconsin, in the United States on December 6, 2006, also resulted in a detonation. Initially a fire got formed which after a certain time transited to a detonation due to the blockages. The initiation of a detonation in propane–air mixture requires energies of about 10 kJ, which is obviously difficult to obtain from accidental ignition sources. However, the interaction of the flame and the flow in the presence of the blockages from machines in the manufacturing plant would have resulted in the formation to a detonation.

Explosions could be similarly caused in flammable mixtures in partially confined areas such as tunnels. Adequate safety measures need to be incorporated while designing them.

9.6   Sequence of Events in Typical Unconfined and Confined Explosions

We consider the mechanisms and sequence of events in the following four gas-phase explosions listed in Chap. 1.

9.6.1   Largest Man-made Unconfined NG Explosion: Ural Mountains

A pipeline 0.7 m in diameter carrying compressed natural gas (methane) from Siberia to Uja at a pressure of 2.5 MPa split over a length of about 1.5–2 km about 1400 km downstream of the supply station. Huge quantities of NG got spilled forming an enormous cloud of methane. The cloud of methane traversed through open land.

The Trans-Siberian railway line between Moscow and Vladivostok was in the region of the traversing methane cloud. Two electric trains traveling in opposite directions entered the cloud simultaneously. The turbulence generated by the passage of the two trains helped in mixing the methane with air and the availability of an ignition source in the form of electric spark from the overhead electric lines caused the explosion. The explosion was so severe that 600 people were killed. Trees were flattened to a distance of 4 km from the site of the explosion.

The drifting methane cloud did not find a suitable ignition source till the passage of the train through it. It mixed well with air to form a turbulent flame due to the turbulence from the movement of the adjacent trains. The blockages, comprising the carriages and other obstructions, led to the transition of the flame to a detonation.

9.6.2 Propane Vapor Explosion and Fireball at Port Hudson, Missouri

This is a widely quoted example of both confined and unconfined explosion. A 0.2 m diameter pipeline conveying liquid propane at a pressure of 6.5 MPa burst at a weld forming a huge fountain of liquid propane of height between 15 and 25 m above the spill. Liquid propane boils at a temperature of −42 °C at atmospheric pressure and a cloud of propane and air comprising the fog and mist of propane got formed. The ambient atmosphere was cold being a winter night on December 9, 1970. A wind at about 2.8 m/s blew the fog down a valley and in the process a mixture of propane and air got formed. Since no ignition source was available, the cloud of propane and air continued to drift along the ground in the direction of the wind without burning.

A concrete warehouse with six deep-freeze units was located at a distance of about 300 m downwind and about 6–9 m below the point of release of the propane. The propane–air permeated into the warehouse through crevices in a sliding door. A motor in the refrigeration unit provided the necessary ignition energy and ignited the mixture. The blockages in the freeze unit helped in the transition of the flame to a detonation. The detonation took place about 13 min after the burst of the pipeline. The explosion in the warehouse corresponded to a confined explosion.

The shock wave front of the detonation from the warehouse broke the glass windows and doors and propagated out and detonated the cloud in the unconfined space. The detonation of the unconfined cloud was heard subsequent to the detonation in the warehouse. The blast wave generated from the unconfined explosion was so strong that windows of buildings 18 km away from the site got ruptured. Tree trunks were uprooted and telephone poles were broken. Houses within a radius of about 3.2 km were extensively damaged. This was followed by a fire from liquid propane accumulated in the spill.

9.6.3 Semi-confined Explosion at Chemical Plant at Flixborough, England

Cyclo-hexane used in the manufacture of caprolactum for nylon leaked when a feedline, $\frac{1}{2}$ m in diameter, bypassing a reactor snapped. 40t of cyclohexane leaked out forming a huge vapor cloud of diameter between 100 and 200 m. The cloud got ignited on coming into contact with a furnace and the obstructions in the plant accelerated the flame to form a detonation. The delay in the ignition after the spill of about a minute helped in the formation of a well-mixed cloud of cyclo-hexane and air. The blast from the detonation killed about 28 persons and injured about 89. Buildings numbering about 1800 within a radius of 1.8 km were destroyed.

9.6.4 Hydrogen Explosion and Rupture of Confinement In Loss of Coolant Accident (LOCA) at Fukushima

The Fukushima Daiichi nuclear power plant that met with a disaster in March 2011, dealt with in Chapter 1, used fission reactions to generate high-pressure and high temperature steam in boiling water reactors. The nuclear reactors used nuclear fuel rods cladded in zirconium or cadmium. These generated the neutrons in the nuclear reaction and heated the high-pressure water in the reactor. The reactors were housed in large containment vessels unlike in the case of the Chernobyl nuclear power plant in which also a LOCA took place in 1986. These large containment vessels could sustain a pressure of about 5 atmospheres (500 kPa).

The power plant had six reactors of which three were working on the day of the accident. A severe earthquake caused the nuclear power plant to trip and as per the design, the three reactors shut down automatically. The shut down of a fission reactor does not stop the nuclear reactions immediately and there is a gradual decay in the reactions lasting over a prolonged period. With the nuclear reactions occurring during this decaying period, there is continual heat generation known as decaying heat. The reactor thus needs to be cooled after its shutdown.

There was a back up in place for the high-pressure water cooling with redundancies during the period of decaying heat. However, following the earthquake, a tsunami struck and the cooling system got flooded and the standby power driving the redundant pumps also failed. An auxiliary battery system kept the power on and supplied the required coolant for about 8 h. But this also failed, and there was a loss of coolant supply to the reactor. It was therefore not possible to get rid of the decaying heat from the nuclear reactions in the primary reactors and the temperature of the water and steam in it shot up. The metal cladding of the nuclear rods melted. The zirconium metal reacts with water at high temperature to form hydrogen as per the chemical reaction

$$Zr + 2H_2O = ZrO_2 + 2H_2$$

The formation of hydrogen gas further increased the pressure in the reactor, and the reactor had to be vented to decrease its pressure to acceptable levels.

The venting of the high-pressure steam and hydrogen from the reactor into the containment vessel carried along with it the radioactivity from the nuclear reactions. Upon expansion, the steam condensed while the hydrogen mixed with the air in the containment vessel to form the explosive hydrogen–air mixture. The explosion of this mixture ruptured the containment vessel and the radioactivity spread to the ambient. The radioactivity of the condensed water and the open spent-fuel ponds also contributed to the spread of radioactivity to the township and the ocean nearby.

We should be able to determine whether the hydrogen–air explosion did indeed cause the rupture of the containment vessel and damaged the building housing the reactors. Further, if at all this was true, was it a constant volume explosion discussed in this chapter or was it a detonation as discussed in Chap. 5? We address these points in the following.

The maximum possible pressure in a constant volume explosion in the containment vessel can be estimated assuming a stoichiometric mixture of hydrogen and air to be formed in it. The reaction can be written as

$$2H_2 + 1(O_2 + 3.76N_2) = 2H_2O + 3.76N_2$$

The heat generated in the above stoichiometric reaction is $-[2\Delta H^0_{fH_2O}]$ where $\Delta H^0_{fH_2O}$ is the standard heat of formation of water. The standard heats of formation for all other quantities are zero since they are elements at the standard state.

The value of the standard heat of formation of water is -286.7 kJ/mole. The heat generated in the reaction is therefore $-[-2 \times 286.7] = 573.7$ kJ.

The number of moles taking part in the reaction is $(2 + 1 + 3.76) = 6.76$ moles. The reactants are all ideal gases, and the volume of 6.76 mol is given by the ideal gas equation of state $pV = nR_0T$ or

$$V = \frac{nR_0T}{p} = \frac{6.76 \times 8.314 \times 298}{10^5} = 0.167\,m^3$$

Here it is assumed that the gas mixture is at a pressure of 100 kPa and a temperature of 25 °C since the initial air in it is at 100 kPa and 25 °C.

The heat released per unit volume

$$\frac{\Delta Q}{V} = \frac{573.7}{0.167} kJ/m^3 = 3435.3\ kJ/m^3$$

We have derived the maximum pressure p_m for a constant volume explosion as

$$p_m - p_i = (\gamma - 1)\frac{\Delta Q}{V}$$

where p_i is the initial pressure. The value of the ratio of the specific heats is 1.4 since all constituents are diatomic. Hence the value of p_m is given by

$$p_m = p_i + 0.4 \times 3435.3 \times 1000\ (J/m^3 = N/m^2 = Pa)$$

With $p_i = 10^5$ Pa, the maximum pressure becomes $100,000 + 1374000 = 1474$ kPa. The above pressure of 1474 kPa exceeds the pressure of about 500 kPa for which the the containment vessel is designed. It could therefore explode by the combustion of hydrogen with air at constant volume in it.

Since the pressure generated is much higher than what the vessel can endure, it is certain that the containment vessel explodes due to the hydrogen gas vented into it from the primary containment of the reactor. Once it explodes the building is also damaged and radioactivity is dispersed into the surroundings.

On examination of the vessels and the buildings, it appeared that two of the three reactors may have exploded at constant volume as the above numbers suggest. However, in case of one reactor, the explosion seems to be a detonation based on the flying off of the roof and the nature of damage to the walls of the building. Based on the Table given in Chap. 5, the Chapman–Jouguet detonation pressure is 1508 kPa for the stoichiometric hydrogen air mixture. This is very near to that of the pressure in the constant volume explosion. The vigorous mixing of the vented steam and hydrogen with air in the containing vessel could have led to severe turbulence and the formation of a turbulent flame brush which in the confinement became a Chapman–Jouguet detonation or a quasi-detonation. The reflection of the detonation wave from the walls would further increase the pressure loading on the wall resulting in a more severe explosion.

The analysis brings out the important role of hydrogen explosion in LOCA at the Fukushima disaster.

9.6.5   Confined Explosion; Fuel Tank of Aircraft

A near-empty fuel tank of an aircraft of a TWA 800 flight exploded 11 min after take off from New York at an altitude of about 4.5 km. At the reduced ambient pressure of 4.5 km, the kerosene vapor from the vaporization of liquid kerosene in the near-empty tank formed a mixture of kerosene vapor–air mixture within the limits of flammability. This was possible due to the reduced mass of air in the tank as air escaped at higher altitudes from the breathing vent holes provided in the tanks. An electrostatic spark is presumed to have caused the ignition of the low-pressure kerosene vapor–air mixture. The pressure, generated by the explosion, could not be withstood by the tank and it ruptured. The blast waves from it caused the adjacent fuel tanks to rupture, spilling the kerosene and formed a fireball, which engulfed the aircraft killing all the passengers and the crew. Details of the conditions for which explosion takes place are given in Chap. 7 under examples.

Examples

Maximum Rate of Pressure Rise in the deflagration of the Kerosene Vapor–Air mixture of an Aircraft Fuel Tank considered in Section 9.6.4 on confined explosion:

Determine the maximum rate of pressure rise in an aircraft fuel tank of volume 0.5 m³ assuming the vent area to be negligibly small when the lean kerosene–vapor mixture in it deflagrates. The value of K_G for the lean kerosene–vapor mixture can be assumed to be 45 bar-m/s.

Solution: The maximum rate of pressure rise is given as

$$\left.\frac{dp}{dt}\right|_m V^{1/3} = K_G$$

Hence

$$\left.\frac{\mathrm{d}p}{\mathrm{d}t}\right|_{m} = K_G/V^{1/3} = 45/(0.5)^{1/3} = 56.7 \text{ bar/s}$$

The maximum rate of pressure rise is 5.67 MPa/s.

Nomenclature

C_V Specific heat at constant volume [kJ/(kg K)]
d Diameter (m)
f Fraction
K_G Constant correlating maximum rate of pressure rise and volume (bar m/s)
p Pressure (Pa)
p_v Vapor pressure (Pa)
Q Energy release (kJ)
R Specific gas constant [kJ/(kg K)]
T Temperature (K, °C)
U Internal Energy (kJ)
V Volume (m^3)
Δ Incremental value
γ Specific heat ratio

Subscripts

a Ambient
f Final; flame
i Initial
m, max   Maximum
R Regulated
v Vapor

Additional Reading

1. Bartknecht, W., *Explosions*, Springer Verlag, Berlin, 1981.
2. Boddurtha, F. T., *Industrial Explosion Prevention and Protection*, McGraw, Hill, New York, 1980.
3. Croft, W. M., Fires involving explosions – a literature review, *Fire Safety Journal*, 3, 3–24, 1980.
4. Crowl, D. A. and Louvar, J.F., *Chemical Safety: Fundamentals with Applications*, Prentice Hall, Englewood Cliffs, NJ, 2002.
5. *Gas Explosion Handbook*, Gexcon, Bergen, Norway, 2007.
6. Murthy Kanury. A., Limiting case of fire arising from fuel tank/pipeline ruptures, *Fire Safety Journal*, 3, 215–226, 1980.
7. Quintiere, J.G., *Fundamentals of Fire Phenomenon*, Wiley, Sussex, 2006.
8. Strehlow, R.A., Unconfined Vapuor–Cloud Explosions – An Overview, Proc. 14th Int. Symposium on Combustion, The Combustion Institute, Pittsburg, PA, 1973, pp. 1189–1200.

9. Strehlow, R. A. and Baker, W.E., The Characterization and Evaluation of Accidental Explosions, NASA CR-134779, National Aeronautics and Space Administration, Cleveland, OH, June 1965.

10. Stull, D. R., *Fundamentals of Fire and Explosion*, AIChE Monograph Series, Vol. 73, No. 10, 1977.

Exercises

9.1 (a) A stoichiometric mixture of propane (C_3H_8)–air mixture is contained in a confined volume of $2\,m^3$ at a pressure of 1 bar and temperature of 298 K. Determine the heat released during the explosion of the mixture. You can assume no dissociation of the fully oxidized combustion products, viz. CO_2 and $H_2O(l)$. Air contains 21% by volume of oxygen and 79% by volume of nitrogen. The heat of formation of propane, carbon dioxide and water are given below:

$$\Delta H^0_{f,C_3H_8} = -104\,\frac{kJ}{mol},$$

$$\Delta H^0_{f,CO_2} = -394\,\frac{kJ}{mol}, \quad \text{and}$$

$$\Delta H^0_{f,H_2O(l)} = -286\,\frac{kJ}{mol}.$$

(b) Determine the value of maximum pressure in the above explosion. You can assume the average specific heats of CO_2, $H_2O(g)$ and N_2 as 0.054, 0.04, and 0.033 kJ/(mol K) respectively. The latent heat of vaporization of water is 40.6 kJ/mol.

9.2 In the problem given above, if the K_G value of the stoichiometric mixture of propane–air is 75 bar-m/s, determine the maximum rate of pressure rise during the explosion.

9.3 Two m^3 of butane–air mixture, contained in a spherical vessel, catches fire and burns at constant volume. Determine the maximum rate of pressure rise in the spherical vessel, given that the value of $K_G = 80$ bar m/s. Would the maximum rate of pressure rise be higher or lower than the above value if the gaseous mixture were detonated than if burnt at constant volume? Give reasons.

9.4 Figure shows the aftermath of the explosion of a diesel tank of an old heavy vehicle in a scrap yard in Chennai on April 29, 2010 in which the blast killed a worker and injured three others. The explosion occurred when the tank was being dismantled from the vehicle using a welding torch. The spark from the welding ignited the left over diesel in the tank and caused the explosion.

Explosion of diesel tank of a lorry.

Assuming the volume of the diesel tank as 130 L and the small volume of diesel in it to produce a flammable mixture of diesel vapor and air, determine the maximum pressure and the maximum rate of pressure rise in the tank.

You can assume the properties of diesel to be the same as that of kerosene. The lower calorific value of kerosene is given 44,000 kJ/kg. The density of the diesel-air mixture could be assumed as $1.2 \, kg/m^3$. The value of K_G for the mixture can be assumed as 50 bar-m/s. The molecular mass of the products of combustion is 35 kg/kmole, and the ratio of specific heats is 1.35. State any other assumptions made.

Chapter 10
Dust Explosions

Large quantities of agricultural dust are involved during the handling of foodgrains for bulk transport and storage. The grains are also ground to fine powder for making food products. These include wheat flour, powdered corn, maize, and barley among the different grains. Sugar is also finely ground and used. Wood and cellulose dust are routinely generated during the manufacture of wooden furniture and products. Inorganic dusts are employed in the paint, pigment, propellant, and explosion-manufacturing industries. The pharmaceutical industry makes use of powders of organic and inorganic substances in the manufacture of medicines. Pulverized coal is used for power plants. Organic dust particles, when heated, volatilize, and the vapor on mixing with air forms an explosive gaseous mixture of fuel vapor and air.

The size of the particulates of dust is an important parameter influencing the formation of an explosive vapor–air mixture. The surface area of a given particle increases substantially compared to its volume when it is ground to small dimensions and the heat transferred to it at its surface gets enhanced compared to its thermal capacity. If the mean diameter of a spherical dust particulate is denoted by d, the temperature increase ΔT from an energy transfer ΔE to the dust particulate is given by

$$\Delta T = \frac{\Delta E}{\rho C(\pi d^3/6)} \tag{10.1}$$

where ρ is the density and C is the specific heat with the product ρC being the thermal capacity. Higher temperatures are obtained for the smaller dust particles and result in copious release of vapor. There exists a threshold value of diameter below which significant vaporization takes place and forms an explosive mixture of the vapor and air. In this chapter, we examine the conditions leading to dust explosions, the characteristic features of these explosions and methods of avoiding the occurrence of these explosions.

K. Ramamurthi, *Modeling Explosions and Blast Waves*,
https://doi.org/10.1007/978-3-030-74338-3_10

10.1   Organic Dust; Lower and Upper Limits of Concentration

Organic dusts are encountered in agriculture, food, pharmaceutical, and coal industry. They release on heating volatile substances, which mix with air to form explosive vapor–air mixtures. The combustion process of the flames and detonations in the flammable vapor–air mixture, generated by the dust, is similar to that of gaseous mixtures dealt with earlier.

The number density of dust particles suspended in the air decides the volume of flammable vapor in the vapor–air mixture. The dust–air mixtures will burn or explode only when the number density of the dust particle dispersed in air is able to form a fuel vapor–air mixture within the limits of flammability or limits of detonation.

The number density is specified as mass of dust particles in a given volume of air in which it is suspended and is known as concentration. The units of concentration are g/cm^3 or kg/m^3. For most organic dust particles, the lower limit of concentration, corresponding to the lower flammability limit, is between 20 and $60\,g/m^3$, while the upper limit of concentration corresponding to the upper flammability limit is about $2–6\,kg/m^3$.

The upper limit of concentration is unlikely to be encountered in practice with dust–air mixtures as very large number density of dust particles cannot be realized in the ambient atmosphere. Even if large amounts of dust is present, any stratification in the distribution of the dust particles can lead to lower levels of concentration. We shall therefore deal only with the lower limit of concentration corresponding to the lean flammability limit. The low limit of concentration is also known as minimum explosive concentration (MEC).

Variations of the size of the dust particles influence the value of MEC. Dust particles of size greater than about 400 μm cannot support a flame or detonation even when high values of ignition energy are provided. This is because volatilization of the large dust particles is not significant, and a flammable vapor–air mixture does not get formed. Fine dust particles less than about 40 μm readily form flammable mixtures. The change of MEC with the size of dust particles is shown in Fig. 10.1. For particle sizes less than about 20 μm, no change in the value of MEC is observed. A rapid increase of MEC is generally observed for particle sizes exceeding 40 μm.

A mixture of fine and coarse dust containing as little as 10–15% fines has been shown to explode. Hence the specification of an average dust particle size larger than certain threshold value does not imply that explosion will not take place. A small content of finer dust can as well lead to burning and an explosion. It is therefore difficult to specify a particle size distribution for avoiding explosions.

In addition to the size of the dust particles, humidity, nature of the dust particle, turbulence, and homogeneity of the dust–air mixture and the magnitude of ignition energy influence the value of MEC. Dry dust is easier to volatilize. Humidity increases the value MEC. Moisture content greater than 60% does not lead to the formation of a dust explosion. The nature of the dust particle and the ease of volatilization or sublimation and charring to form a flammable mixture are also important. The

Fig. 10.1 Variation of minimum explosive concentration with size of dust particles

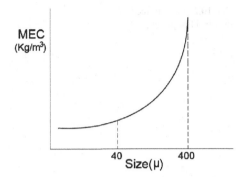

turbulence in the dust–air mixture can lead to segregation of the different sizes of dust particles and to concentration gradients. If the energy release from the ignition source is very rapid, it could blow off the dust particles in its vicinity adversely influencing the concentration. The shape of the dust particles would also influence the MEC with the shapes that give a larger surface area per unit volume being more combustible.

Organic dusts have ignition temperatures between 500 and 800 K. This temperature is near to the auto-ignition temperature of the corresponding vapor.

The minimum ignition energy required for the explosion of dust–air mixtures is higher than for gaseous fuel–air mixtures since part of the ignition energy goes into forming the vapor. The general practice has been to derive the data of minimum ignition energies from experiments considering that a large number of parameters influence the explosion process. The dust is sprayed as uniformly as possible in the experiments in a cylindrical vessel (volume about 1 m^3) and is ignited at the center of the vessel. The diameter of the vessel is about the same as its height. The minimum explosive concentration, the minimum ignition energy, the maximum pressure, and the maximum rate of pressure rise are measured in the experiments.

10.2 Estimation of Concentration

Foodgrains are handled in silos, which are essentially elongated vertical structures with height to diameter ratio greater than about 5. They are fed into the silos using hoppers (Fig. 10.2). Organic substances, after being ground are also transported by pipes or chutes after the sieving process is completed. The following gives a simple procedure for estimating the concentration of dust in a dust–air mixture when the dust is either fed by gravity or is entrained in a flowing gaseous medium.

Fig. 10.2 Gravity feed of
dust in a hopper/chute

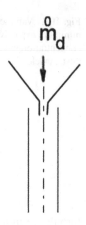

10.2.1 Gravity Feed

If the dust is conveyed by gravity, as in the vertical chute shown in Fig. 10.2, it
soon reaches a constant terminal velocity for which the buoyancy force and drag
force equals its weight. The concentration of the dust particles can be assessed by
determining its terminal velocity in the chute. If the diameter and density of the dust
particles are d m and ρ kg/m^3, respectively, and the ambient density is ρ_a kg/m^3, the
weight W and the buoyancy force B are

$$W = mg = \frac{1}{6}\pi d^3 \rho g \tag{10.2}$$

$$B = \frac{1}{6}\pi d^3 \rho_a g \tag{10.3}$$

The drag force on the dust particle is

$$\frac{\pi d^2}{4} C_D \frac{\rho_a V_t^2}{2} \tag{10.4}$$

where C_D is the drag coefficient and V_t is the terminal velocity in m/s.

The value of the drag coefficient C_D for small spherical particles falling at small
values of velocities is given as a function of the non-dimensional Reynolds number
Re as $C_D = 24/Re$ for $Re < 1$. The Reynolds number is defined with the diameter of
the dust particle d as the characteristic dimension, the terminal velocity V_t and the
viscosity μ of the air medium:

$$Re = \frac{\rho_a V_t d}{\mu} \tag{10.5}$$

Fig. 10.3 Drag coefficient variations

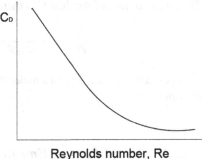

The value of C_D over a wide range of Reynolds number till about 10^3 is given by the empirical equation:

$$C_D = \frac{1}{4} + \frac{24}{Re} + \frac{6}{1 + Re^{1/2}} \tag{10.6}$$

The trend of the variation of C_D with changes of Reynolds number is given in Fig. 10.3.

The condition of achieving terminal velocity is that the buoyancy and drag forces on the dust particle are balanced by its weight to give

$$\frac{\pi d^2}{4} C_D \frac{\rho_a V_t^2}{2} + \frac{1}{6}\pi d^3 \rho_a g = \frac{1}{6}\pi d^3 \rho g \tag{10.7}$$

Solving for the terminal velocity V_t, we get

$$V_t^2 = \frac{4gd}{3C_D}\left(\frac{\rho - \rho_a}{\rho_a}\right) \tag{10.8}$$

Since the density of the dust particle ρ is very much greater than the density of the atmospheric air ρ_a, Eq. 10.8 simplifies to give

$$V_t = \sqrt{\frac{4gd}{3C_D}\frac{\rho}{\rho_a}} \tag{10.9}$$

If the diameter of the chute or silo is D (m) and the mass flow rate of the dust is $\dot{m}_d$, then the amount of air that the dust encounters in one second is

$$\dot{Q}_a = \frac{\pi}{4}D^2 V_t = \frac{\pi}{4}D^2\sqrt{\frac{4gd}{3C_D}\frac{\rho}{\rho_a}} \; m^3/s \tag{10.10}$$

The concentration of the dust C is therefore

$$C = \frac{\dot{m}_d}{\dot{Q}_a} = \frac{\dot{m}_d}{(\pi/4)D^2\sqrt{(4gD/3C_D)(\rho/\rho_a)}} \, \text{kg/m}^3 \qquad (10.11)$$

A large diameter chute D or a smaller throughput of the dust $\dot{m}_d$ will bring down the concentration.

10.2.2 Forced Feed by Entraining with Air

If the dust is transported by entrainment with air such as in a pipe or a chimney or by suction as in a vacuum pump, the concentration is determined as

$$C = \frac{\dot{m}_d}{\dot{Q}_a} \qquad (10.12)$$

where $\dot{m}_d$ is the mass flow rate of dust (kg/s) in the volume flow rate of air $\dot{Q}_a$ (m^3/s). If the velocity of air is V_a m/s, the concentration is

$$C = \frac{\dot{m}_d}{\pi D^2 V_a/4} \qquad (10.13)$$

where D is the diameter of the pipe or chimney conveying the dust.

10.3 Detonation, Smoldering, And Secondary Explosions

The thickness of a dust flame will be much larger than a flame in gaseous mixture. This is due to the additional process of volatilization, char, and mixing of the volatiles with air. These processes lead to additional delays before the gas phase chemical reactions and do not generally favor the formation of a detonation. However, detonations and quasi-detonations have been observed in dust–air mixtures and catastrophic dust explosions have been reported. Organic dust containing oats are shown to support a detonation in the range of concentration between 0.22 and 0.275 kg/m^3. Wheat dust detonates for concentrations between 0.25 and 0.375 kg/m^3. Blockages present in silos contribute to the transition of a deflagration to detonation as in gaseous fuel–air mixtures. Mixtures of coal dust and methane–air mixtures are detonable.

The collection of dust over heated surfaces such as an electric bulb or a furnace wall can cause it to smolder (Fig. 10.4). Smoldering is the thermal degradation from pyrolysis and charring of the material. A flame is not formed due to insufficient

Fig. 10.4 Smoldering from collection of dust on hot surfaces

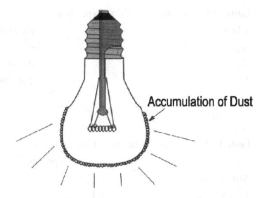

Accumulation of Dust

amount of the air and a glow is often observed. The smoldering can lead to flames being formed in the later stages when the temperature increases releasing copious amounts of vapor. A flashover from smoldering to secondary explosion takes place.

10.4 Characterization of Dust Explosions

10.4.1 Severity, K_{St} and St Classification

Dust explosions take place in confined spaces as the dust particles disperse in an unconfined or open environment. The minimum concentration of dust necessary to form a flammable vapor–air mixture cannot be achieved in an open environment. As in the case of confined explosions, the maximum rate of pressure rise $\frac{dp}{dt}|_m$ depends on the volume of the confinement and is given as

$$\frac{dp}{dt}\bigg|_m V^{1/3} = \text{constant}(= K_{St}) \qquad (10.14)$$

The constant has a specific value for a given dust–air mixture and is denoted by K_{St}. It has units of bar-m/s as for K_G in the gas mixtures. The subscript St in K_{St} denotes the word 'Staub' in German which means dust.

Considering the difficulty in achieving an uniform mixture of dust and air and the associated scatter in the properties, dust explosions are qualitatively classified in four broad categories as St_0, St_1, St_2 and St_3. St_0 implies that $K_{St} = 0$ and that the dust mixture is incapable of achieving an explosion. St_1 refers to K_{St} between 0 and 200. The majority of dust explosions fall in this category. St_2 implies values of K_{St} between 201 and 300 while St_3 denotes very high rates of pressure rise with values of K_{St} greater than 301. High values of K_{St} are obtained with energetic metal powders. The above classification of dust explosions is summarized in Table 10.1.

Table 10.1 Classification of dust explosions

Class	K_{St} (bar-m/s)
St_0	0
St_1	0–200
St_2	201–300
St_3	> 301

Table 10.2 Maximum pressure and K_{St} values for typical dust–air mixtures

Dust	Maximum pressure (bar)	K_{St} (bar-m/s)
Milk powder	8.1–9.7	58–180
Polyethylene	7.4–8.8	54–130
Sugar	8.2–9.4	59–165
Brown coal	8.1–10	93–176
Wood dust	7.7–10.5	83–211
Cellulose	8–9.8	50–229
Pigments	6.5–10.7	28–344
Aluminium powder	5.4–12.9	16–740

An explosion is unlikely to occur in an unconfined dust mixture. This is due to the larger characteristic times of energy release and the absence of a spurt in energy release rate. A fire or a fireball can, however, be formed in the unconfined environment.

Dust explosions are characterized by minimum explosive concentration (MEC), minimum ignition energy (MIE), ignition temperature T_I, maximum pressure p_m, and maximum rate of pressure rise $\frac{dp}{dt}|_m$. The value of K_{St}, which characterizes the maximum rate of pressure rise and the maximum value of pressure p_m, is given in Table 10.2 for some typical dust–air mixtures. The particulate size of dust considered is 200 mesh, viz., 75 μm when the dust–air mixture is of stoichiometric proportion.

The rate of pressure rise, which indicates the violence of the explosion, is observed to be equal if not higher than for gas phase explosions in confined environment.

10.5   Ignition Sensitivity, Explosion Severity, and Index of Explosibility

The St classification, given in the last section, does not provide any information whether an explosion is likely to occur. A relative grading of the different dust mixtures is done by the US Bureau of Mines by comparing the particular dust with a standard Pittsburgh coal dust of 74 μm size. This coal dust has the following properties:

Ignition temperature $T_I = 610\,°C$
Minimum ignition energy (MIE) = 60 mJ
Minimum explosive concentration (MEC) = $0.055\,kg/m^3$
Maximum pressure $p_m = 6.65$ bar
Maximum rate of pressure rise $\frac{dp}{dt}|_m = 156.5$ bar/s

A given dust is more likely to be ignited when the ignition temperature T_I is low, the MIE is small and also when MEC is small. Overall, the sensitivity to ignition can be said to be inversely proportional to these three parameters, viz.

$$\frac{1}{T_I \times \text{MIE} \times \text{MEC}}$$

Based on the given reference coal as the standard, the relative ignition sensitivity (IS) of a given dust is given as

$$IS = \left(\frac{\frac{1}{T_I \times \text{MIE} \times \text{MEC}}|_{\text{dust}}}{\frac{1}{T_I \times \text{MIE} \times \text{MEC}}|_{\text{std.coal}}} \right) = \frac{[T_I \times \text{MIE} \times \text{MEC}]_{\text{std coal}}}{[T_I \times \text{MIE} \times \text{MEC}]_{\text{dust}}} \qquad (10.15)$$

The severity of an explosion depends on both the maximum pressure p_m and maximum rate of pressure rise $\frac{dp}{dt}|_m$. The explosion severity (ES) is therefore defined as being proportional to

$$p_m \times \left(\frac{dp}{dt} \right)_m$$

Based on the reference coal, the explosion severity of a given dust is expressed as

$$ES = \frac{\left(p_m \frac{dp}{dt}|_m \right)\Big|_{\text{dust}}}{\left(p_m \frac{dp}{dt}|_m \right)\Big|_{\text{std.coal}}} \qquad (10.16)$$

While ignition sensitivity implies likelihood or chances for an explosion to occur, explosion severity indicates the consequence of the explosion. An index of explosibility (IE) is therefore defined as the product of ignition sensitivity and explosion severity. It is stated as

$$IE = IS \times ES = \frac{\left(p_m \frac{dp}{dt}|_m \right)}{(T_I \times \text{MIE} \times \text{MEC})}\Big|_{\text{dust}} \div \frac{\left(p_m \frac{dp}{dt}|_m \right)}{(T_I \times \text{MIE} \times \text{MEC})}\Big|_{\text{std.coal}} \qquad (10.17)$$

The dust is said to constitute a weak explosion hazard when $IE < 0.1$. The dust in this case would not be readily amenable to an explosion, and the hazard is small. The hazard becomes moderate when IE is between 0.1 and 1 and is strong when IE is between 1 and 10. The likelihood of an explosion and its severity are severe when $IE > 10$.

10.6   Nonvolatile Dusts

The propagation of flames and explosions in metal dusts and other nonvolatile substances such as diamond are not likely in the vapor phase as in organic dusts. Combustion takes place at the surface somewhat like the smoldering combustion. Zirconium metal dust burns rather violently. Gas phase reactions take place only when flammable vapors are formed from the burning at the surface.

10.6.1   Ignition Sources

The value of MIE depends on the concentration. A typical trend of the variation of MIE with concentration is shown in Fig. 10.5. Below a threshold value of MEC, the dust mixture cannot be ignited. There is a range of concentration over which only the dust is readily ignitable.

Generally the MIE is about one to two orders of magnitude higher than for gaseous fuels. The higher energy requirements arise from the additional energy required for vaporizing and the resulting increase of the flame thickness.

Sparks from grinding machinery, electrostatic sparks from improper earthing of equipment, hot surfaces, overhead bearings, etc., could supply the necessary energy for ignition. Provision of electrically conducting surfaces for chutes and pipes and proper earthing would help to reduce the incidence of electrostatic sparks. Welding carried out in a dusty atmosphere may provide the ignition source for the dust explosion.

Ignition sources cannot always be avoided and the practice for ensuring safe operation would be to ensure the concentration of the dust is less than MEC. This is done by introducing an inert gas such as nitrogen. The direct admission of dust at a level below the top of the chute, so as to reduce the volume and residence time and thus prevent the formation of dust–air mixture will also enhance the safety. The other alternative is to use a slurry of dust with a liquid and handle the slurry instead

Fig. 10.5 Minimum ignition energy changes with concentration

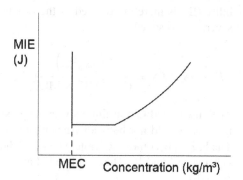

of the dust and air. Inert dusts such as Fuller's earth or extinguishing agents like ammonium phosphate are mixed with the given dust to reduce its sensitivity.

The provision of vents in confined spaces relieves the pressure as discussed in Chap. 9. It is important that dust does not accumulate over warm surfaces and lead to smoldering, flashover and secondary explosions.

Examples

10.1 *Gravity Feed Rate of Sugar to Avoid Possibility of Explosion in a Confectionery Unit*: Finely powdered sugar of 10 μm mean diameter is smeared over chocolate candy in the final operation of preparing the candy in a mechanized plant making chocolates. The sugar powder is fed by gravity from a hopper through a long vertically mounted cylindrical chute of 0.2 m diameter onto a tray where the chocolates are kept. Determine the terminal velocity of the sugar particles in the chute and the upper limit of the feed rate in kg/hour of the finely powdered sugar so as to avoid the possibility of a dust explosion. The minimum explosive concentration (MEC) of the 10 μm sugar powder is 0.08 kg/m^3. The density of sugar is 1.3 g/cc, and the density of air is 0.001 g/cc. The sugar particles can be assumed to be spherical and its drag coefficient C_D under the flow condition to be 0.95. The value of gravitational constant (g) is 9.81 m/s^2.

Solution

(a) *Terminal velocity of the sugar dust particles*: Consider the force balance on a single sugar particle of diameter d given as 10 μm: The weight is balanced by the air resistance and buoyancy when it achieves its terminal velocity. Denoting mass of a sugar particle as m kg and the terminal velocity by V m/s, we have the force balance:

$$mg = C_D A \frac{1}{2} \rho_a V^2 + m_a g$$

Here m_a is the mass of air displaced by the dust particle = $\frac{\pi}{6} d^3 \rho_a$, where ρ_a is the density of air. A is the area over which the drag force acts and is $\frac{\pi}{4} d^2$. Substituting the values, we get

$$\frac{\pi}{6} d^3 \rho_d g = \frac{1}{2} C_D \rho_a \frac{\pi}{4} d^2 V^2 + \frac{\pi}{6} d^3 \rho_a g$$

This gives

$$\frac{\pi}{6} \times (10 \times 10^{-6})^3 \times 1300 \times 9.81 - \frac{\pi}{6} \times (10 \times 10^{-6})^3 \times 1 \times 9.81$$
$$= \frac{0.95}{2} \times 1 \times \frac{\pi}{4} \times (10 \times 10^{-6})^2 V^2$$

or

$$6.6675 \times 10^{-12} - 5.1365 \times 10^{-15} = 3.73064 \times 10^{-11} V^2$$

Therefore
$$V = 0.423 \text{ m/s}$$

The terminal velocity of sugar dust is 0.423 m/s.

(b) *Upper limit of feed rate*: The MEC is given as 0.08 kg/m³. The volume of air swept per second relative to the sugar dust $= A \times V$. Here A is the cross-sectional area of the chute $= \frac{\pi}{4}D^2$, where D is the diameter of the chute. If the mass flow rate of the sugar is $\dot{m}$, the concentration of sugar dust is $\frac{\dot{m}}{AV}$ kg/m³, with $\dot{m}$ measured in kg/s, and AV the volume, measured in m³/s. The maximum flow rate of sugar should not exceed the MEC value of 0.08 kg/m³. Hence the maximum mass flow rate of sugar dust is

$$\dot{m} = MEC \times AV = 0.08 \times \frac{\pi}{4} \times 0.2^2 \times 0.423$$

Here 0.2 m refers to the diameter of the chute. Therefore,

$$\dot{m} = 1.06315 \times 10^{-3} \text{ kg/s}$$

This gives the maximum feed rate of sugar as $1.06315 \times 10^{-3} \times 3600$ kg/h = 3.83 kg/h.

10.2 *Coal Dust Explosion in Chimney*: The efflux of fine coal particles and air escaping from the chimney is catalogued as a Class II explosive dust cloud of strong explosive rating with a K_{St} value of 200 bar-m/s. If the minimum explosive concentration (MEC) of the coal dust in the chimney of a thermal power plant is 0.08 kg/m³ and the environment surrounding the coal dust in the chimney can be assumed to be air, determine the maximum flow rate of the coal dust in kg/h, which can be tolerated so that no explosion would take place. The chimney is of diameter 2 m and is 15 m long. The mean draft velocity of air through the chimney is 1.2 m/s.

Solution: Mean air velocity of the medium through the chimney = 1.2 m/s
Volumetric flow rate of air $= \frac{\pi}{4} \times 2^2 \times 1.2 = 1.2\pi$ m³/s.
If the mass flow rate of the coal dust through the chimney is $\dot{m}$ (kg/s), concentration of the coal dust is

$$C = \frac{\dot{m}}{1.2\pi} \text{ kg/m}^3$$

The minimum explosive concentration (MEC) is given as 0.08 kg/m³. Hence

$$\frac{\dot{m}}{1.2\pi} = 0.08$$

This gives the maximum flow rate of $\dot{m}$.
The maximum flow rate of the coal dust in kg/h $= 0.08 \times 1.2\pi \times 3600 = 34.56$ kg/h.

10.3 *Ignition Sensitivity and Index of Explosibilty of Confined Cloud of Cellulose Dust*

(a) If the maximum pressure in the explosion of a confined cloud of fine cellulose dust of volume $2\,m^3$ is 0.95 MPa and the maximum rate of pressure rise of the cloud is 90 bar/s, determine the explosion severity (ES) of the cloud. The reference coal dust–air mixture can be assumed to have a maximum explosion pressure of 0.8 MPa and a KSt value of 120 bar-m/s.

(b) Experiments give the minimum ignition energy (MIE) to be 0.5 J and the minimum ignition temperature (T_I) as 400 °C for the given cellulose dust cloud. The minimum explosive concentration (MEC) of the mixture is $0.12\,kg/m^3$. Determine the ignition sensitivity (IS) of the cloud. (You can assume for the standard coal dust mixture, the MIE to be 1 J, the ignition temperature to be 800K and MEC to be $0.2\,kg/m^3$.)

(c) What is the index of explosibility of the cellulose–dust mixture?

Solution

(a) The maximum pressure of the cloud is independent of the volume $= 0.95$ MPa $= 9.5$ bar. The maximum rate of pressure rise of the cloud $= 90$ bar/s. K_{St} value of the reference coal dust mixture $= 120$ bar-m/s. However, the maximum rate of pressure rise depends on the volume V and is related to the K_{St} value by

$$\left.\frac{dp}{dt}\right|_m V^{1/3} = K_{St}$$

The maximum rate of pressure rise for the given volume in the reference coal dust is

$$\left.\frac{dp}{dt}\right|_m = \frac{K_{St}}{V^{1/3}} = \frac{120}{2^{1/3}} = 95.24 \text{ bar/s}$$

The maximum explosion pressure of the standard coal dust mixture $= 0.8$ MPa $= 8$ bar. Explosion severity of the cloud

$$\frac{9.5 \times 90}{8 \times 95.24} = 1.122$$

The fine cellulose dust cloud has a higher explosion severity than the reference coal dust.

(b) Ignition sensitivity (IS) of the given cloud is given by

$$IS = \left(\frac{\left.\frac{1}{T_I \times MIE \times MEC}\right|_{dust}}{\left.\frac{1}{T_I \times MIE \times MEC}\right|_{std.coal}}\right) \qquad (10.18)$$

Substituting the values, we get

$$IS = \frac{800 \times 1 \times 0.2}{673 \times 0.5 \times 0.12} = 3.96$$

The cellulose dust has a much higher sensitivity to ignition than the standard coal dust.

(c) The index of explosibility (IE) is the product of explosion severity and ignition sensitivity. It equals $1.122 \times 3.96 = 4.44$. The confined cloud of fine cellulose dust is therefore a strong explosion hazard.

Nomenclature

B	Buoyant force (N)
C	Concentration (kg/m^3); Specific heat [kJ/(kg K)]
C_D	Drag coefficient
d	Diameter of dust particle (m, μm)
D	Diameter of conveyor, chute (m)
E	Energy (kJ)
ES	Explosion severity defined by Eq. 10.16
K_{St}	Constant correlating maximum rate of pressure rise & volume (bar-m/s)
g	Gravitational field (m/s^2)
IS	Ignition sensitivity defined by Eq. 10.15
IE	Index of explosibility defined by Eq. 10.17
m	Mass (kg)
$\dot{m}$	Mass flow rate (kg/s)
MIE	Minimum ignition energy (J)
MEC	Minimum explosive concentration (g/m^3)
p	Pressure (Pa)
$\dot{Q}$	Volumetric flow rate (m^3/s)
Re	Reynolds number defined by Eq. 10.5
T	Temperature (K, °C)
V	Velocity (m/s)
V_t	Terminal velocity (m/s)
ρ	Density (kg/m^3)
ρ_a	Density of air (kg/m^3)

Subscripts

a	Air
d	Dust
I	Ignition
m	Maximum
t	Terminal

Further Reading

1. Bartknecht, W., *Explosions*, Springer Verlag, Berlin, 1981.
2. Beever, P. F., Fire and explosion hazards in the spray drying of milk, *J. Food Technology*, 20, 637–645, 1985.
3. Boddurtha, F. T., *Industrial Explosion Prevention and Protection*, McGraw Hill, New York, 1980.
4. Chen, X. D., *Fire and Explosion Protection in Food Drier*, Chap. 10 of *Drying Technologies in Food Processing*, eds., Chen, X. D. and Majumdar, A. S., Blackwell–Wiley, West Sussex, UK, 2008.
5. Crowl, D. A. and Louvar, J.F., *Chemical Safety: Fundamentals with Applications*, Prentice Hall, Englewood Cliffs, NJ, 2002.
6. Eckhoff, R. K., *Dust Explosions in Process Industries*, Butterworth- Heinemann, Oxford, 1991.
7. Eckhoff, R. K. and Enstad, G., Why are long electric sparks more effective as dust explosion initiators than short ones? *Combustion and Flame*, 27, 129–131, 1976.
8. Hertzberg, M., Zlochower, I. A., Cashdollar, K. L., Metal Dust Combustion, Explosion Limits, Pressures and Temperatures, Proc. 24th International Symposium on Combustion, The Combustion Institute, Pittsburg, pp. 1827–1835, 1992.
9. Proust, C. and Veyssiere, B., Fundamental properties of flames propagating in starch–dust–air mixtures, *Combustion Science and Technology*, 62, 149–172, 1988.
10. Quintiere, J.G., *Fundamentals of Fire Phenomenon*, Wiley, Susex, 2006.
11. Stull, D. R., *Fundamentals of Fire and Explosion*, AIChE Monograph Series, Vol. 73, No. 10, 1977.

Exercises

10.1 (a) A sugar candy coating unit of an automated confectionary handles fine sugar powder at a rate of 5 kg/h. The fine sugar powder is transported in a cylindrical pipe of length 5 m by blowing it with dry air at a speed of 1.5 m/s. Determine the minimum diameter of the cylindrical pipe such that the explosion of the sugar powder will not take place in the pipe. The minimum explosive concentration for the given fineness of the sugar powder in air is 0.08 kg/m^3.

   (b) Due to a fault in the air blower, the concentration of the sugar powder increases to 0.095 kg/m^3. If the sugar–air mixture encounters an electric spark and explosion takes place, determine the maximum rate of pressure rise in the pipe due to the explosion? The value of K_{St} for the given sugar powder at the concentration of 0.095 kg/m^3 is 220 bar-m/s.

10.2 Dry wood dust of mean size of 25 μm is removed from a carpentry shop by a vacuum cleaner. The hose of the vacuum cleaner has a diameter of 200 mm. The suction pressure available across the hose is 25 Pa. Given that the minimum explosive concentration (MEC) of the given wood dust in air is 0.06 kg/m^3, determine the maximum rate at which the wood dust can be sucked along with air in order that an explosion will not occur in the hose.

10.3 (a) Corn flour is conveyed through a vertical cylindrical chute (tube) 1.5 m in diameter in a large bakery. The corn flour is introduced at the top of the chute by a hopper. The length of the vertical tube is long enough for the corn dust to attain terminal velocity. Determine the maximum permissible rate at which the corn flour can be supplied by the hopper so that no explosion is possible. You can assume the following:

- The corn flour has been sieved in a 200 mesh screen so that the particle diameter can be assumed to be 74 μm. The density of the corn flour can be taken as 1200 kg/m^3.
- The powdered particles of corn flour can be assumed be spherical. The drag coefficient C_D can be taken as unity.
- The minimum explosive concentration (MEC) of the corn flour for the given particle size of 74 μm is 0.0794 kg/m^3.

(b) Considering the demand of products from the bakery, the management wishes to increase the throughput of the corn flour through the 1.5 m diameter pipe to augment the production. They, however, would not like to violate any safety aspect and decide to increase the particle size of the flour to 85 μm from the existing 74 μm. Assuming that the dependence of MEC on mean particle size is given by

$$MEC(\text{kg/m}^3) = 0.035 + 0.0006d$$

where d is the diameter of the flour dust in μm, determine the permissible increase in the rate of supply of the corn flour to the chute.

10.4 The index K_{St}, describing the maximum rate of pressure rise in a particular dust mixture of milk powder and air, is 75 bar-m/s when the concentration of the milk powder is 100 g/m^3. A concentration of 100 g/m^3 is obtained while conveying the milk powder in a cylindrical duct 1.2 m diameter and 6 m long. Determine the maximum rate of pressure rise if the dust mixture in the above column explodes?

10.5 (a) Find the maximum rate of pressure rise in an explosion of fine cellulose dust of mass 0.15 kg in a confined volume of 0.5 m^3 given that the K_{St} value for the dust mixture is 200 bar-m/s. The dust does not explode for concentrations below this value.

(b) If the maximum explosion pressure in the above question is 0.95 MPa, determine the explosion severity. The reference coal–air dust mixture has an explosion pressure of 0.8 MPa and a maximum pressure rise of 120 bar/s.

(c) Experiments for the given dust cloud give the minimum ignition energy to be 5 mJ and the minimum ignition temperature as 400 °C. Determine the ignition sensitivity of the given dust mixture. The minimum ignition energy for the standard coal dust mixture is 10 mJ, the ignition temperature is 800 K and the minimum explosive concentration is 02 kg/m^3.

(d) What is the index of explosibility of the given dust–air mixture?

(e) Would you rate the dust–air mixture as a weak hazard, moderate hazard, strong hazard or a severe hazard and why?

Chapter 11
Physical Explosions and Rupture of Pressure Vessels

Chemical reactions are not to the only means of obtaining rapid release of energy for the occurrence of explosions. A liquid maintained at metastable conditions such as at temperatures much above its boiling temperature corresponding to the ambient conditions of pressure could behave as an explosive under the ambient conditions. The spontaneous heating of water from the large mass of molten lava resulted in the flash or impulsive formation of steam and a strong blast wave in the volcanic eruption at Krakatau on August 27, 1883 (discussed in Chap. 1). The sudden increase in volume is similar to the increase in the volume of gaseous products in the condensed phase explosions. The explosion is termed as hydrovolcanic explosion, and the cause for the explosion is flash vaporization. The sudden heating of water, like in the explosion at the copper smelter in Flin Flon, also discussed in Chap. 1, was from flash vaporization though at a very much reduced scale compared to the Krakatau explosion. Release of liquids in the ambient from containers at temperatures above their atmospheric boiling point can also lead to explosion from the rapid and spontaneous vaporization of a significant portion of the liquid. Explosions resulting from the sudden change of liquid phase to vapor phase are called physical explosions.

The spontaneous release of the potential energy of a pressurized gas, contained in a pressure vessel, due to busting of the vessel or otherwise also leads to an explosion. Accidental explosions from the rupture of high-pressure gas vessels have been frequently encountered. Concrete tanks used as digesters for biogas and sewage lines are known to explode when inadvertent accumulation of gas takes place and adequate venting is not provided. A newly commissioned concrete tank of a biogas plant exploded at Edathala in the state of Kerala in India on August 26, 2009, due to high pressure built up within it killing three people and injuring many. Physical explosions involving flash vaporization and bursting of vessels are dealt with in this chapter.

K. Ramamurthi, *Modeling Explosions and Blast Waves*,
https://doi.org/10.1007/978-3-030-74338-3_11

11.1 Flash Vaporization

11.1.1 Principle

Consider a liquid existing as a saturated liquid at a high value of pressure p_2. This is shown by point B having a temperature T_2 on the saturated liquid line in the temperature T vs. the specific volume v diagram in Fig. 11.1. This figure shows the equilibrium states of the liquid, liquid and vapor, and the superheated vapor in the $T - v$ diagram. The wet region, comprising the liquid and vapor phase, is bounded on the left side by the saturated liquid line and on the right side by the saturated vapor line as shown. To the left of the saturated liquid line, only liquid state exists, while to the right of the saturated vapor line, only the vapor exists. The constant pressure lines at high pressure p_2 and at low pressure (atmospheric) p_1 are shown by $ABCD$ and $A'B'C'D'$, respectively. AB in Fig. 11.1 denotes the liquid state, BC is the wet region, and CD is the superheat corresponding to pressure p_2. At the lower value of pressure p_1, $A'B'$, $B'C'$, and $C'D'$ denote the liquid, wet, and superheated states.

If the liquid at the saturated liquid state B at pressure p_2 is suddenly exposed to a lower pressure p_1, for which the boiling temperature is T_1 as shown at B' (Fig. 11.1), it partly gets converted to vapor since it cannot remain in equilibrium at the higher value of boiling point. The specific enthalpy corresponding to state B is h_{f2} and is higher than the specific enthalpy at B' of h_{f1}. It cannot therefore exist as a saturated liquid at B'. The excess liquid enthalpy $(h_{f2} - h_{f1})$, locked within it, vaporizes part of the liquid. If the latent heat of vaporization at the reduced value of pressure p_1 is denoted by h_{fgl}, the fraction of the liquid f converted to vapor is

$$f = \frac{h_{f2} - h_{f1}}{h_{fgl}} \tag{11.1}$$

The entire liquid flashes to vapor when $f = 1$. A value of $f > 1$ implies that superheated vapor would be formed.

Fig. 11.1 Flash vaporization process

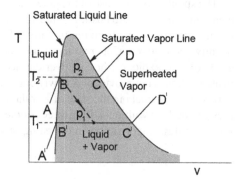

If we consider the initial state of a liquid to be along the liquid line at pressure p_1 but is maintained at a temperature T_2 corresponding to pressure p_2, its state cannot be stable, i.e., in equilibrium since for the ambient pressure p_1 it can only be in equilibrium at temperature T_1. The liquid at the metastable state at T_2 flashes into vapor, the fraction of the vapor formed being given by Eq. 11.1. Vapor has larger volume compared to a liquid, and spontaneous release of vapor having much larger value of the specific volume compared to the liquid causes a sudden expansion and gives rise to the explosion.

11.1.2 Spillage of Volatile Liquids Stored at High Pressures

The normal boiling point temperature is defined as the temperature at which the liquid boils at the ambient pressure of 1 atmosphere. At this temperature, the vapor pressure of the liquid is the same as the ambient pressure of 1 atmosphere. When the pressure at which a liquid is stored is increased to values greater than the ambient, its boiling temperature (at the higher pressure) would be greater than the normal boiling temperature. It can therefore be stored at higher pressures greater than the ambient pressure as a liquid, while at ambient temperature and ambient pressure it would be a vapor.

As an example, the variation of boiling temperature of Butane at four different pressures is given in Table 11.1. It is seen that butane, which is normally a gas at the ambient conditions of 1 atm. and 32 °C, can be stored as liquid butane at 3 atm. pressure and the same temperature of 32 °C.

Consider a spill of liquid butane from a cylinder or a pipeline wherein it is at 3 atm. pressure into the ambient atmosphere at 1 atm. pressure and a temperature of 32 °C. The liquid, when spilled, will be at 1 atm. pressure for which its equilibrium temperature is −0.5 °C. It cannot exist at its temperature of 32 °C. We call this state of 32 °C and 1 atm. pressure as a metastable state, and it contains excess heat corresponding to product of the mass spilled, the specific heat of liquid butane, and the difference in temperature between 32 and −0.5 °C. This heat flashes part of the mass spilled into vapor.

The leakage of liquefied gases at pressurized conditions from pipelines or release of liquefied fuels, contained at high pressure in containers, thus leads to a rapid

Table 11.1 Variation of boiling temperature with pressure for liquid butane

S. No.	Pressure (atmospheres)	Boiling temperature (°C)
1	1	−0.5
2	1.5	10
3	2	21
4	3	32

formation of vapor. The efflux of the pressurized liquids at ambient temperature from storage vessels or pipelines into the ambient results in copious formation of vapor. The quantity of vapor m_V formed from a spill of liquid of mass m is

$$m_V = f \times m \tag{11.2}$$

Here the value of f is the fraction determined from the liquid enthalpy at the storage pressure, the saturation liquid enthalpy at the ambient pressure, and the latent heat of vaporization at the ambient pressure following Eq. 11.1.

If the leakage is over a prolonged duration from a constant storage pressure p to the ambient pressure p_a, the mass flow rate of the liquid $\dot{m}$ is

$$\dot{m} = C_d A_0 \sqrt{2\rho(p - p_a)} \tag{11.3}$$

when the flow through the opening is not choked. C_d in the above equation is the discharge coefficient, A_0 is the opening, and ρ is the density of the liquid. When the flow is choked, the mass flow rate is proportional to the pressure p in the vessel or pipeline.

The mass flow rate $\dot{m}$ in Eq. 11.3 is at a temperature higher than the saturation temperature corresponding to the ambient pressure p_a. Denoting the enthalpy and the temperature corresponding to the liquid at the release as h_2 and T_2 and the saturation temperature corresponding to ambient pressure as T_1, we have

$$\dot{m}(h_2 - h_{f1}) = \dot{m}C_m(T_2 - T_1) \tag{11.4}$$

C_m in the above equation denotes the mean specific heat of the liquid, while h_{f1} denotes the enthalpy of the saturated liquid at the ambient pressure. Equation 11.4 gives the excess enthalpy contributing to the vaporization, and the rate of vaporization $\dot{m}_V$ is being

$$\dot{m}_V = \frac{\dot{m}C_m(T_2 - T_1)}{h_{fg1}} \tag{11.5}$$

where h_{fg1} is the latent heat of vaporization of the liquid corresponding to the ambient pressure.

11.1.3 General Formulation for Flash Vaporization

The flash evaporation takes place when the metastable liquid at the ambient pressure is at a temperature greater than the normal boiling point temperature. Let us presume that a mass of metastable liquid is m, its specific heat is C and its temperature is a small amount dT higher than the normal boiling temperature. Let the latent heat of vaporization at the ambient pressure be h_{fg}. If a mass that flashes to vapor is dm, the energy balance gives

$$dmh_{fg} = mCdT$$

or

$$\frac{dm}{m} = \frac{CdT}{h_{fg}} \qquad (11.6)$$

If the normal boiling temperature is T_B and the metastable liquid is at temperature T_0, Eq. 11.6 can be integrated between limits of initial mass m_i of the spilled liquid and the final mass of the spilled liquid, which is $m_i - m_V$ where m_V is the mass that flashes into vapor. The limits of temperature accordingly are between T_0 and T_B. The integral becomes

$$\int\limits_{m_i}^{m_i-m_V} \left(\frac{dm}{m}\right) = \int\limits_{T_0}^{T_B} \left(\frac{CdT}{h_{fg}}\right) \qquad (11.7)$$

Solving Eq. 11.7, we have

$$\ln\left(\frac{m_i - m_V}{m_i}\right) = C(T_B - T_0)/h_{fg} = -C(T_0 - T_B)/h_{fg}$$

and simplifying we have

$$1 - \frac{m_V}{m_i} = e^{-C(T_0-T_B)/h_{fg}}$$

We therefore get the fraction of the mass that flashes to vapor as

$$\frac{m_V}{m_i} = 1 - e^{-C(T_0-T_B)/h_{fg}} \qquad (11.8)$$

If the specific volume of the vapor is v, the volume of V vapor formed is $v \times m_V$.

$$V = v \times m_V$$

11.1.4 Energy Release in Flash Vaporization

The energy (E) associated in the flash vaporization process is the product of the mass of vapor formed and the latent heat of vaporization corresponding to the ambient pressure. Similarly the rate of energy release (dE/dt) is the product of the rate of formation of the vapor and the latent heat of formation. These are given below:

$$E = m_V h_{fg1}$$
$$\frac{dE}{dt} = \dot{m}_V h_{fg1} \qquad (11.9)$$

11.1.5   Flash Vaporization of Water Thrown in A Hot Furnace as an Example

Consider the example of 50 kg of water being thrown into a furnace, which is at a temperature of 300 °C. The thermal capacity of the furnace can be assumed to be very large, and the water would spontaneously reach a metastable condition of 300 °C. This is shown by point A in the $T - v$ diagram in Fig. 11.2.

The excess enthalpy of water at A, over that for the equilibrium saturated condition of water given by B at 0.1 MPa, goes to spontaneously vaporize a part of the water corresponding to the saturated vapor conditions at C. The fraction of the water vaporized is given by

$$f = \frac{h_A - h_B}{h_{fg}} \tag{11.10}$$

In the above expression h_A and h_B are the enthalpies at A and B, while h_{fg} is the latent heat of vaporization corresponding to line BC (Fig. 11.2). From steam tables we have

$$h_A = 1344 \, kJ/kg$$
$$h_B = 419 \, kJ/kg$$
$$h_C = 2258 \, kJ/kg$$

Substituting the above values in Eq. 11.10 gives

$$f = 0.41$$

The total quantity of water that flashes into vapor is therefore

$$m_V = 50 \times 0.41 = 20.5 kg$$

Fig. 11.2 Metastable condition A and vaporization

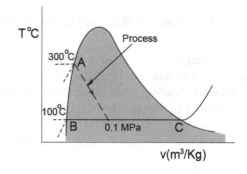

The volume of water, which flashes to vapor, is

$$V = 20.5 v_B$$

where v_B is the specific volume of water corresponding to the initial metastable condition B. The difference in the volume of water would not be very different between conditions A and B. The volume of water vapor at the final condition is however much larger and equals 20.5 v_C where v_C is the specific volume of the saturated vapor at the ambient pressure (condition C in Fig. 11.2). Pressure is built up spontaneously from the large increase of volume. The high pressure drives the blast wave.

Liquids stored or handled under pressure above their normal boiling point, corresponding to the ambient conditions, are hazardous.

11.1.6 Rollover and Explosion in Cryogenic Storage Vessels

Cryogenic liquids such as LNG are stored in insulated containers. LNG is not pure methane. It contains about 95% methane and small quantities of methane, ethane, propane, and gasoline. The normal boiling temperatures of methane and the other constituents in LNG are given in Table 11.2.

Liquid ethane, propane and gasoline are heavier than liquid methane and would settle at the bottom of the storage vessel of LNG. Liquids having the normal boiling temperatures less than $-150\,°C$ (123 K) are referred to as cryogenic liquids or cryogens. They need to be stored in well-insulated storage vessels if at atmospheric pressure. LNG is a cryogen and is stored in insulated vessels.

Though the storage vessel containing LNG is insulated, the small amounts of heat leaking into the liquid from the walls of the container result in a decrease of the density of the liquid adjacent to the wall. The warm liquid moves up due to its buoyancy and collects as a stratified warm layer at the liquid surface. This is schematically indicated in Fig. 11.3.

The warm stratified liquid methane at the top layer evaporates, and the evaporation of the warm liquid at the surface reduces its temperature. The pressure increase, due to the enhanced evaporation of the warm stratified liquid, causes an increase in pressure of the vapor, and some vapor is vented out. The expansion during the venting results

Table 11.2 Variation of boiling temperature with pressure for LNG

S. No	Liquid	Normal boiling temperature (°C)
1	Liquid methane	-162
2	Liquid ethane	-85
3	Liquid propane	-42.4
4	Gasoline	$+69$

Fig. 11.3 Stratification and rollover

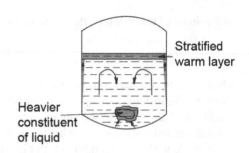

Stratified warm layer

Heavier constituent of liquid

in a drop in pressure, and hence temperature of the stratified layer decreases. An equilibrium pressure is reached with the venting and cooling of the warm stratified layers from the evaporation.

However, the heat in-leak at the bottom of the container, which contains the denser fraction, also generates warm liquid. This warm liquid, though at a temperature higher than the mean temperature, cannot convect up as its density, in spite of it being heated, is still high. It can only conduct heat through the liquid. The warm liquid, if it was methane, would normally have risen to the top because of its buoyancy. But not these denser fractions. The bottom layers, though warmer, therefore exist in equilibrium with the lower temperatures above them since the density of the heavier constituent at the higher temperature is more than the density of the lighter constituent at the lower temperature. The warm dense layer at the bottom is known as thermal overfill.

When the temperature of this warm and relatively heavier constituent (thermal overfill) further increases, it becomes less dense and convects to the surface. The temperature of the thermal overfill is much higher than the equilibrium value corresponding to the liquid at the surface. It has excess internal energy. The mixing of this very warm thermal overfill and the lower temperature stratified layer at the surface leads to a somewhat large increase in the temperature at the surface. The liquid phase at the surface is no longer in thermodynamic equilibrium with the vapor above it. Rapid evaporation as in flash vaporization takes place, causing an increase of the gas pressure above the liquid. The phenomenon is known as 'rollover'. The rapid flash vaporization at the surface from the enhanced surface temperature of the liquid leads to abnormal pressure rise and rupture of the LNG storage vessel.

11.1.7 Rollover Explosion at Cleveland, Ohio and Intense Damage of Sewage Lines, Homes and Skyrocketing of Manhole Covers

On October 20, 1944, a huge LNG tank containing about 2.8 million cubic meters of LNG at Cleveland, Ohio developed pressure due to rollover, and the tank ruptured at the dome. Huge quantities of natural gas escaped from the tank. The escaping natural gas, viz. methane, was at the temperature of LNG, i.e., −161 °C. The molecular mass of methane is 16 g/mole, and its density at the ambient temperature of 25 °C and 100 kPa pressure is 0.64 kg/m^3 [$10^5 \times 0.016/(8.314 \times 298)$]. This is about half the density of air, and it will therefore readily disperse upwards. However at the low temperature of -161 °C, the density of methane is $0.64 \times 298/112 = 1.8$ kg/m^3. It is therefore very much denser than air, and consequently it remains on the ground and cools the ground.

The cold methane gas spilled over the streets. There was a sewage unit nearby, and the heavy cold methane reached it hugging to the ground. At the sewage plant, there were perforations on the ground through which sewage gas could escape from the sewage lines. The dense methane gas entered the sewage lines through the perforations and mixed with air and sewage gas in the lines. The movement of the gas created static electricity, and the energy release from the electrical spark ignited the mixture of sewage gas, methane, and air. A flame was thus formed.

The roughness of the sewage lines and blockages in it contributed to accelerate the flame and form shock waves and a detonation. The heavy manhole covers of the sewage lines sky-rocketed upwards from the high pressure of the detonation. The high-temperature gases from the sewage lines connected to the houses entered the homes and caused fire and a lot of discomfort. Part of the township got destroyed. The explosion from the rollover is referred to as East Ohio Gas Disaster and killed about 130 people.

11.1.8 Condensation Shock and Rupture of Cryogenic Pressure Vessels

Thermal overfill and rollover cause the explosion of pressure vessels when the cryogenic fluid stored in it had some impurities such as denser fractions. Such a situation is unlikely with an insulated vessel contains pure cryogenic liquid such as say liquid hydrogen, liquid oxygen, or liquid helium. However, a phenomenon known as condensation shock seen during the expansion of humid air in wind tunnels and in the expansion of steam in turbines, given below, could cause shock waves and the rupture.

A cryogenic storage vessel is inherently provided with a vent valve to prevent overpressurization of the vessel from the vapors formed by the stratification process. The vent valve operates for a finite pressure difference between the vapor in the vessel and the ambient pressure. During the venting process, a flow of vapor takes place in the vapor above the cryogenic liquid as its pressure decreases. Wave motion is induced in the volume of the vapor with expansion waves originating at the vent valve.

The fall in pressure in the vapor column above the liquid, associated with the opening of the vent valve, causes a metastable vapor to be formed in the zones of convection of the vapor. This is due to the abrupt decrease in pressure communicated by the expansion waves. The temperature of the vapor, however, retains the saturation value corresponding to the higher pressure before the venting operation as it takes time for it to relax to the lower equilibrium value. Hence a metastable layer of

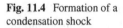

Fig. 11.4 Formation of a condensation shock

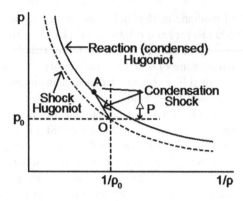

vapor is formed with its temperature higher than its equilibrium temperature. It flash-condenses to the liquid phase. In doing so, it releases a spurt of energy.

The energy release in flash condensation occurs in a similar way to energy absorption and formation of vapor with flash evaporation. The energy release forms compression disturbances, and these intensify to form shocks very much like the energy of a moving accelerating piston driving shock waves. Such shocks driven by the condensation of the metastable vapor are known as condensation shocks. The Mach number of the condensation shocks is generally small, and the maximum is of the order of about Mach 1.4. The pressure jump Δp across the condensation shock is shown in Fig. 11.4 in a plot of the shock Hugoniot before vent and a reaction Hugoniot formed after vent in the lower pressure medium with the energy release from condensation. The condensation shock is shown by the line OA with the point O on the shock Hugoniot denoting the state of the vapor when the vent valve opens and the point A lying on the reaction Hugoniot. The energy release in the condensation Hugoniot corresponds to the energy liberated in the flash condensation process.

Though the magnitudes of the pressure rise across the condensation shock would be small, the reflection of the condensation shock at the walls of the vessel would give rise to higher values of pressure. Such pressures can cause the vessels that are designed to hold the cryogenic liquids at ambient pressures to rupture. It is necessary to ensure that venting of the vapor is done at low pressure drops across the vent to avoid the formation of the condensation shock.

11.2   Burst of Pressure Vessels

When a vessel containing a high-pressure gas ruptures, the energy contained in the gas is released to the ambient. The energy released in the bursting process is determined in the following.

Consider a vessel of volume V_0 containing a gas at temperature T_0 and high pressure p_0 (Fig. 11.5). After the vessel ruptures, the volume V_0 contains gases at the ambient pressure p_∞ and temperature T_∞.

Fig. 11.5 Conditions in pressure vessel before rupture

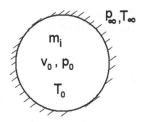

Denoting the initial mass of the gas contained in the vessel as m_i in the volume V_0 and the mass in this volume after the bursting as m_f, we have the energy released during the rupture E as

$$E = m_i u_i - m_f u_f \qquad (11.11)$$

Here u_i and u_f are the specific internal energies of the gas at the initial state and after the rupture.

The specific internal energy $u = C_V T$, where C_V is the specific heat at constant volume, and T is the temperature. Writing C_V as $R/(\gamma - 1)$, where R is the specific gas constant in kJ/(kg K) and γ is the specific heat ratio of the gas, we have

$$E = m_i C_V T_0 - m_f C_V T_\infty = \frac{m_i R T_0}{(\gamma - 1)} - \frac{m_f R T_\infty}{(\gamma - 1)} \qquad (11.12)$$

Assuming perfect gas and noting that $pV = mRT$ from the equation of state for an ideal gas, we get

$$\begin{aligned}
E &= \frac{p_0 V_0}{R T_0} \frac{R T_0}{\gamma - 1} - \frac{p_\infty V_0}{R T_\infty} \frac{R T_\infty}{\gamma - 1} \\
&= \frac{(p_0 - p_\infty) V_0}{\gamma - 1}
\end{aligned} \qquad (11.13)$$

The energy released in the bursting is directly proportional to the difference between the initial pressure p_0 and ambient pressure p_∞ and the volume of the vessel V_0. The specific heat ratio γ is also seen to influence the energy release in the burst of the vessel.

11.2.1 Influence of Specific Heat Ratio

Consider the bursting of a vessel having a volume of $1\,\mathrm{m}^3$ and containing air at a pressure of 10 MPa. The energy release from Eq. 11.13 is

$$E = \frac{(10^7 - 10^5) \times 1}{(1.4 - 1)} = 24.75\,\mathrm{MJ} \qquad (11.14)$$

Fig. 11.6 Energy release
due to changes in γ

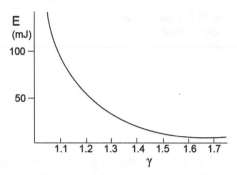

If instead of air for which γ is 1.4, helium having γ of 1.67 is used, the energy release
is

$$E = \frac{(10^7 - 10^5) \times 1}{(1.67 - 1)} = 10.8\,\text{MJ} \tag{11.15}$$

Gases having complex molecular structure have smaller values of the specific heat
ratio. This is due to the higher values of the internal energy stored in the complex
molecules. Gases such as the refrigerant trichlorofluoromethane CCl_3F and dichloro
difluoro methane CCl_2F_2 have values of γ near to 1.1. This gives the energy from
the bursting of a volume of $1\,\text{m}^3$ at an initial pressure of 10 MPa to be

$$E = \frac{(10^7 - 10^5) \times 1}{(1.1 - 1)} = 99\,\text{MJ} \tag{11.16}$$

The energy is seen to be considerably enhanced when the specific heat ratio of the
gas is smaller. Figure 11.6 shows the variation of energy release with specific heat
ratio changes for the particular example that we considered.

A smaller value of γ implies larger molecular or larger internal energies. Since
the change in internal energy contributes to the released energy, gases with smaller
γ release more energy during the burst of the pressure vessel.

11.2.2   Heat Addition to a Gas at Constant Pressure

The volume of a gas in an unconfined geometry increases when heat is added to it
at constant pressure. Let us consider the volume to change from an initial value V_i
to a final volume V_f at a pressure p_0. This is a slow process, and the work done or
energy release is

$$\begin{aligned} E &= p_0(V_f - V_i) \\ &= mR(T_f - T_i) \end{aligned} \tag{11.17}$$

Here m is the mass of gas in the given volume, and R is the specific gas constant. The temperature of the gas increases from initial value T_i to the final value T_f. Since $C_p = \gamma R/(\gamma - 1)$, the above equation can be written as

$$E = \frac{\gamma - 1}{\gamma} m C_P (T_f - T_i) \tag{11.18}$$

The heat transferred to the gas during the expansion denoted by Q is

$$Q = m C_P (T_f - T_i) \tag{11.19}$$

Combining Eqs. 11.18 and 11.19, we have

$$\frac{E}{Q} = \frac{\gamma - 1}{\gamma} \tag{11.20}$$

In the case of a slow equilibrium process for which the time tends to be large, the ratio of the energy available to heat release is $(\gamma - 1)/\gamma$. However, when no expansion work takes place as in a constant volume process having small characteristic times, the heat added is

$$Q = m C_V (T_f - T_i) = E \tag{11.21}$$

This gives the value of $E/Q = 1$. For short durations of heat release, more of the heat is available for energy as shown in Fig. 11.7.

Examples

11.1 *Flash Vaporization of Water*: Consider the physical explosion involving flash vaporization of water in a reverberatory furnace. 20 l of water initially at a temperature of 15 °C is suddenly thrown into the hot furnace with the purpose of cooling it down. The flash vaporization of the water causes a very sudden release of vapor. Assuming the thermal capacity of the furnace to be large and the temperature of the furnace to be 350 °C before the 20 l of water were poured into it, calculate the fraction of the water which flashes to vapor and the energy release in the process.

Fig. 11.7 Heat addition available as energy release

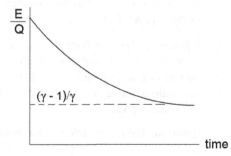

Solution: The water is momentarily subject to a temperature of 350 °C when it comes in contact with the furnace. The enthalpy of water corresponding to saturated water conditions at 350 °C using steam tables is 1671.2 kJ/kg (corresponds to point A in Fig. 11.2).

The saturated water conditions corresponding to the ambient pressure of 0.1 MPa are 417.51 kJ/kg (point B in Fig. 11.2). The latent heat of vaporization at the ambient pressure of 0.1 MPa is 2257.5 kJ/kg. The fraction of the water, which flashes to vapor, based on Fig. 11.2, is

$$f = \frac{h_{f,350\,°C} - h_{f,100\,kPa}}{h_{fg,100\,kPa}}$$
$$= \frac{1671.2 - 417.5}{2257.5}$$
$$= 0.555$$

Total mass of water that flashes to steam $= 20 \times 0.555 = 11.1$ kg. The energy associated with the above flashing of steam $= 11.1 \times h_{fg,100\,kPa} = 11.1 \times 2257.5 = 25,058$ kJ. The energy available in the process is 25.06 MJ.

11.2 *Pressure Increase from Stratification*: An insulated tank contains liquid hydrogen at a bulk temperature of −253 °C. Due to the small amounts of heat that leaks into the insulated tank from the ambient, a stratified layer is formed at a temperature 2° above the bulk temperature. Determine the increase of pressure in the tank.

Solution: The variation of the vapor pressure with temperature of liquid hydrogen is given in the table below:

S. No	Temperature (K)	Vapor pressure (atm)
1	20	1
2	21	1.2
3	22	1.6
4	23	2
5	24	2.6
6	25	3.2

The surface temperature of liquid hydrogen from the stratification is −251 °C, i.e., 22 K. The vapor pressure is therefore 1.6 atm. Hence the pressure increase in the tank is 0.6 atm.

11.3 *Influence of Specific Heat Ratio on Energy Release in an Explosion*: A high-pressure storage vessel of volume 0.5 m³ contains a refrigerant trichlorotrifluoromethane at a pressure of 20 MPa and ambient temperature. If the specific heat ratio of the refrigerant vapor is 1.1, determine the energy released in case the vessel ruptures.

Solution: The energy released has been derived as

$$\frac{(p_0 - p_\infty)V_0}{\gamma - 1}$$

Substituting the values of pressures and volume we get the energy released as

$$\frac{19.9 \times 6 \times 0.5}{0.1} = 99.5 \times 10^6 \text{ J} = 99.5 \text{ MJ}.$$

11.4 *Steady State Flash Vaporization*: A pipeline carrying liquid butane at a pressure of 2.13 kPa and room temperature of 25 °C spurts liquid butane through a small opening of 0.1 cm². If the discharge coefficient of the opening through which the liquid butane escapes is 0.7 and the flow is not choked, determine the rate at which the butane flashes to vapor. You can neglect the cooling of ambient air from the vaporization process. The pressure in the pipeline can be assumed to be a constant at 2.13 kPa.

Solution: The saturated liquid enthalpy of liquid butane at 2.13 kPa is 355 kJ/kg, while the saturated liquid enthalpy at atmospheric pressure is 300 kJ/kg. The enthalpy of the saturated vapor at atmospheric pressure is 680 kJ/kg. The density of liquid butane is 610 kg/m³.

The rate at which liquid butane squirts out is given by $C_D A_0 \sqrt{2\Delta p \rho}$ where Δp is the pressure drop across the opening, ρ is the density of liquid butane, and A_0 is the area of the opening. C_D is the discharge coefficient. The above formula comes from Bernoulli equation. Substituting the values given in the problem, we have the mass flow rate in kg/s as $0.7 \times 0.1 \times 10^{-4} \times \sqrt{2 \times (2.13 - 1.03) \times 10^5 \times 610} = 0.11$ kg/s.

The liquid is not in equilibrium with the ambient. It has an excess heat equal to the value of the saturated enthalpy as a liquid at 2.13 kPa, viz. $h_f = 355$ kJ/kg as compared to the saturated enthalpy at atmospheric pressure $h_f = 300$ kJ/kg. The excess heat is 55 kJ/kg. The latent heat of vaporization of liquid butane at atmospheric pressure is the difference between its enthalpy as saturated vapor and its enthalpy as saturated liquid and is $680 - 300 = 380$ kJ/kg. Hence the fraction that vaporizes is $55/380 = 0.145$.

Hence of the liquid spurting out at the rate of 0.11 kg/s, the quantity that flash vaporizes is $0.145 \times 0.11 = 0.016$ kg/s.

Nomenclature

A_0 Area of opening (m²)
C_d Discharge coefficient
C_m Specific heat of liquid [kJ/(kg °C)]
C_p Specific heat at constant pressure [kJ/(kg °C)]
C_V Specific heat at constant volume [kJ/(kg °C)]
E Energy (kJ)
f Fraction
h Specific enthalpy (kJ/kg)
h_f Specific enthalpy of saturated liquid (kJ/kg)

h_g Specific enthalpy of saturated vapor (kJ/kg)
h_{fg} Latent heat (kJ/kg)
m Mass (kg)
m_V Mass of vapor (kg)
$\dot{m}$ Mass flow rate (kg/s)
p Pressure (Pa)
Q Heat transfer (kJ)
R Specific gas constant [J/(kg K)]
t Time (s)
T Temperature (°C)
u Specific internal energy (kJ/kg)
v Specific volume (m³/kg)
V Volume (m³)
γ Specific heat ratio
ρ Density (kg/m³)

Subscripts

f Final
u Initial
∞ Ambient

Further Reading

1. Crowl, D. A. and Louvar, J. F., *Chemical Safety: Fundamentals with Applications*, Prentice Hall, Englewood Cliffs, NJ, 2002.
2. Irvine, T. F., Jr., and Hartnett, J. P., (eds.) *Steam and Air Tables in SI Units*, Hemisphere, Washington, DC, 1976.
3. Landau, L. D. and Lifshitz, E. K., *Fluid Mechanics*, Pergamon Press, London, 1966.
4. Liepman, H.W., and Roshko, A., *Elements of Gas Dynamics*, Wiley, New York, 1957.
5. Moron, J. M. and Shapiro, H. N., *Fundamentals of Engineering Thermodynamics*, Wiley, New York, 2000.
6. Strehlow, R. A. and Baker, W. E., *The Characterization and Evaluation of Accidental Explosions*, NASA CR-134779, National Aeronautics and Space Administration, Cleveland, OH, June 1965.

Exercises

11.1 (a) A spherical vessel of volume 0.5 m³ capacity bursts while being filled with compressed air. The pressure in the vessel at the instant of the explosion is 15 MPa. Determine the energy released during the rupture of the pressure vessel.

(b) If instead of air, Freon gas was being charged into the same spherical vessel and it exploded when the pressure was 15 MPa as given earlier, what is the energy release in the explosion?

You can assume that the specific heat ratio for air as 1.4 and for Freon gas as 1.1.

11.2 A high-pressure air-storage tank of $10\,m^3$ capacity contains air at the ambient temperature of 27°C and pressure of 5 MPa ruptures. The ambient corresponds to sea level conditions. Determine the energy released in the explosion.

11.3 (a) The failure of a weld joint in a pipeline conveying liquid propane at a pressure of 4 MPa causes liquid propane to be released in the atmosphere at the rate of 5 kg/s. Determine the rate at which the liquid propane would flash into vapor. You can assume the initial state of liquid propane in the pipeline to be a saturated liquid, the saturated liquid enthalpy at 4 MPa being 515 kJ/kg. At the ambient pressure of 0.1 MPa, the value of saturated liquid enthalpy is 99 kJ/kg, and the saturated enthalpy of the vapor is 526 kJ/kg. The boiling temperature of liquid propane at the ambient pressure is 231 K.

(b) Determine the dryness of the propane vapor formed.

(c) Show the process of the flash vaporization on a temperature vs. specific volume diagram by a dotted line.

11.4 (a) Propane is stored as a liquid at room temperature at a pressure of 4.1 MPa in a large diameter cylindrical pressure vessel. If the dome of the pressure vessel gives way due to defective weld and the propane is subjected to the ambient pressure of 0.1 MPa, determine the fraction of the liquid propane that would flash into vapor.

The boiling point of propane is 231 K at the ambient pressure of 0.1 MPa. At this, temperature and pressure, the value of saturated liquid enthalpy is 99 kJ/kg, and the value of the saturated enthalpy of the vapor is 526 kJ/kg. At 4.1 MPa at which the liquid was contained, the saturated liquid enthalpy is 515 kJ/kg.

(b) Determine the energy release per kg of liquid propane.

11.5 LPG is stored in a huge tank at atmospheric pressure but at a cryogenic temperature corresponding to the boiling temperature of liquid methane at −162°C. Due to the presence of small quantities of propane, petrol, and trace quantities of other hydrocarbons, a region of thermal overfill at a temperature of −156°C is formed at the base of the liquid. Rollover of the thermal overfill region abruptly takes place due to density equalization, and a layer of a warm liquid methane is formed at the surface. Determine the sudden increase in pressure in the tank due to the rollover.

The vapor pressure of liquid methane with temperature in the region of interest is given in the following table:

S. No	Temperature of liquid methane in K	Vapor pressure in mm of mercury
1	111	760
2	113	847.5
3	115	993
4	117	1157
5	119	1340
6	121	1545
7	123	1772
8	125	2023
9	127	2300

760 mm mercury corresponds to 1 atm. pressure.

11.6 A condensation shock is formed at a Mach number of 1.5 from the super-saturated liquid hydrogen due to the venting of gaseous hydrogen in a liquid hydrogen storage tank. Determine using the pressure in the condensation shock and the increase in overpressure when the shock is reflected at the wall, the pressure experienced at the dome of the tank. The pressure at which the vent valve operates is 120 kPa, and the pressure after venting is 105 kPa. The temperature of liquid hydrogen is 20 K, and the molecular mass of hydrogen is 2 g/mole.

Chapter 12
TNT Equivalence and Yield from Explosions

The overpressure and impulse at a distance R_s from an explosion were determined in Chap. 2 by assuming the energy to be released instantaneously at a point. A length .scale for the energy release called as 'explosion length R_0' was defined, and a non-dimensional parameter R_s/R_0 was used to find the values of overpressure and impulse following Sach's scaling law. In practice, all sources of energy release would have a finite volume and the release of energy cannot be instantaneous. The overpressure and impulse for such non-ideal sources would deviate from the predictions made assuming ideal conditions. The asymmetrical geometry of the energy source would also form non-spherical waves. We address the explosive TNT as a near-ideal source and discuss the role of non-idealities in influencing the blast wave in this chapter.

12.1 TNT as an Idealized Energy Source

12.1.1 Energy Release from TNT

Consider the energy release from the detonation of a given mass of TNT. TNT is a fuel-rich explosive as detailed in Chap. 8. The chemical reaction for evaluating the energy release of TNT was given by

$$CH_3C_6H_2(NO_2)_3 \rightarrow 2\frac{1}{2}H_2O + 3\frac{1}{2}CO + 3\frac{1}{2}C + 1\frac{1}{2}N_2 \qquad (12.1)$$

The energy release in the above reaction is determined by considering the products and the TNT to be at the standard state of 25 °C and is given as

$$-\left[2\frac{1}{2}\Delta H_{f,H_2O}^0 + 3\frac{1}{2}\Delta H_{f,CO}^0 - \Delta H_{f,TNT}^0\right] \qquad (12.2)$$

© The Author(s), under exclusive license to Springer Nature Switzerland AG 2021
K. Ramamurthi, *Modeling Explosions and Blast Waves*,
https://doi.org/10.1007/978-3-030-74338-3_12

The standard heat of formation of TNT ($\Delta H^0_{f,\text{TNT}}$) is -54.4 kJ/mole while the values of $\Delta H^0_{f,\text{H}_2\text{O}}$ and $\Delta H^0_{f,\text{CO}}$ are -286.7 and -110.5 kJ/mole, respectively. The molecular mass (M) of TNT is 227 g/mole. The energy liberated per kg of TNT is therefore

$$\frac{-[-2.5 \times 286.7 - 3.5 \times 110.5 + 54.4]}{0.227} = 4621.6 \text{ kJ/kg} \qquad (12.3)$$

In the above calculations, we assumed H_2O in the products as water for the standard conditions of 25 °C while in practice it would be in the vapor phase at the high temperatures. We also neglected the dissociation of the products at high temperature of the combustion products and assumed carbon C to be formed because of insufficient oxygen. The actual value of energy release is actually somewhat smaller and is 4520 kJ/kg.

12.1.2   Energy Liberated at Constant Volume

The energy of TNT could be released almost at constant volume. In Sect. 12.1.1, we computed the energy using the heats of formation that are at constant pressure. For constant volume heat release, we need to use the internal energy of formation rather than the enthalpy or heat of formation.

The internal energy per mole is related to enthalpy per mole through the relation

$$u = h - pv \qquad (12.4)$$

where u and h are internal energy and enthalpy per mole, p is the pressure, and v is the specific volume per mole. Assuming the gaseous products formed in the reaction as ideal gas, we write using the ideal gas equation

$$v = \frac{R_0 T}{p} \qquad (12.5)$$

and for the gaseous species, we get the standard internal energy of formation as

$$\Delta u^0_f = \Delta h^0_f - R_0 T \qquad (12.6)$$

For the condensed species, however, the volume is negligible and we get from Eq. 12.4

$$\Delta u^0_f = \Delta h^0_f \qquad (12.7)$$

Substituting the values of the standard internal energy of formation in place of the standard heat of formation in Eq. 12.2 we have the heat liberated at constant volume in TNT as:

$$- \left[2.5(\Delta H^0_{f,H_2O}) + 3.5(\Delta H^0_{f,CO} - R_0 T) - \Delta H^0_{f,TNT} \right] \qquad (12.8)$$

With $R_0 = 8.314$ J/(mole K) and $T = 298$ K, in the above equation, we get the energy release at constant volume as 4659.8 kJ/kg of TNT. The value is marginally higher than the value of 4621.6 kJ/kg obtained in Eq. 12.3. The higher value is to be expected since there is no expansion work done for constant volume heat release.

12.1.3   Rate of Energy Release in TNT

The density of TNT is 1640 kg/m^3, and the volume of 1 kg mass of TNT is 6.098×10^{-04} m^3. If the volume is considered as spherical, the radius of 1 kg mass of TNT is 0.0526 m.

The detonation velocity (Chapman–Jouguet velocity) in TNT is 6940 m/s (Chap. 8). The time taken for the energy release in the detonation of a spherical charge of TNT of 1 kg is therefore $0.0526/6940 = 7.58 \times 10^{-06}$ s. Here, it is assumed that the detonation is initiated at the center of the spherical volume of TNT, the time taken for energy release being the radius of the charge divided by the detonation velocity. The time taken for the energy release is seen to be small of the order of a few microseconds.

12.1.4   Approximation of TNT as a Point Source

If we consider the energy release E_0 in a spherical volume of TNT of radius R_E, as shown in Fig. 12.1, we have

$$E_0 = m C_V T \qquad (12.9)$$

where m is the mass of TNT in the spherical volume of TNT of radius R_E, C_V the specific heat of the combustion products at constant volume, and T the final temperature which is very much higher than the initial temperature of the explosive $T_0 (T \gg T_0)$. Assuming the ideal gas equation to be valid for the combustion products, we get the maximum pressure p_e as

$$p_e \frac{4}{3} \pi R_E^3 = mRT \qquad (12.10)$$

where R is the specific gas constant. Noting that for a perfect gas

$$C_V = \frac{R}{\gamma - 1} \qquad (12.11)$$

Fig. 12.1 Energy release in
a spherical volume of TNT
of radius R_E

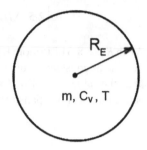

we get from Eqs. 12.9 and 12.10

$$E_0 = \frac{4}{3}\pi R_E^3 \frac{p_e}{\gamma - 1} \tag{12.12}$$

The explosion length was defined in Chap. 2 as $R_0 = (E_0/p_0)^{1/3}$, where p_0 is the ambient pressure. Substituting in Eq. 12.12 gives

$$\left(\frac{R_E}{R_0}\right)^3 = \frac{3(\gamma - 1)}{4\pi}\frac{p_0}{p_e} \tag{12.13}$$

The pressure generated in the detonation of TNT is about 190 MPa (Chap. 8). Assuming the value of specific heat ratio of the combustion products as 1.35, we have

$$\frac{R_E}{R_0} = \left(\frac{3 \times 0.35}{4\pi}\frac{0.1}{190}\right)^{1/3} = 0.035 \tag{12.14}$$

The radius of the source is seen to be a very small fraction of the explosion length for the explosive TNT.

If R_S denotes the radius of the blast wave at any time t, the value of R_S is very much greater than R_E. The properties of a blast wave were derived earlier as a function of the non-dimensional parameter R_S/R_0. Except in the very early stages of a blast wave, for which $R_S \approx R_E$, the value of R_S is very much larger than R_E. The dimensions of the source can therefore be neglected compared to R_S in the case of TNT. This can also be seen from the fact that the time involved in the energy release in TNT explosion is negligibly small compared to the time t at which a blast wave arrives at the distance R_S. This is schematically indicated in Fig. 12.2, which shows the blast wave trajectory with the blast wave at distance R_S from the source at time t and $\tau = \tau_E$ is the duration of energy release in the source of radius R_E.

Fig. 12.2 Blast wave formed from energy release E_0 over duration τ

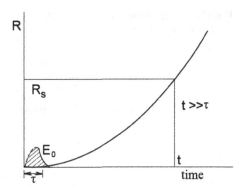

12.1.5 Characterization of Ideal Explosion by TNT

The explosion of TNT tends to be an instantaneous point source and the blast wave formed from it can be considered as ideal except for the region very near the source. In the region near to the charge of TNT, depending on the shape of the charge of TNT, the blast wave could follow the geometry of the charge and would be asymmetrical. However, as the wave propagates outwards, the local curvatures get smoothened out due to the high velocity of sound behind the wave and a spherical blast wave is formed. The explosion of TNT can therefore be taken to conform to an ideal blast wave.

Instead of expressing the energy release from the explosion of TNT in Joules, a common practice is to use the energy release in terms of mass of TNT. We have seen earlier that 1 kg of TNT releases 4520 kJ/kg of energy so that the equivalent mass of TNT for energy release E_0 Joules is

$$m_{\text{TNT,E}} = \frac{E_0}{4520} \text{ kg} \qquad (12.15)$$

The overpressure from the explosion of TNT would follow the ideal case given in Chap. 2. The scaled distance is directly given in terms of mass of TNT as $R_S/m_{\text{TNT}}^{1/3}$ instead of R_S/R_0. Here, R_S is in m as shown and m_{TNT} is in kg. Figure 12.3, derived from Fig. 2.24, shows the overpressure as a function of mass of TNT used.

12.2 Role of Nonidealities and Yield

A slower rate of energy release and a larger volume in which energy gets released produces different characteristics of blast wave than one formed from an ideal source such as TNT. A larger value of energy release under non-ideal conditions (neither instantaneous nor at a point) would be required compared to energy under ideal

Fig. 12.3 Overpressure as a
function of mass of TNT

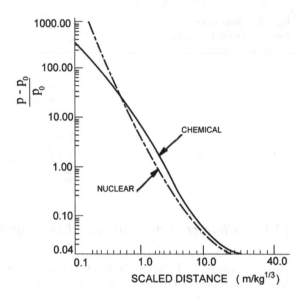

SCALED DISTANCE $(m/kg^{1/3})$

conditions to generate a blast wave having the same overpressure and impulse at a
given distance R_S from the center of the explosion. This is due to the dissipation of
part of the energy released as heat rather than the entire energy contributing to the
formation of the blast wave. The reduced effectiveness of energy in forming a blast
wave is known as the yield of the explosion.

The yield of an explosion is determined in the following manner. The mass of
TNT, which generates the same overpressure or impulse as that observed in a non-
ideal explosion at the particular distance of interest is determined from the effect of
the explosion. This is denoted by $m_{TNT,B}$. Here the suffix B denotes that the mass of
TNT is derived on blast wave basis. The mass of TNT is also determined based on
the energy release criterion from Eq. 12.15 and is denoted as $m_{TNT,E}$. The yield is
given by

$$\text{Yield} = \frac{m_{TNT,B}}{m_{TNT,E}} \tag{12.16}$$

The TNT equivalence is generally based on the observed blast wave damage from
an explosion, which is characterized by $m_{TNT,B}$ since it considers the actual effect
of the explosion, viz. the overpressure and impulse of the blast wave so generated.
This is determined from the overpressure (or impulse) curve such as the one given
in Fig. 12.3. The value of overpressure (or impulse) is based on the actual damage
at the particular distance. As an example, the TNT equivalence of the explosion
due to impact of a meteor or comet in 1908 over Tunguska was estimated from the
observed destruction to be between 10 and 15 Mt (megatons). The TNT equivalence
of the atomic bomb dropped over Hiroshima in 1945 was 15 kt (kilotons). The TNT
equivalence of the explosion in the metro rail in Moscow on 29 March 2010 was
estimated from the observed blast damage as 4 kg.

12.3 Influence of Source and Surroundings

Spherical blast waves are considered in this chapter. A planar source of energy release generates a planar wave, which in the far-field transforms to a spherical wave. A line source of energy generates a cylindrical blast wave near the source, which again in the far field becomes spherical. The blast wave formed in the near region of the source retains the geometrical character of the source of energy release.

The interaction of the blast wave with surfaces such as the ground or buildings forms reflected waves and complex shock and flow patterns. These cause the overpressures and impulses to be different from the values determined assuming a spherical blast wave to propagate freely in the ambient atmosphere. These have been dealt with in Chap. 3.

Examples

12.1 *TNT Equivalence and Yield due to Burst of a Pressure Vessel:*

(a) A pressure vessel of volume $1 \, m^3$ is designed to withstand a pressure of 500 bar. However, after prolonged use and corrosion at a weld, the pressure vessel fails at a pressure of 450 bar when it is being charged with compressed air. Determine the TNT equivalence for the explosion based on energy release in the rupture.

(b) If in the above accidental explosion, an overpressure of 75 kPa was observed at a distance of 10 m from the pressure vessel, find the percentage yield.

Solution:

(a) **TNT Equivalence**: The burst corresponds to a constant volume energy release. The energy release is $mC_V(T - T_0)$ which for an ideal gas whose equation of state is $pV = mRT$ and $C_V = R/(\gamma - 1)$ becomes

$$E = \frac{(p - p_0)}{\gamma - 1} V_0$$

Here V_0 is the volume of the pressure vessel and p the burst pressure. p_0 is the ambient value of pressure. Taking the specific heat ratio γ as 1.4 and substituting the values of the burst pressure, volume, and ambient pressure, we get

$$E = \frac{(450 - 1) \times 10^5 \times 1}{1.4 - 1} = 112.25 \, MJ$$

The energy release from TNT is 4520 kJ/kg. This gives equivalent mass of TNT as 112.25×10^3 (kJ)/4520 (kJ/kg) = 24.83 kg of TNT.

(b) **Percent Yield**: The overpressure of 75 kPa corresponds to the non-dimensional overpressure ratio of 75 (kPa)/100 (kPa) = 0.75. The mass of TNT, which causes the above value of overpressure, is determined from Fig. 12.3. We observe that the overpressure ratio of 0.75 is obtained at the value of $R_s/M_{\text{TNT}}^{1/3} = 9$. The mass of TNT is

$$m_{\text{TNT}} = \left(\frac{10}{9}\right)^3 = 1.37\,\text{kg}$$

$$\text{Percentage yield of the explosion} = \frac{1.37}{24.83} \times 100 = 5.52\%$$

12.2 *Overpressure from a given TNT Equivalent Explosion*: The TNT equivalence of a given explosion is given as 1.68 t. Find the distance from the site of the explosion over which people are knocked down by the blast. You can assume an average overpressure of 17 kPa to knock down people.

Solution: The overpressure of 17 kPa corresponds to the dimensionless overpressure of $17/100 = 0.17$. Here the ambient pressure is taken to be 100 kPa.
From charts, such as the one given in Fig. 12.3 derived from scaling law, the above value of dimensionless overpressure corresponds to a value of $R_S/M_{\text{TNT}}^{1/3} = 5$, where M_{TNT} is in kg. The equivalent mass of TNT is given as 1680 kg. Hence the distance over which the overpressure is greater than 17 kPa is $5 \times 1680^{1/3} = 59.4$ m.
People get knocked down upto a distance of about 60 m from the site of explosion.

12.3 *Boiler Explosion*: A high-pressure water boiler, used in a process plant for supplying hot water at a saturation temperature of 250 °C and a pressure of 4 MPa, bursts. The amount of hot water in the boiler at the moment of bursting is 10,000 kg.

(a) Find the mass of water that flashes into vapor?
  The ambient pressure is 100 kPa. You can use the following data from the steam tables: At 4 MPa:

$$\text{Saturation temperature} = 250\,^\circ\text{C}$$

$$\text{Enthalpy of saturated liquid} = 1087.31\frac{\text{kJ}}{\text{kg}}$$

$$\text{Enthalpy of saturated vapor} = 2801.4\frac{\text{kJ}}{\text{kg}}$$

At 100 MPa:

$$\text{Saturation temperature} = 100\,^\circ\text{C}$$

$$\text{Enthalpy of saturated liquid} = 417.46\frac{\text{kJ}}{\text{kg}}$$

$$\text{Enthalpy of saturated vapor} = 2675.3\frac{\text{kJ}}{\text{kg}}$$

(b) What is the energy released in the flash vaporization process and what is the equivalent mass of TNT on energy basis?

(c) If the yield from the above explosion is 2%, determine the overpressure at a distance of 100 m from the burst, given that for an ideal point explosion the relation between the non-dimensional overpressure ($\bar{p}$) and non-dimensional distance ($\bar{R}$) is given by

$$\bar{p} = e^{-0.24 \times \bar{R}}$$

where $\bar{p}$ = (overpressure/ambient pressure) and $\bar{R}$ = (distance R_S/Explosion length R_0).

Solution:

(a) **Mass of water flashing into vapor:**
Fraction of water which flashes to vapor

$$f = \frac{h_{f,4\text{MPa}} - h_{f,100\text{kPa}}}{h_{g,100\text{kPa}} - h_{f,100\text{kPa}}}$$

$$f = \frac{1087.31 - 417.46}{2675.3 - 417.46} = \frac{669.85}{2257.84} = 0.2967$$

Amount of water that flashes into vapor $= 0.2967 \times 10,000 = 2967$ kg.

(b) **Energy released in the flash vaporization process:**

$$\text{Energy released} = \text{Mass of flash vapor}$$
$$\times \text{Latent heat at the ambient pressure}$$
$$= 2967(h_{fg\,100\text{kPa}})$$
$$= 2967 \times 2257.84$$
$$= 6.699 \times 10^6 \text{ kJ}$$

$$\text{Equivalent mass of TNT} = \frac{6.699 \times 10^6 \text{ kJ}}{(4520\, kJ/kg)}$$
$$= 1482 \text{ kg of TNT}$$

We have taken the energy release per kg of TNT as 4520 kJ/kg.

(c) **Overpressure from the Explosion:** Since the yield of the flash vapor explosion is given to be 0.02, the energy which drives the blast wave is only 0.02 of the total energy release. The energy contributing to blast wave $= 0.02 \times 6.699 \times 10^6$ kJ $= 133.98 \times 10^3$ kJ

$$\text{The explosion length } R_0 = \left(\frac{E_0}{p_0}\right)^{1/3}$$
$$= \left(\frac{133.98 \times 10^6}{1 \times 10^5}\right)^{1/3} = 11.02 \text{ m}$$

The scaled distance $\bar{R} = R_S/R_0 = 100/11.02 = 9.07$. The non-dimensional overpressure from the given equation is

$$\bar{p} = e^{-0.24 \times \bar{R}} = e^{-0.24 \times 9.07} = 0.113$$

For the ambient pressure of 0.1 MPa the overpressure at a distance of 100 m is 0.0113 MPa, which is 11.3 kPa.

Nomenclature

C_V	Specific heat at constant volume [kJ/(kg K)]
D	Detonation velocity (m/s)
E_0	Source energy (J)
m	Mass (kg)
m_{TNT}	Mass of TNT (kg)
p	Pressure (Pa)
p_e	Pressure of explosive source (Pa)
p_0	Ambient pressure (Pa)
R	Specific gas constant [kJ/(kg K)]
R_E	Radius of charge (m)
R_S	Radius of shock wave (m)
R_0	Explosion length (m)
t	Time (s)
T	Temperature (°C)
γ	Specific heat ratio
τ	Duration of energy release (s)

Subscripts

B	Blast
E	Energy

Further Reading

1. Baker, W. E., *Explosions in Air*, University of Texas Press, Austin, 1973.
2. Boddurtha, F. T., *Industrial Explosion Prevention and Protection*, McGraw Hill, New York, 1980.
3. Crowl, D. A. and Louvar, J.F., *Chemical Safety: Fundamentals with Applications*, Prentice Hall, Englewood Cliffs, NJ, 2002.
4. Strehlow, R. A. and Baker, W. E., The Characterization and Evaluation of Accidental Explosions, NASA CR-134779., National Aeronautics and Space Administration, Cleveland, OH, June 1965.
5. Stull, D. R., *Fundamentals of Fire and Explosion*, AIChE Monograph Series, Vol. 73, No. 10, 1977.

Exercises

12.1 The explosion of an improvised explosive device (IED) consisting of a mixture of ammonium nitrate and fuel oil (ANFO) causes an overpressure of 10 kPa at a distance of 10 m from the site of the explosion. The energy release per kg of the mixture of ANFO is 2800 kJ/kg, and the yield of the explosion is 10%. Determine the mass of the mixture of ammonium nitrate and fuel oil in the explosive device.

You can assume the dependence of the dimensionless overpressure for an instantaneous point explosion, as obtainable from TNT, to be given in the range of the overpressure of interest by the expression

$$\bar{p} = e^{-1.2\frac{R_S}{R_0}}$$

In the above expression, R_S is the distance from the site of the explosion and R_0 the explosion length. $\bar{p}$ is the dimensionless overpressure. The energy release in the explosion of TNT is 4520 kJ/kg. The ambient pressure can be taken as 100 kPa.

12.2 An improvised explosive device (IED) comprising a 5 kg mixture of ammonium nitrate and a heavy fuel oil is detonated in a marketplace. The yield of the explosion is 0.25. Determine the TNT equivalence of the IED. You can assume the heat of combustion of TNT as 4600 kJ/kg while the heat of combustion of the particular mixture (ANFO) used is 6000 kJ/kg.

The ambient pressure can be taken as 100 kPa.

12.3 Copious amounts of propane vapor leak from a high pressure pipeline at a rate of 0.5 kg/s for 10 min before the leak is arrested. A flammable cloud of propane–air mixture is formed, and this flammable cloud encounters an ignition source and explodes about 15 min after the start of the leak. Assuming that the yield of the explosion is 0. 06 and the heat of combustion of propane to be 50,000 kJ/kg, calculate the equivalent mass of TNT from the explosion.

12.4 (a) A high-pressure storage vessel of volume 0.5 m³, containing a refrigerant at ambient temperature and a pressure of 20 MPa explodes at the ambient pressure. If the specific heat of the refrigerant is 1.1 and the yield of the explosion is 2%, determine the equivalent mass of TNT of the explosion? Energy release from 1 kg of TNT is 4520 kJ.

   (b) What would be the overpressure at a distance of 10 m from the center of the explosion given that for an idealized point explosion the relation between the non-dimensional overpressure $\bar{p}$ and non-dimensional distance $\bar{R}$ is given by

$$\bar{p} = e^{-0.24 \times \bar{R}}$$

Here $\bar{p}$ = overpressure/ambient pressure and $\bar{R}$ = distance R_S from energy source/explosion length R_0.

Chapter 13
Atmospheric Dispersion of Flammable Gases and Pollutants

Explosions involving gaseous fuels, vapors, dust, and condensed phase materials have been discussed in the earlier chapters. The release of fuel gas or vapor into the ambient from leaks in pipelines conveying gaseous fuel and volatile liquid fuels at high pressures and from spills of gaseous and liquid fuels from fuel storages vessels, oil rigs and refineries, and chemical process industries could form explosive gas clouds in the atmosphere. The rapid release of gases from pressurized containers and the bursting of high-pressure storage vessels, such as in a BLEVE explosion, would also release copious amounts fuel vapor, which disperse in the atmosphere. The fuel–air clouds in the atmosphere could either burn as a fireball or explode when they encounter an ignition source provided they are flammable or detonable.

The pollutants, and toxic products formed from uncontrolled chemical reactions in reactors, storage vessels, and in the confined and unconfined explosions would also disperse in the atmosphere, causing damage to the environment and living organisms. Glaring examples are the Bhopal gas tragedy in 1984, the Visakhapatnam gas leak in 2020, and the Great SMOG of London in 1952. The dispersion depends on the conditions of the atmosphere and is dealt with in this chapter.

13.1 Dispersion of Gases Released in the Atmosphere

Dispersion of gases in the atmosphere is strongly governed by the temperature distribution in it. The temperature distribution is governed by the radiant heating of the earth by the sun, the winds, and cloud cover in the different seasons. A basic understanding of the temperature distribution in the atmosphere as influenced by the different parameters is therefore first considered followed by the dispersion of gases in it.

13.1.1 Temperature Distribution Above the Surface of Earth

The variation of temperature with height above the surface of the Earth under normal atmospheric conditions is shown in Fig. 13.1. Here the temperature of air is shown in °C on the X-axis while the altitude in km is shown on the Y-axis. The temperature initially decreases as the altitude increases. The decrease in temperature of the air above the earth is due to the surface of the earth getting directly heated by the radiation from the sun and thereafter the heated earth's surface conducting away the heat to the air above it. The region above the Earth in which the temperature decreases with height is known as troposphere and extends to an altitude between 10 and 15 km as shown in Fig. 13.1.

The incoming solar radiation is known as insolation (INcoming SOLAr radia-TION). Insolation is higher when the sun shines vertically above such as during mid-day. It decreases with increasing inclination of the sun from the vertical. The inclination of the sun changes during the different seasons. The rate of temperature decrease (dT/dz) with height, where T is the temperature and z is the height is known as the lapse rate. The lapse rate varies with the moisture content in the air. For dry air, the lapse rate is as high as 9.8 °C/km, while for moist air the lapse rate is about 5 °C/km. The lower rate for moist air results from the condensation of the vapor as the latent heat is evolved in the process of condensation.

Above the troposphere is stratosphere, in which the temperature increases from the lower value of about −50 °C to about 7 °C (Fig. 13.1). The stratosphere extends from 10–15 km to about 50 km. The increase of temperature in the stratosphere is from the

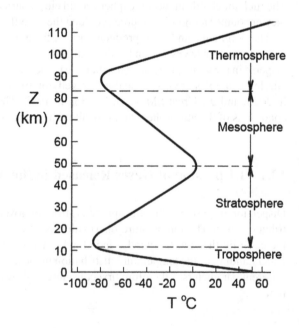

Fig. 13.1 Variation of temperature with altitude

absorption of solar radiation by ozone present in it. The heating of ozone provides circulation in the stratosphere in a manner similar to the wind in the troposphere.

Over a height of 25–40 km above the stratosphere, the temperature drops again to a minimum value of about −90 °C due to the non-availability of gas molecules, which can absorb solar radiation. This region of decreasing temperatures is known as mesosphere. The mesosphere is followed by the thermosphere in which the temperature monotonically increases from the photochemical recombination reaction of oxygen molecules with oxygen atom ($O_2 + O$), which liberates heat. The thermosphere extends up to about 500 km and reaches temperatures as high as 1200 °C. The high temperatures from the recombination reactions produce a display of colors known as 'aurora.' The aurora is visible in countries in northern latitudes.

The ions formed in the thermosphere reflect electromagnetic waves and are useful for radio communication. The layer of these ions is known as 'ionosphere.' The extent of ionosphere depends on the solar activity.

The region between the troposphere and stratosphere is known as tropo-pause, while the region between the stratosphere and mesosphere is known as strato-pause. The low temperature in the tropo-pause condenses water vapor to ice.

13.1.2 Wind, Air Currents, and Clouds

The differences in the rate of heating of air above the earth's surface causes the air to move horizontally. The horizontally moving air is called as wind. Wind speeds over the surface of earth vary. Wind speeds generally increase with height. The movement of air in the vertical direction is known as air current.

The presence of clouds in the near region in the troposphere influences the heating of the earth. Clouds absorb heat radiation and also emit radiation. The wind over the surface of the earth and the presence of clouds influence the heating of earth.

Wind is not a steady one-dimensional flow. It is attended with gusts and comprises of eddies. The wind is therefore generally treated to be turbulent, with turbulence arising from interaction with objects on the ground and the thermal drive.

13.1.3 Temperature Inversion

Instead of temperature above the surface of the earth decreasing with height as shown in Fig. 13.1, the temperature sometimes increases with altitude in the region near the surface. This is indicated in a small region of the troposphere over a height of about 500 m, above the surface of the earth in Fig. 13.2. An increase of temperature with height occurs when the ground cools down more rapidly such as at night and during winters when the air above it is still warm. The increase of temperature with height is known as temperature inversion.

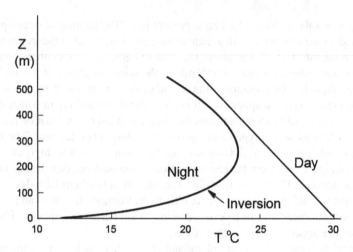

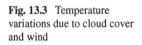

Fig. 13.2 Temperature variations with height and inversion

Fig. 13.3 Temperature
variations due to cloud cover
and wind

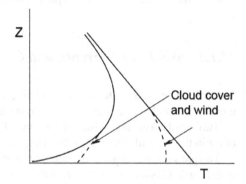

The effect of the wind is to bring down the magnitude of the gradient dT/dz near the surface whether temperature inversion is present or not. Similarly, the existence of a cloud is to reduce the insolation during the day and thereby decreases the value of dT/dz. The thermal radiation from the cloud during the night would enhance the surface temperature and reduce the positive gradient dT/dz associated with temperature inversion. Figure 13.3 illustrates the shift of the temperature profile due to wind and cloud cover for both positive and negative gradients of dT/dz. As the wind velocity and cloud cover increase, the temperature inversion could cease altogether.

13.2 Escape of Gas Released in the Atmosphere: Atmospheric Stability

The stratification of temperature and hence the change of density in the atmospheric layer above the ground causes the gas released into it to either escape as a vertical current or else accumulate and disperse over the ground. If we consider a parcel of gas of volume V, density ρ_p in atmosphere of density ρ_a, the net buoyancy force experience by the parcel is $(\rho_a - \rho_p)Vg$ (Fig. 13.4).

The buoyant acceleration α of the parcel therefore is

$$\alpha = \frac{(\rho_a - \rho_p)Vg}{\rho_p V} \tag{13.1}$$

where $\rho_p V$ is the mass of the parcel of gas of volume V. The acceleration is

$$\alpha = \frac{(\rho_a - \rho_p)g}{\rho_p} \tag{13.2}$$

When the parcel of gas has the same molecular mass as the air but is at a different temperature, the acceleration is given by $[(T_p - T_a)/T_a]g$. Here T_a is the temperature of air and T_p is the temperature of the parcel in Kelvin.

A parcel of gas encounters a lower temperature as it moves up when the temperature gradient is negative. The upward acceleration is positive with the buoyancy force further accelerating the parcel upwards. When the temperature gradient is positive, as under the condition of temperature inversion, the parcel of gas encounters a higher temperature, which causes a negative acceleration and the parcel is pushed back vertically downwards.

The atmosphere is said to be *stable* when the buoyant acceleration is negative and causes the parcel of gases vertically displaced back to its original point. However, if the acceleration is positive, the parcel keeps getting displaced further vertically upwards and the parcel escapes from the ground. This condition of the atmosphere, which gives the positive acceleration to the parcel of gas is called as *unstable* atmosphere. When there is no acceleration, i.e., buoyancy force equals the weight of the parcel, the stability of atmosphere is said to be *neutral*. In summary, the atmosphere is stable, neutral, or unstable according to whether the buoyant acceleration is

Fig. 13.4 Buoyant force $\rho_a Vg$ and weight of parcel $\rho_p Vg$

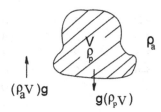

negative, zero, or positive, respectively. A positive lapse rate dT/dz is associated with the stable condition, a negative dT/dz with the unstable and $dT/dz = 0$ for the neutral stability.

13.2.1 Pasquill's Classification of Atmospheric Stability

The insolation is a maximum when the sun is vertically up and gives a negative lapse rate dT/dZ. In the absence of insolation, such on a clear winter night the lapse rate dT/dZ is positive. The value of dT/dZ, as discussed earlier, is influenced not only by the level of insolation but also by the wind velocity, turbulence in the atmosphere, humidity, and the cloud cover over the ground. A pioneering meteorologist, Pasquill proposed six discrete classes of atmospheric stability ranging from class A to class F depending on the insolation, wind speed, and cloud cover to determine the dispersion of gas release in the atmosphere. Class A represents about the most unstable condition with very strong insolation (sun shining vertically on a clear summer day) and a very small wind speed. Class F represents the most stable condition such as a winter night with clear sky and low wind speed. Class C is near to neutral. The classification of the stability based on insolation, wind speed, and cloud cover is given in Table 13.1.

Strong insolation refers to mid-day on a clear sunny day in summer while moderate refers to the winter sun. Night is considered for the period an hour before sunset and one hour after sunrise.

Within the troposphere the region of interest for the dispersion is the lowest part adjacent to the ground. This region extends to a height about one km over the ground and is known as the atmospheric boundary layer (ABL). The atmospheric stability conditions in ABL would govern the escape or retention of a gaseous release. The dispersion of gases in the atmosphere would depend on its stability that is governed by multiple parameters like the insolation, wind, cloud cover, temperature inversion, etc. However, before proceeding with the above, we need to address the dispersion or diffusion of a gas in a given medium.

Table 13.1 Pasquill's stability classification

Surface wind (m/s)	Insolation			Night	
	Strong	Moderate	Slight	Overcast with cloud cover >1/2	Cloud cover <3/8
<2	A	A-B	B	E	F
2–3	A-B	B	C	E	F
3–5	B	B-C	C	D	E
5–6	C	C-D	D	D	D
>6	C	D	D	D	D

13.3 Dispersion

13.3.1 Diffusion of Gas in One Dimension

Consider the dispersion of gas A in a single dimension along the X-axis in a medium of gas B. The rate of diffusion of mass is given by the Fick's law of diffusion, which states that the diffusing mass flux is proportional to its concentration gradient and takes place in the direction of decreasing gradient. Denoting the mass flux by $\dot{m}''$ and the concentration by C, we have

$$\dot{m}''_A = -D_{AB} \frac{\partial C_A}{\partial x} \tag{13.3}$$

The mass flux of gas $A(\dot{m}''_A)$ has units of kg/(m^{2}s), and the concentration is given as kg/(m^3). D_{AB} is the diffusion coefficient of gas A in gas B and has units of m^2/s. If the concentration is specified in moles/m^3 as in Chap. 4, the diffusion coefficient has to be multiplied by the molecular mass M kg/mole. Due to the diffusion, the concentration changes with time, and for one-dimensional diffusion, the variation of concentration is given by the partial differential equation

$$\frac{\partial C}{\partial t} = D \frac{\partial^2 C}{\partial x^2} \tag{13.4}$$

Here the suffix for the diffusing gas A and for the gas B in which it diffuses is not given for generality. If the concentration C at time $t = 0$ and $x = 0$ is C_i, the concentration C at any time t and distance x is obtained from solution of Eq. 13.4 subject to the initial and boundary conditions as (given in Appendix G):

$$C(x, t) = C_i - \frac{2C_i}{\sqrt{\pi}} \int\limits_{0}^{x/\sqrt{Dt}} \exp\left(-\frac{x^2}{4Dt}\right) d\left(\frac{x}{\sqrt{Dt}}\right) \tag{13.5}$$

where $x^2/(Dt)$ is a non-dimensional parameter characterizing the diffusion process with time and distance x. The variation of the concentration as a function of the non-dimensional parameter $x^2/(Dt)$ is shown in Fig. 13.5.

13.3.2 Standard Gaussian Distribution

The standard Gaussian distribution is given by

$$p(w) = \frac{1}{\sqrt{2\pi}\sigma_w} \exp\left(-\frac{w^2}{2\sigma_w^2}\right) \tag{13.6}$$

Fig. 13.5 Variation of
concentration with distance
and time

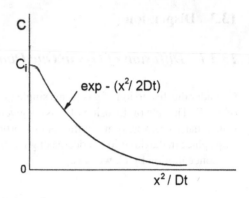

Fig. 13.6 Standard
Gaussian distribution

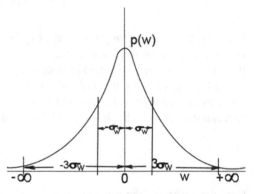

where p is the probability density function, w is a dummy variable, and σ_w is the standard deviation. The probability density function $p(w)$ is a real positive value, whose integral over the variable w in the entire range is 1, i.e.,

$$\int_{-\infty}^{\infty} p(w)\mathrm{d}w = 1 \qquad (13.7)$$

13.3.3   Significance of Standard Deviation in the Standard Gaussian Distribution

The standard deviation σ_w is a measure of the width of the probability density function. The square of standard deviation denotes the variance σ_w^2 and is given by

Fig. 13.7 Small and large
values of standard deviation
σ

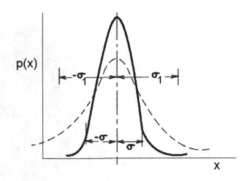

$$\sigma_w^2 = \overline{w'^2} = \int\limits_{-\infty}^{\infty} w'^2 p(w)dw \tag{13.8}$$

Here, w' is the fluctuation in the value of the variable w. If $\bar{w}$ is the mean, $w = \bar{w} + w'$.

The standard deviation σ_w is a measure of the variation from the mean and is shown in Fig. 13.6. Values of the standard deviation of three times the value of σ_w are also shown in the figure.

When the standard deviation σ is small, the dispersion of the values about the mean is small. Figure 13.7 illustrates small and large values of the standard deviation. Here the dummy variable is x and the probability distribution function is denoted by $p(x)$. A small value of σ, shown for the distribution indicated by the dark line, suggests the deviations to be small and near to the mean. When the standard deviation is larger, as shown by σ_1 for the wider distribution indicated by dotted line, the dispersion about the mean is larger.

The area between $-\sigma$ to $+\sigma$ is 68.3% of the total area under the probability distribution, which is unity. If the deviation of three times the standard deviation, viz. 3σ is considered, as shown in Fig. 13.6, the area between -3σ to $+3\sigma$ is 99.7% of the total area. This is shown in Fig. 13.8 for values of σ, 2σ and 3σ.

13.3.4 Comparison of the Diffusion of Concentration in One Dimension with the Standard Gaussian Distribution

The equations describing the diffusion of the concentration given by Eq. 13.5 (Fig. 13.5) and the Gaussian distribution of Eq. 13.6 (Fig. 13.6) are similar. As the mass gets diffused, its concentration decreases just as the probability function decreases with increase of the standard deviation σ. Comparing Eqs. 13.5 and 13.6 also implies that $\sigma^2 = Dt$. The concentration at a distance x from a source can therefore be expressed as

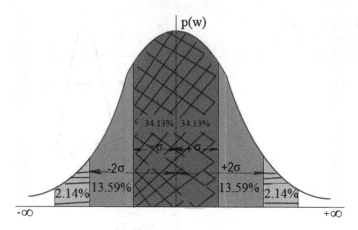

Fig. 13.8 Standard deviation σ, 2σ and 3σ

$$C(x) = \frac{Q}{\sqrt{2\pi}\sigma} \exp\left(-\frac{x^2}{2\sigma^2}\right) \tag{13.9}$$

Here Q is the initial mass release at $t = 0$ and $x = 0$, and for the one-dimensional diffusion, is equal to

$$\int_{-\infty}^{\infty} C(x)\mathrm{d}x = Q \tag{13.10}$$

This is in analogy to $\int_{-\infty}^{\infty} p(w)\mathrm{d}w = 1$ given by Eq. 13.7. The initial peak in concentration $C(0)$ or C_i at $x = 0$ and $t = 0$ gets increasingly diffused with larger values of σ as x and t increases.

Initially, the dispersion is small and the concentration is high and localized at the source $x = 0$. As time increases, the concentration decreases and gets diffused over a larger distance x. The dispersion is therefore similar to the standard distribution as was observed with increase of σ.

The fluctuations in the case of atmospheric dispersion originate from the turbulent fluctuations in the atmosphere. Over the mean wind velocity $\bar{u}$, there is an unsteady component u', giving $u = \bar{u} + u'$. The root mean square value of the fluctuation $\sqrt{\bar{u'^2}}$ is the standard deviation. The initial peak in concentration $C(0)$ or C_i at $x = 0$ and $t = 0$ gets increasingly diffused with larger values of σ as x and t increases.

Standard deviations can be readily measured in practice unlike the diffusion coefficient D. Predictions are therefore made for the concentration $C(x)$ using Eq. 13.9 and the measured values of standard deviations.

13.4 Dispersion in Three Dimensions of a Mass of Gas Release

Extrapolating to three dimensions, in which the mass would spread along x, y, and z, and denoting the concentration at a point x, y, and z from the source as $\chi(x, y, z)$, instead of C in Eq. 13.9, we get

$$\chi(x, y, z) = \frac{Q}{(2\pi)^{3/2}\sigma_x\sigma_y\sigma_z} \exp\left[-\frac{x^2}{2\sigma_x^2} - \frac{y^2}{2\sigma_y^2} - \frac{z^2}{2\sigma_z^2}\right] \qquad (13.11)$$

Here the Cartesian coordinate system is used. σ_x, σ_y, and σ_z are the standard deviations in the $x-$, y-, and $z-$ directions. The instantaneous release of Q kg forms a cloud or puff, which expands out with time and distance and disperses by atmospheric turbulence as shown in Fig. 13.9. However, if there is wind blowing in the X-direction with a velocity u, we have in the frame of reference of the wind the concentration at x, y, z and at time t as (Fig. 13.9)

$$\chi(x, y, z, t) = \frac{Q}{(2\pi)^{3/2}\sigma_x\sigma_y\sigma_z} \exp\left[-\frac{(x-ut)^2}{2\sigma_x^2} - \frac{y^2}{2\sigma_y^2} - \frac{z^2}{2\sigma_z^2}\right] \qquad (13.12)$$

Based on the mass conservation Eq. 13.10, we get at any time t

$$\int_{\infty}^{\infty}\int_{\infty}^{\infty}\int_{\infty}^{\infty} \chi(x, y, z, t)\mathrm{d}x\mathrm{d}y\mathrm{d}z = Q \qquad (13.13)$$

If the mass release is expressed in kg (Q kg), the unit of concentration is seen to be χ kg/m^3 as is to be expected.

The concentration at the given distance and at a given time can therefore be determined provided the standard deviation of the concentration is known. If the release of mass Q kg is at a height h above the ground (such as from a chimney) and the earth reflects the vertical component of mass incident on it vertically upwards (i.e., along z), Eq. 13.12 gets modified for the incident and reflected concentration as

Fig. 13.9 Instantaneous mass release forming a puff

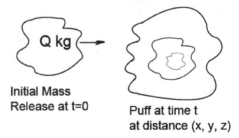

Initial Mass Release at t=0

Puff at time t at distance (x, y, z)

$$\chi(x, y, z, t) = \frac{Q}{(2\pi)^{3/2}\sigma_x\sigma_y\sigma_z} \exp\left[-\frac{(x-ut)^2}{2\sigma_x^2} - \frac{y^2}{2\sigma_y^2}\right]$$

$$\times \exp\left[-\frac{(z-h)^2}{2\sigma_z^2} - \frac{(z+h)^2}{2\sigma_z^2}\right] \qquad (13.14)$$

Here, the distance traveled vertically along z by the release is $(z - h)$. The reflection of the release from the ground travels the distance $[(z - H) + H + H] = z + H$, and the contribution from both are added as shown in Eq. 13.14.

13.5 Dispersion of Steady Release: Plume

A steady release at the rate of $\dot{Q}_S$ kg/s in the atmosphere having wind velocity u (m/s) is different from the instantaneous release of Q kg, considered in Eqs. 13.12 and 13.14. In a steady release, the time over which the release takes place is much larger than the time taken for the release to reach the given location. The release is transported along the wind as a plume as shown in Fig. 13.10. The plume spreads out in the lateral and vertical directions and is illustrated for mass release from a chimney or a stack in Fig. 13.11. The dimension of the plume increases in the shape of a diverging cone. Here the height at which the steady mass flow rate is released is denoted by h. H denotes the maximum height reached by it due to its vertical momentum before the wind velocity begins to diffuse it. As the plume is blown by the wind in the direction x, it diffuses in the direction normal to the wind (y) and the vertical (z-direction) as illustrated in Fig. 13.11. The wind direction is along x. The diffusion in the $y-$ and z-directions is mainly by atmospheric turbulence, and we can assume a Gaussian distribution in the y- and z-directions. The atmospheric fluctuations in the $y-$ and z-directions are given by v' and w' with standard deviations being $\sqrt{\overline{v'^2}}$ and $\sqrt{\overline{w'^2}}$.

The value of the steady mass release rate $\dot{Q}_S$ kg/s from the continuous source is conserved along the plume in the direction of the wind blowing at speed u along x (Fig. 13.11) to give

Fig. 13.10 Steady mass release forming a plume

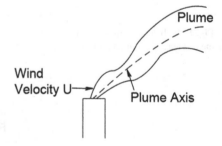

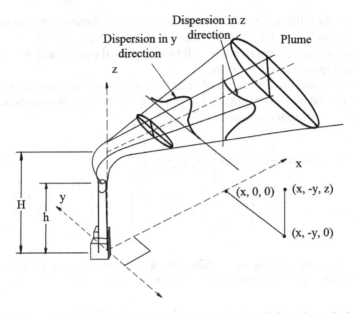

Fig. 13.11 Concentration profiles lateral to the direction of wind and along the vertical

$$\int\limits_{\infty}^{\infty}\int\limits_{\infty}^{\infty} \chi u \, dy dz = \dot{Q}_S \qquad (13.15)$$

Here χ is the concentration at (x, y, z) in kg/m^3. This can be seen if we were to consider the frame of reference at (x, y, z) as moving at the wind speed u. Being a steady mass release, the concentration at a given point does not change with time.

The effect of the wind speed is to reduce the concentration along the wind. The wind also provides a conical shape to the plume.

For constant wind speed u, Eq. 13.15 reduces to

$$\int\limits_{\infty}^{\infty}\int\limits_{\infty}^{\infty} \chi \, dy dz = \dot{Q}_S/u \qquad (13.16)$$

The concentration χ at any distance (x, y, z) is determined by considering the wind direction to be along the x-axis and the diffusion to be in the lateral y- and the vertical z-directions. Following Eqs. 13.11 and 13.13, it becomes

$$\chi(x, y, z) = \frac{\dot{Q}_S}{2\pi u \sigma_y \sigma_z} \exp\left[-\frac{y^2}{2\sigma_y^2} - \frac{z^2}{2\sigma_z^2}\right] \qquad (13.17)$$

In this case, the diffusion in the y-direction is lateral to the direction of the wind while the z-direction is vertically upwards. The release takes place at $x = y = z = 0$.

The above equation would not be valid for very small wind speeds for which the expanding plume may not be formed.

When the release is from a stack of effective height H and the plume gets reflected from the ground, we have following Eq. 13.14 the steady value of concentration at a location (x, y, z) from the source as

$$
\chi(x, y, z; H) = \frac{\dot{Q}s}{2\pi u} \frac{1}{\sigma_y} \exp\left[-\frac{1}{2}\left(\frac{y}{\sigma_y}\right)^2\right]
$$

$$
\times \frac{1}{\sigma_z}\left\{\exp\left[-\frac{1}{2}\left(\frac{z-H}{\sigma_z}\right)^2\right] + \exp\left[-\frac{1}{2}\left(\frac{z+H}{\sigma_z}\right)^2\right]\right\}
$$

(13.18)

The steady concentration along the centerline of the plume (i.e., $y = 0$) at the surface of the ground ($z = 0$) from the above equation becomes

$$
\chi(x, 0, 0, H) = \frac{\dot{Q}s}{\pi u \sigma_y \sigma_z} \exp\left[-\frac{H^2}{2\sigma_z^2}\right]
$$

(13.19)

13.6 Dispersion Coefficients

We have been able to determine the concentrations as a function of the standard deviations σ_x, σ_y, and σ_z. The standard deviations are known as dispersion coefficients. They must be known for predicting the concentrations for both the steady and spontaneous releases. They are determined by measurements in field experiments for steady plumes and puffs as function of distance (x) along the wind for different conditions of atmospheric stability. In this way, the parameters of the atmosphere such as insolation, wind, and cloud cover are taken care of in the predictions. Planned experiments are done, and based on the measurements of concentration in the lateral and vertical directions at different downstream locations along the wind for a given arbitrary release, the distribution of concentration along the centre line and the cross wind locations are found and the values of the standard deviations σ_y and σ_z are determined. These coefficients depend strongly on the atmospheric stability conditions. The values of the dispersion coefficients σ_x and σ_y for instantaneous release are the same. In the case of steady plume, the transport in the x-direction (along the wind) is by the wind and the dispersion coefficient σ_x is not required.

The site, whether it an open space or a place with tall multistoried buildings, would additionally influence the dispersion coefficients since the buildings introduce blockages and control the turbulence and mixing. The dispersion coefficients are therefore given separately for rural and urban areas. Rural settings presume an open space while urban imply considerable blockages. The sampling periods used for the measurement of the concentration for the steady-state releases are typically the average values taken between 10 and 90 min. The instantaneous concentration would be higher, and a factor of 2 is assumed for determining the instantaneous concentration. The instantaneous concentration along the center-line of a plume from Eq. 13.19 is therefore given as

$$\chi_i(x, 0, 0, H) = \frac{2\dot{Q}_s}{\pi u \sigma_y \sigma_z} \exp\left[-\frac{H^2}{2\sigma_z^2}\right] \tag{13.20}$$

Values of diffusion coefficients σ_y and σ_z for continuous release (plume) and σ_x, σ_y, and σ_z for instantaneous release (puff) are available in the form of charts and equations in books on atmospheric dispersions. The variations σ_y and σ_z as a function of x are given separately for rural and urban conditions for the six different conditions of atmospheric stability A to F. Figure 13.12 illustrates the typical trends of the dispersion coefficients of σ_y and σ_z as a function of downwind distance x for a steady release as the atmospheric stability conditions change between class A and class F. Typical values of the dispersion coefficients for different atmospheric stability conditions in open and urban environments are given in Tables 13.2 and 13.3, respectively, as a function of the distance downwind (x) in meters.

Fig. 13.12 Trend of variation of σ_y and σ_z with distance downwind from release

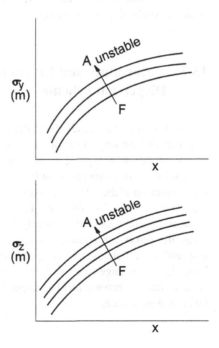

Table 13.2 Dispersion coefficients for continuous release (rural area)

Atmospheric stability	σ_y (m)	σ_z (m)
A	$0.22x(1 + 0.0001x)^{-1/2}$	$0.20x$
B	$0.16x(1 + 0.0001x)^{-1/2}$	$0.12x$
C	$0.11x(1 + 0.0001x)^{-1/2}$	$0.08x(1 + 0.0002x)^{-1/2}$
D	$0.08x(1 + 0.0001x)^{-1/2}$	$0.06x(1 + 0.00055x)^{-1/2}$
E	$0.06x(1 + 0.0001x)^{-1/2}$	$0.03x(1 + 0.0003x)^{-1/2}$
F	$0.04x(1 + 0.0001x)^{-1/2}$	$0.016x(1 + 0.0003x)^{-1/2}$

Table 13.3 Dispersion coefficients for continuous release (Urban area)

Atmospheric stability	σ_y (m)	σ_z (m)
A-B	$0.32x(1 + 0.0004x)^{-1/2}$	$0.24x(1 + 0.001x)^{-1/2}$
C	$0.22x(1 + 0.0004x)^{-1/2}$	$0.20x$
D	$0.16x(1 + 0.0004x)^{-1/2}$	$0.14x(1 + 0.0003x)^{-1/2}$
E-F	$0.11x(1 + 0.0004x)^{-1/2}$	$0.08x(1 + 0.0005x)^{-1/2}$

A workbook on use of the scheme along with the dispersion plots has been issued by the Environmental Protection Agency in the USA. These are also available in standard textbooks on atmospheric dispersion and pollution.

13.7 Uncertainty and Errors in the Gaussian Dispersion Scheme

The direction of the wind would decide the x-axis, and small errors in the wind direction and the magnitude of the wind will lead to changes in the predicted values of concentration at a given point. The assumption of a constant wind speed and constant value of atmospheric turbulence and stability is not expected to hold over a given duration of the release. Losses such as due to washout or removal by reactions are not accounted for. Vertical and crosswinds, which lead to lateral and vertical convection, are neglected. Similarly in the presence of buildings, such as in urban areas, the wakes and eddies, which cause mixing are not adequately included. The dispersion coefficients are averages determined between 10 min and more than an hour. As the average time of the measurements increases, the measured values of concentrations decrease and the dispersion coefficients, so determined, are likely to be an underestimate.

13.8 Buoyant and Heavy Gases

The above discussions on dispersion are valid when the gases dispersed by the wind have the same density as the atmospheric air in which it is released. When a buoyant gas, such as hydrogen, is released from a stack velocity with a velocity w, the initial motion of the gas is determined by its momentum and buoyancy. When the effect of the initial momentum ceases, the buoyancy force further causes the gas release to rise.

The rise of the gas before the dispersion can be segregated into two phases–initial momentum and buoyancy. The extent of these momentum- and buoyancy-related rise depends on the release velocity w (m/s) and the density of the buoyant gas (ρ_g) and the stack diameter (d). A densimetric Froude number (Fr_D) is defined as

$$Fr_D = \frac{w}{\left(dg(|\rho_g - \rho_a|)/\rho_a\right)^{1/2}} \tag{13.21}$$

Here w is the vertical release velocity in m/s from the stack of diameter d, ρ_g is the density of the gas, ρ_a is the density of the ambient, and g is the acceleration due to gravity. The buoyancy term in the denominator has dimensions of m/s.

For heights above the stack (z_h) such that

$$\frac{z_h}{d} \leq 1.1 Fr_D \tag{13.22}$$

the initial momentum from velocity w dominates the vertical rise. When the value of z_h/d is much larger than the densimetric Froude number, i.e.,

$$\frac{z_h}{d} > 5.6 Fr_D \tag{13.23}$$

the buoyancy forces from the low gas density governs its motion. The interaction of the vertical rise velocity (from the initial momentum and the buoyancy force) with the wind velocity increases the effective height of the stack from h to H as shown in Fig. 13.13. The location of the effective height of release is also shifted downwind by distance x_f as shown in Fig. 13.13.

The dispersion in the vertical direction and in the lateral direction with respect to the wind is initiated after the plume reaches the effective release height H. The buoyancy effects thereafter are not generally influential, and the cloud gets

Fig. 13.13 Effective location of release

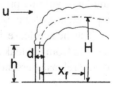

sufficiently diluted so as to be treated as from an ambient air cloud. When the density of the gas release from the stack is heavier than the surroundings, the initial rise from momentum is followed by a decrease in the effective height of the release. The reader is referred to standard texts on atmospheric dispersion and air pollution for details on the dispersion of buoyant and heavy gases.

In spite of the different sources of error and the approximate nature of the above prediction procedure, the method is very useful. This is because the different parameters of the atmosphere, though qualitative, are considered in the predictions, and this would be difficult in numerical computations.

13.9 Major Explosions and Pollutions Involving Atmospheric Dispersion

13.9.1 Dispersion of Propane in the Explosion at Port Hudson in December 1970

The dispersion of propane due to the burst of a pipeline conveying liquid propane at a pressure of 6.5 MPa caused an explosion. The accident happened as detailed in the following.

A weld joint of a 20 cm diameter pipeline, 40 years old, failed and high-pressure liquid propane at the rate of about 3.5 m³ per minute spilled from the pipeline, resulting in the formation of a fountain with a height of 25 m of propane. The incident occurred at 10:20 P.M. on a cold winter day (December 9, 1970) at Franklin, Port Hudson, Missouri, USA.

The pipeline was on a hillside, and there was a slight wind downhill at a speed of about 2.7 m/s. Being a cold winder night with clear sky, the low-wind speed provided a stable atmosphere at condition F. The propane fog and vapor were sheltered in the region above the ground. The higher density of propane compared to air also contributed to it being at the ground level. The dispersion of propane downwind mixed with air, forming a flammable concentration.

At a distance of 330 m downwind from the spill, the propane–air mixture entered a building. Here the mixture encountered an ignition source (hot compressor/spark in a refrigerating unit) and was ignited. The presence of blockages within the confined room brought about the transition of the flame to a detonation.

The detonation within the building got transmitted to the propane–air cloud outside the building by shattering the glass panes and doors. It triggered the detonation of the large volume propane–air cloud (about 30,000 m³) in the open space. The TNT equivalence of the explosion based on the observed damages was about 23,000 kg.

The heat release in the explosion caused the burning of the fuel-rich propane issuing from the leak and formed a huge fireball about 30 m high. The fire at the zone of the release persisted for about 12 h.

13.9.2 Dispersion of Pollutant in the Great Smog of London, December 1952

December of 1952 was particularly cold period in London. The residents burnt more coal than usual to keep them warm. A fog comprising of smoke, known as smog, got formed.

Being winter, the insolation was small. There was almost no wind during the period December 3 to 9. This gave rise to a very stable atmospheric condition (about 'F' in the Pasquill's classification). The pollutants from the burning of coal could not therefore escape upwards and dispersed over the ground and diffused into the buildings. The use of low-grade coal containing sulfur increased the pollution. A total of about 12,000 people died by inhaling the poisonous fumes. Their death was attributed to a toxic content exceeding about 10 parts per million of sulfur dioxide and the smoke. We shall consider the dose for quantifying damages in the next chapter.

13.9.3 Dispersion of Poisonous Gases from Runaway Chemical Reaction in a Chemical Plant at Bhopal (December 2–3, 1984)

Poisonous methyl isocyanide (MIC) was an intermediate in the manufacture of a pesticide 'carbonyl' in a chemical plant at Bhopal, India. About 43 tons of liquid MIC was stored in a tank. The inadvertent ingress of water from cleaning of pipelines and valves introduced water in the tank. The hydrolysis of MIC is exothermic as per the chemical reaction.

$$CH_3CNO + H_2O \rightarrow \quad CH_3NH_2 \quad + \quad CO_2 \qquad (13.24)$$
$$\text{(MIC)} \qquad \text{(water)} \quad \text{(monomethyl amine)} \quad \text{(carbon dioxide)}$$

In this reaction, monomethyl amine and CO_2 are formed. The CO_2 increased the pressure in the tank, while the exothermicity of the reaction enhanced the temperature. The increase of temperature further enhanced the speed of the reaction. A diaphragm-motorized valve did not relieve this initial pressure buildup. The tank had the provision for active cooling using a refrigerating unit. However, this was not working at the time of the accident. The rate of temperature rise could therefore not be controlled, and the rise in temperature was not noticed by the operating staff as the temperature alarm was not reset.

The monomethyl amine formed by reaction given in Eq. 13.23 reacted with MIC to form dimethyl urea (DMU) and the dimethyl urea further reacted with MIC to form trimethyl buret(TMB). The reactions are given below:

$$CH_3CH_2 \quad + CH_3CNO \rightarrow CH_3NHCONHCH_3 \qquad (13.25)$$
$$\text{(monomethyl amine)} \quad \text{(MIC)} \qquad \text{(DMU)}$$

$$CH_3NHCONHCH_3 + CH_3CNO \rightarrow CH_3NHCOCNONHCH_3 \qquad (13.26)$$

$$\text{(DMU)} \qquad \text{(MIC)} \qquad\qquad ||$$

$$CH_3\text{(TMB)}$$

The above two reactions, being exothermic, further increased the temperature and the rate of the hydrolysis reaction (Eq. 13.24).

It was suspected that due to inadequate purification during the distillation process, the MIC in the tank was probably contaminated with a chlorine containing compound such as chloroform ($CHCl_3$). About 9 kg of metal, found in the residue of the tank, suggested substantial corrosion of the tank caused by chlorine. The presence of iron catalyzes the reaction (termerization) of MIC to give MIC trimer.

$$3CH_3CNO \rightarrow MIC\,Trimer \qquad (13.27)$$

The initial hydrolysis (Eq. 13.24) followed by the formation of DMU and TMB (Eqs. 13.25 and 13.26) and the termerization (Eq. 13.27) led to a runaway in temperature and pressure. The high pressure activated the safety relief valve and the rupture disk, and a two-phase flow of the boiling solution containing about 40–50% of MIC at a pressure of about 1.35 MPa and 200 °C got released at the stack. The gas scrubber meant for washing out the poisonous MIC gas with sodium hydroxide and the flare tower to burn the release were not functioning. The last safety feature of a water curtain could not be sufficient considering the large release of the gas. About 40 tons of MIC got released over the duration of $1^{1}/_{2}$ hours starting from 00:30 hrs.

The ambient atmosphere at midnight of December was cold with a clear sky and low-wind speed of 2.9 m/s. This provided a condition of relatively stable atmosphere. The heavier-than-air MIC traversed along the wind and dispersed laterally. The concentration of the poisonous MIC was sufficiently high to kill people till a downwind distance of about 3 to 4 km. The precise nature of the release was not immediately available to suggest a corrective remedy and save people.

13.9.4 Dispersion of Styrene Vapor in Atmosphere in the Visakhapatnam Gas Leak, India (May 7, 2020)

Styrene vapor leaked from a storage tank at a plastic and polymer plant in Visakhapatnam at 2 a.m. in the morning on May 7, 2020. Styrene at ambient temperatures is a volatile liquid. It is based on benzene with a molecular formula C_6H_6-C_2H_2 and is used in the manufacture of polystyrene, plastics, and high molecular mass polymers.

1800t of liquid styrene was contained in two storage tanks at the plant. Liquid styrene has a density of 900 kg/m^3 and is lighter than water. Its vapor pressure at the standard temperature of 25 °C is 1.08 kPa, which is about 1/100 of the atmospheric pressure of 100 kPa. The vapor of styrene is flammable with a minimum oxygen concentration (MOC) of 8% by volume, viz. 8% volume styrene by the volume of the mixture with air. Based on discussions of MOC and flammability limits in

Chap. 7, the lean flammability limit would be MOC/ν, where ν is the stoichiometric coefficient. The equation for the stoichiometric reaction of styrene with oxygen is $C_6H_6C_2H_2 + 10O_2 = 8CO_2 + 4H_2O$ giving $\nu = 10$. The lean limit of flammability of styrene vapor in air is therefore 0.8% volume styrene in the styrene–air mixture.

The concentration of styrene in air formed from the leak at the ambient temperature of 25 °C is $1.08/100 = 1.08\%$, where 1.08 is the vapor pressure in kPa. The leak is near to the threshold of lean flammability limit of 0.8 and could therefore form a flame if an ignition source is present. Since it was 2 a.m. in the morning, there was no activity at the plant or near by with no ignition source being present and no explosion took place.

The boiling temperature of liquid styrene at atmospheric pressure is 145 °C. If we assume the ambient temperature to be about 30 °C, and the liquid styrene to be at the ambient temperature, it cannot boil. But liquid styrene polymerizes when its temperature exceeds 25 °C, especially when the dissolved oxygen concentration in it is less than about 15 to 20 PPM (parts per million). Control of temperature of liquid styrene to less than about 20 °C and ensuring more than 20 PPM of dissolved oxygen in it, and also, addition of a retardant chemical are required to prevent the polymerization. The polymerization generates heat leading further increase of the temperature and to a runaway polymerization.

Unfortunately, the cooling or refrigeration system was not working. The amount of retardant chemical nor the dissolved oxygen in the liquid styrene was being monitored nor controlled. Polymerization is believed to have taken place and the temperature of liquid styrene in one of the storage tanks went up. The increase of the bulk temperature of liquid styrene increase its vapor pressure from 0.8 kPa at 20 °C to 1.08 kPa at 30 °C and 1.45 kPa at 35 °C. The increase of vapor pressure caused the vent valve to open, and styrene vapor was vented out into the ambient atmosphere.

If we assume the venting took place when the bulk temperature of the liquid styrene was 35 °C, and the ambient temperature as 30 °C, some of the styrene vapor would condense and form a fog. Indeed a whitish fog was noticed.

The early morning of 2 a.m. had no insolation. The sky was clear, and a very slow wind was blowing due north. The atmospheric stability condition on the Pasquil classification would have been about 'C.' The slow wind carried the styrene vapor due north. The maximum concentration of the styrene vapor at the ambient temperature of 30 °C was the pressure of vapor at 30 °C divided by the ambient pressure which gives the concentration as $1.08/100 = 0.011$ volume/volume and is 11,000 PPM. This value is very much greater than the immediately dangerous to life and health (IDLH) value of 700 PPM for styrene. The acceptable ceiling for styrene vapor over a prolonged period is 200 PPM. People therefore became unconscious with about 13 persons loosing their lives, and many required hospitalization.

An assessment of the dispersion of concentration could be made using the dispersion relations in the atmosphere. The atmospheric stability is near class 'C.' The area of spread was essentially rural. The release from the vent is a steady one. The values for the dispersion coefficients for the steady release at condition 'C' are

$$\sigma_y = 0.11x(1 + 0.0001x)^{-1/2}$$

and

$$\sigma_z = 0.08x(1 + 0.0002x)^{-1/2}$$

meters, respectively. The distance along the wind x is in meters.

People were affected to a maximum distance of 3 km. The concentration there would be greater than 700 PPM. Let us assume this value as 1000 PPM. If the release through the vent is $\dot{Q}_S$ m³/s, the average concentration along the wind at a distance of x m from the source of the leak from Eq. 13.19, assuming the release to be at the ground level i.e., $H = 0$, is

$$\chi(3000, 0, 0) = \frac{\dot{Q}_S}{\pi u \sigma_y \sigma_z}$$

Substituting the values, we get $\sigma_y = 289$ m and $\sigma_z = 190$ m. Much of the styrene vapor is carried along the wind with smaller amounts being dispersed lateral to it. The dispersion upward is smaller than that laterally. In fact people, who moved in north-easterly direction compared to the northerly wind, did not face any health issues.

With the assumed wind velocity of 2 m/s and the concentration downwind at 3 km as 1000 PPM, we get $\chi(3000, 0, 0) = 10^{-3}$ volume by volume. This gives the value of $\dot{Q}_S = \chi(3000, 0, 0)\pi u \sigma_y \sigma_z$, and this works out to be 345 m³/s corresponding to 100 kPa pressure and 30 °C. The density of the styrene vapor at the pressure of 100 kPa and 30 °C is $\frac{10^5}{(8.314/0.104) \times 303} = 4.1$ kg/m³. Here 0.104 is the molecular mass of styrene $C_6H_6C_2H_2$ in kg/mole. The mass flow rate of the leak is $345/4.1 = 84$ kg/s.

Examples

13.1 *Efflux from Ground measurements of Concentration*: Determine the efflux of CO_2 from a stack if the maximum concentration of CO_2 measured at a location 1 km downwind (along the direction of the wind) on the ground from the stack is 5 mg/m³. The mean wind velocity is 4 m/s. The height of the stack is small and can be assumed to be at ground level. The dispersion coefficients σ_y and σ_z corresponding to the above wind and atmospheric condition are 70 m and 30 m, respectively.

Solution: Let the efflux of CO_2 from a stack be $\dot{Q}_S$ kg/s. A plume is formed giving rise to steady-state concentrations of 5×10^{-6} kg/m³ 1 km downwind ($x = 1000$ m, $y = 0$). The dispersion coefficients σ_y and σ_z are given as 70 m and 30 m, respectively. The steady-state concentration at x m downwind is obtained from Eq. 13.19 as

$$\chi(x, 0, 0) = \frac{\dot{Q}_S}{\pi u \sigma_y \sigma_z}$$

This concentration is given as 5×10^{-6} kg/m^3. Substituting the values, we get

$$\dot{Q}_S = 5 \times 10^{-6} \pi u \sigma_y \sigma_z = 5 \times 10^{-6} \times \pi \times 4 \times 70 \times 30 = 0.132 \text{ kg/s}$$

The efflux of CO_2 from a stack is $0.132 \times 3600 = 475$ kg/h.

13.2 *Role of Atmospheric Conditions in Dispersion:* The toxic release of methyl isocyanide (MIC) in the Bhopal gas tragedy took place on a cold winter night when
the sky was clear and the wind speed was about 2.9 m/s. Comment whether this
ambient condition was helpful in the spread of the toxic MIC over the ground?
The volume of MIC released can be assumed as 50,000 m^3 over an effective
duration of 90 min around midnight.

Solution: Let us assume that the height of the stack from which MIC was
released is 30 m.

$$\text{Rate of release} = \frac{50,000}{(90 \times 60)} = 9.26 \text{ m}^3/\text{s}$$

On the cold winter night, the wind speed of 2.9 m/s and the clear night provides
atmospheric stability as per Table 13.1 to be moderately stable. The condition
is between E and F. It is therefore difficult for the release to disperse as in the
unstable atmosphere. MIC is also heavier than air. As it mixes with air at the
exit of the stack, its density approaches that of air even though the concentration of MIC in it is significant.

The volumetric rate of MIC release ($\dot{Q}_S$) is 9.26 m^3/s. Considering the long
period of 90 minutes over which the release takes place and spreads downwind
as a plume, the release can be considered as steady. The maximum concentration will be on the centerline of the plume at any downwind distance x from
the stack. The concentration for a plume is given along the downwind direction
by Eq. 13.19 as

$$\chi(x, 0, 0, H) = \frac{\dot{Q}_S}{\pi u \sigma_y \sigma_z} \exp\left[-\frac{H^2}{2\sigma_z^2}\right]$$

The dispersion coefficients σ_y and σ_z are taken from Table 13.2 for continuous
release assuming the release to be in an environment without tall buildings, viz.
a rural area. Let us consider four distances of 1, 2, 3, and 4 km downwind from
the source. The values of dispersion coefficients σ_y and σ_z and the concentration
of χ are given in the table given below. It is to be noted that the units of
concentration χ is kg/m^3 when the release $\dot{Q}_S$ is specified in kg/s. Since in the
present problem, the release is given in m^3/s, the concentration is in volume
fraction.

Dispersion coefficients and concentration

Distance stability parameter	1 km (E-F)	1 km (A)	2 km (E-F)	2 km (A)	3 km (E-F)	3 km (A)	4 km (E-F)	4 km (A)
σ_y m	48	210	91	402	132	579	169	745
σ_z m	23	200	29	400	36	600	42	800
χ vol. fraction	3.13×10^{-4}	2.32×10^{-5}	2.26×10^{-4}	6.3×10^{-6}	1.51×10^{-4}	2.92×10^{-6}	1.11×10^{-4}	1.70×10^{-6}
χ parts per million	313	23.2	226	6.3	151	2.92	111	1.7

In the same table, the values of the dispersion coefficients and concentration are shown for the atmosphere being unstable such as a hot afternoon with the sun shining vertically above. The atmospheric stability is considered to be 'A.' While the concentration of MIC drops off rapidly from 23.2 ppm at a distance of 1 km to 1.7 ppm at a distance of 4 km for unsteady atmosphere with a rating of 'A,' the concentrations are about two orders of magnitude higher for the moderately stable atmosphere between 'E' and 'F.' The rate of drop of the concentration is small with the stable atmosphere. The release of MIC on the cold winter night with clear sky and small wind speed has also contributed to larger concentration of MIC affecting the people.

Nomenclature

C	Concentration (moles/m^3, kg/m^3, ppm)
D	Diffusion coefficient (m^2/s)
g	Gravitational field of Earth (m/s^2)
h	Height of release (m)
H	Effective height of release (m)
$\dot{m}''$	Mass flux [kg/(m^2/s)]
p	Pressure (Pa, bar, mbar); probability density function
Q	Release (kg, m^3)
$\dot{Q}$	Rate of release (kg/s, m^3/s)
T	Temperature (°C)
t	Time (s)
u, U	Steady wind speed (m/s)
u'	Unsteady component of wind speed (m/s)
$\bar{u}$	Mean wind speed (m/s)
V	Volume (m^3)
v'	Unsteady horizontal (lateral) component of wind speed (m/s)
w	Vertical velocity of release from stack or zone of release (m/s); Dummy variable
w'	Unsteady vertical component of wind speed (m/s)
x	Distance downwind from release (m)
y	Lateral distance from centerline along the wind from the release (m)
z	Height of stack above ground level (m)
z_h	Height above the stack (m)
χ	Concentration at any point due to dispersion (ppm, kg/m^3)

ρ Density (kg/m^3)
ρ_a Density of ambient atmosphere (kg/m^3)
ρ_g Density of release (kg/m^3)
ρ_p Density of parcel of gas that disperses (kg/m^3)
σ Standard deviation
σ_x Dispersion coefficient for puff in the x-direction (m)
σ_y Dispersion coefficient lateral to the direction of the wind along the ground (m)
σ_z Dispersion coefficient in the vertical direction (m)
M Molecular mass (g/mole).

Subscripts

a Ambient
i Initial
p Parcel of gas that disperses.

Additional Reading

1. Arya, S. P. *Air Pollution Meteorology and Dispersion*, Oxford University Press, New York, 1999.
2. Boddurtha, F. T., *Industrial Explosion Prevention and Protection*, McGraw Hill, New York, 1980.
3. Pasquill, F., *Atmospheric Diffusion*, D. van Nostrand Company, London, 1962.
4. Ramamurthi, K, Karnam Bhadraiah and Srinivasa Murthy, S, *Influence of Temperature of Hydrogen Release on Formation of Hydrogen–Air Cloud*, Int. J. Hydrogen Energy, vol. 34, pp. 8428–8437, 2009
5. Turner, D.B., *Workbook of Atmospheric Dispersion Estimates*, Office of Air Programs Pub. No. AP-26, US Environmental Protection Agency, Research Triangle Park, NC, 1970.
6. Vallero, D., *Fundamentals of Air Pollution*, 4th Ed., Elsevier/Academic Press, London, 2008.

Exercises

13.1 Carbon monoxide continuously escapes from a stack having effective height of 100 m at noon on a sunny day without clouds when the wind speed is 2 m/s. The concentration of carbon monoxide 200 m directly downwind is measured as 50×10^{-5} g/m^3. Estimate the mass of carbon monoxide released from the stack in g/s. The dispersion coefficients for the sunny afternoon at 2 m/s wind speed corresponds to class A atmospheric stability for which $\sigma_y = 0.22x(1 + 0.0004x)^{-1/2}$ m and $\sigma_z = 0.20x$ m. Here the downwind distance x is in m.

13.2 Acetylene gas leaks at the rate of 30 g/s from a cylinder. Determine the concentration of acetylene downwind at a distance of 200 m on an overcast cold winter night with a wind speed of 7 m/s. Use tables for the dispersion coefficients appropriately. State any assumptions made.

13.3 Hydrogen gas is accidentally released from a stack at the rate of 4 kg/min. The effective height of the release is 5 m and this effective height takes into account the actual height of the stack and the rise due to the initial velocity of release and the buoyancy of hydrogen gas. The hydrogen gas disperses in the atmosphere from this effective height. The concentration along the direction of the wind on the surface of the ground in kg/m^3 is:

$$\chi(x, 0, 0, H) = \frac{\dot{m}_S}{\pi u \sigma_y \sigma_z} \exp\left[-\frac{H^2}{2\sigma_z^2}\right]$$

Here x is the distance downwind from the point of release, $\dot{m}_S$ is the steady release in kg/s, u is the wind velocity in m/s and σ_y and σ_z are the dispersion coefficients in m lateral to the direction of the wind and along the vertical. Determine the concentration of the hydrogen in kg/m^3 and in % volume/volume of the mixture at a distance 100 m downwind for the following two atmospheric conditions.

(a) During noon time when the sun is shining vertically above, the wind velocity is 1.5 m/s and there are no clouds in the sky. The ambient temperature is 37 °C.

(b) During night time when there is heavy wind of 5.5 m/s and there is heavy cloud cover. The ambient temperature is 2 °C.

Also determine the concentration of hydrogen in kg/m^3 at a point on the ground the coordinates of which are (100m, 50m, 0m) relative to the stack for atmospheric conditions given in a. and b. above, respectively. You may note that this point is 100 m downwind and is at a lateral distance of 50m from $y = 0$.

13.4 (a) A farmer burns the stubble after harvesting his crop in his farm. The farm is situated 100 km north and 25 km to the west from the center of a small township near to Delhi in India. He burns the stubble on his farm at night when the mean wind is from north to south with a velocity of 3 m/s. The atmospheric condition is moderately unstable at condition E. The hot particulates and the gaseous emission from the burning rise up by natural convection to a mean height of 50 m before they attain the same density and temperature as the ambient air and thereafter are carried by the wind and the emission get dispersed in the directions lateral and normal to the wind. The township is at a mean height of 750 m above the farmland. If the particulates are very fine with a mean diameter of 1.2 micro-meters and are generated at a mean rate 10^{16} particulates per minute, determine the resulting concentration of these particulates in units of number of particulates per m^3 at the center of the town. You can neglect the ground effects including the friction and reflection and assume the particulates to be carried by the wind without any settling by gravity. The dispersion coefficient for atmospheric stability condition E in the y-direction (lateral to wind) is $\sigma_y = 0.16x(1 + 0.0004x)^{-1/2}$ meters, while the dispersion coefficient along the vertical z-direction (normal to the wind) is $\sigma_z = 0.14x(1 + 0.0003x)^{-1/2}$ meters. Here x denotes the downwind distance in meters from the source.

(b) Would the number density of particulates be more or less than in problem 1a if the stubble were burnt during the day when the insolation was high though the wind velocity was about the same leading to atmospheric stability near A. Give reasons.

13.5 (a) A pipeline carrying propane gas for domestic cooking applications accidentally bursts and releases gaseous propane on the ground in a rural area at a steady rate of 1 kg/s. Determine the concentration of the propane gas at a distance of 0.2 km in the windward direction over the ground from the zone of accidental burst. The day of the burst is a very hot sunny afternoon when the wind velocity is very low at 0.5 m/s. State the unit of concentration. The dispersion constants under different atmospheric stability conditions at a downwind distance of x m for the rural setting are given below:

Dispersion coefficients for continuous release (Rural)

Atmospheric stability	σ_y (m)	σ_z (m)
A	$0.22 \times (1 + 0.001x)^{-1/2}$	$0.20\times$
B	$0.16 \times (1 + 0.001x)^{-1/2}$	$0.12\times$
C	$0.11 \times (1 + 0.001x)^{-1/2}$	$0.08 \times (1 + 0.002x)^{-1/2}$
D	$0.08 \times (1 + 0.001x)^{-1/2}$	$0.06 \times (1 + 0.00055x)^{-1/2}$
E	$0.06 \times (1 + 0.001x)^{-1/2}$	$0.03 \times (1 + 0.000x)^{-1}$
F	$0.04 \times (1 + 0.0001x)^{-1/2}$	$0.016 \times (1 + 0.0003x)^{-1}$

(b) If the above pipeline were to burst on a very cold night under very low wind conditions and severe conditions of temperature inversion, determine the concentration of propane at the same location of 0.2 km in the windward direction.

(c) Would an explosive gas mixture be formed?

(d) What would be the concentration of propane at a house located in the windward direction at a distance of 0.2 km and a lateral distance of 0.15 km and a height of 0.01m from ground level?

Chapter 14
Quantification of Damages

The principles governing explosion of gaseous, dust, and condensed phase substances are discussed in the previous chapters. Confined and unconfined explosions, physical explosions, and explosions of pressure vessels are modeled. Chapman–Jouguet detonations and quasi-detonations are seen to be more destructive than flames in a confined geometry. The destruction is through the blast wave from the explosion and the fire that often follows or precedes the explosions. The overpressure, impulse, and high velocity fragments from a blast wave led to damages. The reflection of blast waves at interfaces also contributed to spall failure. The dispersion of products of combustion of flammable gas, explosion products and pollutants in the atmosphere are also dealt with and are shown to result in damages to life and property. In this chapter, we address the quantification of the damages.

14.1 Probabilistic Nature of Damages

Table 14.1 gives the consequences of overpressure in a blast wave from an explosion. As the level of overpressure increases, the observed consequences are seen to become more catastrophic.

At the same level of overpressure, not all people and environment respond in the same manner. In the case of human beings, the health of the individual, age, and body weight are expected to influence the consequences. If we consider the specific case of rupture of the ear drum of people subject to an overpressure greater than the threshold value of 34 kPa (Table 14.1), we find that a small number have very high resistance and will not get their hearing impaired when the overpressure is much greater than 34 kPa. Similarly a small number have their ear drums ruptured at much lower pressures. A majority of people, however, has their ear drums ruptured at about 34 kPa. Some people are more vulnerable than the others. Since a small fraction responds only for very large stimulus and similarly a small fraction responds for low level of stimulus, with the majority responding at the average value, a probability distribution of the form shown in Fig. 14.1 reasonably represents the consequences of an explosion.

© The Author(s), under exclusive license to Springer Nature Switzerland AG 2021   349
K. Ramamurthi, *Modeling Explosions and Blast Waves*,
https://doi.org/10.1007/978-3-030-74338-3_14

Table 14.1 Consequences of overpressure from an explosion

S. No.	Overpressure greater than (kPa)	Consequences
1	0.7	Window glass panes break
2	34	Ear drum ruptures
3	55	Brick buildings get damaged
4	70	Persons get knocked down
5	210	Lungs get damaged
6	700	Lethal for human beings

Fig. 14.1 Fraction or probability of response

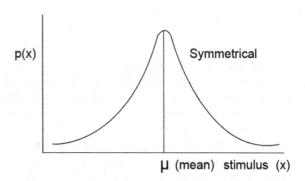

Here the stimulus x is the parameter causing the damage like the overpressure in a blast wave while $p(x)$ represents the probability of people getting affected. This is not only true for living organism but all objects including buildings, bridges, and roads since there will be scatter in their properties.

We had addressed the standard Gaussian distribution in the last chapter on atmospheric dispersion. The distribution was given by

$$p(x) = \frac{1}{\sqrt{2\pi}\sigma} e^{-[(x-\mu)/\sigma]^2} \tag{14.1}$$

where

$p(x)$ is the fraction or probability of a particular response or effect
x is the stimulus
σ is the standard deviation, and
μ is the mean.

A probabilistic procedure of addressing the damages from an explosion is dealt with in the following.

14.2 Dose Signifying the Stimulus

A single parameter, such as the overpressure in the example cited for eardrum rupture, is not always capable of describing a particular consequence or damage to occur. A combination of parameters may be required. As an example, the nature of injury or damage from a cloud of burning fuel–air mixture (fireball) would depend not only on the heat radiated from it but also on the duration of the fire. A fireball, which occurs as a flash, would not be as detrimental as one sustaining over a long period. An effective parameter identifying the stimulus would be a combination of the heat radiation from the fireball and the time. The particular combination needs to be determined based on the observed damage. Similarly, the adverse effect of toxic concentration such as styrene vapor, being dispersed in the atmosphere, depends not only on the magnitude of the concentration but also the duration over which the concentration is present. This effective quantity or parameter, which governs the adverse influence, is called as 'dose.'

Dose influences the response, viz. the degree of damage or the number of people influenced. The extent of the harm or damage increases as the dose increases. Text books on pollution invariably carry a quote of a sixth century scientist Paracelsus, viz. *'Dose alone makes a poison ... All substances are poisons; there is none which is not a poison. The right dose differentiates a poison from a remedy.'*

When a toxic substance is inhaled by a living organism, the adverse effect on the organism, called as receptor, depends on the nature of the substance, its concentration, and the time over which the receptor is subject to it. Based on studies conducted with toxic releases, the fatalities of human receptors from poisonous carbon monoxide or poisonous phosgene gas are seen to depend on the cumulative function of the concentration and time. In the case of toxic ammonia gas or nitrogen dioxide, the level of concentration is more damaging than the duration and the fatality depends on the product of the square of concentration and the time of exposure. The dose for a few different categories of injury and damage from toxic releases, explosion, and fire is given in Table 14.2.

The dose in the above table consists of $C \times t$ and $C^2 t$ for toxic dispersion, Δp, I_{sp} for explosion, and $I_e^{4/3} t_e / 10^4$ for thermal radiation from fire. The dose is defined in a similar way for the different consequences of an explosion by a single parameter or combination of parameters.

14.3 Dose Response Curves

As the dose increases, the adverse influence or response of the receptor increases in a cumulative fashion. After a certain threshold value of dose, a response is observed which progressively increases as the dose is enhanced and thereafter gets saturated. The response is therefore described by the cumulative distribution function $\phi(x)$ obtained by integrating the probability function to the particular value of dose x.

Table 14.2 Dose for toxic release, explosion, and fire

	Dose
Toxic Release	
Fatality from carbon monoxide	$C \times t$
	C = Concentration in ppm
	t = time in seconds
Fatality from ammonia gas	$C^2 t$
	C = Concentration in ppm
	t = time in seconds
Explosion	
Fatality from hemorrhage	Overpressure Δp (N/m^2)
Eardrum rupture	Overpressure Δp (N/m^2)
Injury from impact	Impulse I_{sp} (N-s/m^2)
Glass breakage	Overpressure Δp (N/m^2)
Fire	
Fatality from fire	$I_e^{4/3} t_e / 10^4$
	I_e = Effective radiation intensity (kW/m^2)
	t_e = Effective time duration (s)

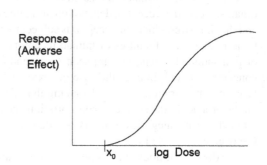

Fig. 14.2 Adverse effect or response as a function of dose

$$\phi(x) = \frac{1}{\sqrt{2\pi}\sigma} \int_{-\infty}^{x} e^{-(1/2)[(x-\mu)/\sigma]^2} dx \qquad (14.2)$$

The cumulative distribution gives increasing adverse influence as dose increases. This is shown in Fig. 14.2 where x_0 denotes the threshold value of the dose. Such curves, based on the cumulative distribution or response, are known as dose–response curves. They are usually plotted as a function of logarithm of the dose in order to expand out the region of small dose and compress the regions of higher dose.

Small levels of dose do not generally lead to major destruction. As an example, in the case of humans, the effect would be reversible in that he would recover. However, if the dose is high, it could be lethal. When the adverse effect or response is lethal,

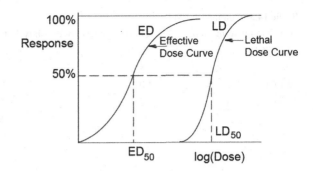

Fig. 14.3 Lethal and effective dose

the dose response curve is called as lethal dose (LD) curve. The adverse effect or response, if minor and if reversible, is called as effective dose (ED) curve. The threshold for lethal dose is much higher than for effective dose as shown in Fig. 14.3.

The dose that leads to 50% of the receptors having minor reversible injury is denoted by ED_{50}, while the dose causing death of 50% of those exposed to the dose is denoted as LD_{50}. The values of ED_{50} and LD_{50} are shown in the dose response curve in Fig. 14.3.

In the case of toxic gas release, if the response of the receptor is not lethal but the effect is not reversible, the response curve is known as toxic dose (TD). The dose corresponding to 50% of the receptors being influenced in this manner is indicated as TD_{50}. It lies in between the effective dose and lethal dose.

The dose–response curves are useful in determining the safe and hazardous values of dose originating from an explosion, fire, spill, and atmospheric dispersion. The dose–response curves have a sigmoid shape following the cumulative distribution curve when plotted as logarithm of the dose.

The standard deviation is higher for low values of dose that results in less amount of damage like effective dose. We already have seen in the chapter on atmospheric dispersion the significance of the standard deviation. The reason for the larger scatter and the associated larger values of standard deviation for small levels of dose is due the insensitivity to small amounts of the stimulus causing the harm.

The effective dose, which results in less harm, will thus have the cumulative response curve with a larger slope compared to the lethal dose as seen in Fig. 14.3. If we could define ED_{20} and LD_{20}, corresponding to 20% of the receptors being influenced, the value of $ED_{50} - ED_{20} > LD_{50} - LD_{20}$. The slope of the toxic dose will be between those of effective dose and lethal dose.

14.4 Standard Deviations σ, 2σ and 3σ

The physical significance of the standard deviation has been considered in the last chapter. The Gaussian distribution $p(x)$ having mean μ and standard deviation σ is given by Eq. 14.1 and shown in Fig. 14.1. The total area bound by the distribution

Table 14.3 Confidence level

Deviation in terms of σ	Confidence level or certainty%	Odds that deviation is greater
$\pm 0.6754\sigma$	50	1 in 2
$\pm\sigma$	68.3	About 1 in 3
$\pm 1.645\sigma$	90	1 in 10
$\pm 2\sigma$	95.4	About 1 in 20
$\pm 3\sigma$	99.7	About 1 in 370
$\pm 4\sigma$	99.994	About 1 in 16,000

is unity. The area under the curve between $(\mu - \sigma)$ and $(\mu + \sigma)$ is 0.6827 of the total area, implying that 68.27% of the values in the given distribution fall within the dispersion band between $(\mu - \sigma)$ and $(\mu + \sigma)$. If we consider the variations between $(\mu - 2\sigma)$ and $(\mu + 2\sigma)$, the area is 95.4 of the total, whereas between $(\mu - 3\sigma)$ and $(\mu + 3\sigma)$ the fraction is 0.997 (Fig. 13.8 in Chap. 13). Stated differently, the confidence of finding an outcome in the interval $(\mu - \sigma)$ and $(\mu + \sigma)$ is 68.27%, while between $(\mu - 3\sigma)$ and $(\mu + 3\sigma)$ the confidence is 99.7%. The confidence level or level of certainty for the different levels of σ for a Gaussian distribution and the odds that the deviation is greater is given in Table 14.3.

For the cumulative distribution function $\phi(x)$ given by Eq. 14.2 and shown as response in Figs. 14.2 and 14.3, $\phi(x)$ or response till a value of -3σ would be $\frac{1}{2}(1 - 0.997) = 0.0035$. Similarly, the response till a value of $+3\sigma$ would be $1 - 0.0035 = 0.9965$. The response till a value of -4σ is $\frac{1}{2}(1 - 0.99994) = 0.00003$ while for -5σ would be almost 0.

14.4.1 Probit

If the cumulative probability distribution function $\phi(x)$ is transformed by taking the inverse of it so that it indicates the width of the distribution along x, viz. the amount of standard deviation σ, we have

$$Y = \phi^{-1}[\phi(x)] \tag{14.3}$$

Here Y is referred to as Probit or probability unit. Since $\phi^{-1}[\phi(x)] = x$, $Y = x$. This is shown in Fig. 14.4.

As an example, for a standard normal distribution

$$\phi(-\sigma) = \frac{1 - \phi(\sigma)}{2} = \frac{1 - 0.683}{2} = 0.158$$

If $\phi(-\sigma) = 0.158$, then taking the inverse of $\phi(-\sigma)$ gives $\phi^{-1}[\phi(-\sigma)] = \phi^{-1}$ $(0.158) = -\sigma$. In other words, Probit of 0.158 is $-\sigma$.

Fig. 14.4 Cumulative response

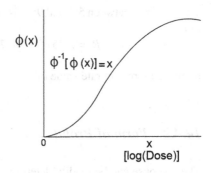

If the value of Probit is to be kept positive, base of the Probit must be about -5σ. Hence the base of the Probit which is the standard deviation σ is set at -5σ, and the response obtained would invariably be positive. We therefore get the cumulative probability distribution or response as

$$\phi = \frac{1}{\sqrt{2\pi}} \int_{-\infty}^{Y-5} e^{-(u^2/2)} du \tag{14.4}$$

Here Y is the Probit and u is a dummy variable. The variation of Y with cumulative response is given in Table 14.4 and Fig. 14.5.

The dependence of the response on Probit is more linear than on the dose. A linear fit for the response versus Probit is given by the equation

$$R = 38.2Y - 141 \tag{14.5}$$

Table 14.4 Variation of cumulative response with Probit

Probit Y	3	3.5	4	4.5	5	5.5	6	6.5	7	7.5
Cumulative function/ Response %	2.3	6.7	15.9	30.9	50	69.1	84.1	93.3	97.7	99.38

Fig. 14.5 Variation of cumulative response with Probit

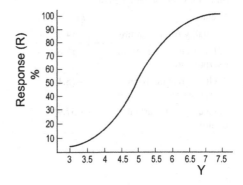

The response between 5% and 95% is sometimes fitted by the following cubic equation

$$R = 3.25Y^3 + 48.76Y^2 - 206.6Y + 270.35 \tag{14.6}$$

to give a more accurate value of the response.

14.4.2 Form of Probit

The response can be readily determined from Fig. 14.5 or Eqs. 14.5 and 14.6 if the value of Probit is quantified for a given consequence or effect of an explosion.

The logarithm of dose is plotted on the x-axis in the dose–response curve (Fig. 14.2). Denoting the dose as D and the threshold value of the dose for the particular influence as D_0 (instead of x_0 in Fig. 14.2), and noting that the dose axis is generally plotted as logarithm of the dose, we get

$$\phi = \frac{1}{\sqrt{2\pi}} \int_{-\infty}^{\ln D - \ln D_0} e^{-(u^2/2)} du = \frac{1}{\sqrt{2\pi}} \int_{-\infty}^{Y-5} e^{-(u^2/2)} du \tag{14.7}$$

This gives the form of Probit Y as

$$Y - 5 = \ln D - \ln D_0 \tag{14.8}$$

Since the threshold value of dose for a given damage is fixed, the equation for the Probit becomes

$$Y = k_1 + k_2 \ln D \tag{14.9}$$

Here k_1 and k_2 are constants and are known as Probit parameters. They are specified based on known damages and response. As an example, the parameters k_1 and k_2 are given in Table 14.5 for some observed damages.

Table 14.5 Probit parameters

Event	k_1	k_2
1. Fatality from flash fire	−14.9	2.56
2. Ear drum rupture from overpressure	−15.6	1.93
3. Glass breaking from overpressure	−18.1	2.79
4. Fatality from carbon monoxide	−37.98	3.7

It may be recalled that the dose for above four examples, based on Table 14.2, is $I_e^{4/3} t_e/10^4$, Δp, Δp, and $C \times t$, respectively.

The value of dose is obtained from methods given in Chaps. 3–13. Using the values of k_1 and k_2, the Probit Y is ascertained. The response or percentage of people affected is thereafter found from the relation between response and Probit.

Examples

14.1 *Fatality from Heat Radiation:* A person is exposed to heat radiation of $15\,\text{kW/m}^2$ for 1 min. The dose as a function of intensity of heat radiation I and time t is given by

$$D = \left[I \left(\frac{W}{m^2} \right) \right]^{1.33} \times t(s)$$

The Probit for fatal injury is given by $Y = k_1 + k_2 \ln D$, where $k_1 = -37.23$ and $k_2 = 2.56$. Determine the probability of fatal injury. You can assume the relation between percent response R and Probit Y to be given by the linear equation $R = 38.2Y - 141$.

Solution

$$Y = k_1 + k_2 \ln D$$
$$D = (15,000)^{1.33} \times 60$$

Hence
$$Y = -37.23 + 2.56 \ln[(15,000)^{1.33} \times 60] = 6$$

The value $Y = 6$ corresponds to $R = 38.2 \times 6 - 141 = 88$, suggesting that 88% of the people would die from the radiated heat.

14.2 *BLEVE Explosion and Fireball:* A huge fireball is formed in a BLEVE explosion of a tanker containing 62.5 tons of liquefied natural gas (LNG). The fireball is of diameter 200 m and is formed for a duration of 25 s. Assuming the lower calorific value (LCV) of LPG as 50,000 kJ/kg, determine:

(i) heat radiation from the fireball in megawatts.
(ii) intensity of radiant heat in W/m^2 for an observer at a distance of 500 m from the fireball
(iii) probability of people at a distance of 500 m getting killed when exposed to the fireball.

The dose for heat radiation is $D = I^{1.33} \times t$ where I is the intensity of heat radiation in W/m^2 and t the duration of the fireball in seconds. The Probit equation is: $Y = k_1 + k_2 \ln(D/10,000)$, $k_1 = -10.7$, $k_2 = 1.99$. You can assume 25% of heat of combustion is radiated and the fraction of radiation from the surface of the fireball reaching the observer 350 m away due to the view factor and atmospheric attenuation is 10%.

Solution

(i) Heat radiation from fireball. Assume that the entire mass of LPG in the tanker forms the fireball. Mass of LPG in the fireball $= 62.5 \times 1000$ kg $= 62,500$ kg. Lower calorific value (LCV) is the heat released by the fuel when the products of combustion contain H_2O in the vapor phase. It is lower than the higher calorific value (HCV), which corresponds to the heat released per unit mass of the fuel when H_2O in the products exists in the liquid phase. The value of LCV $= 50,000$ kJ/kg.

$$\text{Heat release per unit time} = \frac{62,500 \times 50,000 \times 10^3}{25}$$
$$= 1.25 \times 10^5 \text{ MW}$$

Therefore, heat radiation from the fireball $= 0.25 \times 1.25 \times 10^5 = 31,250$ MW.

(ii) Intensity of Radiation at a Distance 500 m away. Radiation intensity at the surface is:

$$\frac{31250}{(\pi \times 200^2)} = 0.249 \text{ MW/m}^2$$

Here $(\pi \times 200^2)$ represents the surface area of the fireball which is assumed to be spherical. Intensity of radiation at a distance 500 m away $= 249 \times 0.1$ kW/m^2 $= 24.9$ kW/m^2.

(iii) Dose $D = I^{1.33} \times t = (24.9 \times 1000)^{1.33} \times 25 = 1.76 \times 10^7$. Probit equation $Y = k_1 + k_2 \ln(D/10,000)$, $k_1 = -10.7$, $k_2 = 1.99$.

$$Y = -10.7 + 1.99 \ln\left(\frac{D}{10,000}\right) = 4.171$$

From the linear equation between response (R) and Probit (Y), the response or probability of the number of people suffering fatal injury is $R = 38.2 \times Y - 141 = 18.3$ for $Y = 4.171$. 18% of the people would therefore suffer fatal injury.

14.3 *Lung Hemorrhage and Loss of Hearing from Overpressure*: A blast from an explosion results in an overpressure of 75,000 Pa at a given distance from the source of the explosion. The overpressure is high enough and could lead to lung hemorrhage and consequent death of people and also rupture of their ear drums. If the Probit for death from lung hemorrhage due to overpressure Δp in Pa is given as

$$Y = [-77.1 + 6.91 \ln(\Delta p)]$$

and the Probit for ear drum rupture is given as

$$Y = [-15.6 + 1.93 \ln(\Delta p)]$$

determine the fraction of people who would die as a result of lung damage and the fraction of people whose ear drums are ruptured.

Solution

The given overpressure is 75,000 Pa. The Probit for death from lung hemorrhage and Probit for ear drum rupture are therefore

$$Y_D = [-77.1 + 6.91 \times \ln(75,000)] = 0.466$$

and

$$Y_E = [-15.6 + 1.93 \times \ln(75,000)] = 6.065$$

respectively. From Table 14.4 for response versus Probit, the values of Probit of 0.466 and 6.065 give the fraction number of persons being killed by lung hemorrhage as 0% and fraction having their ear drum ruptured as 85%.

Nomenclature

C	Concentration (ppm)
D	Dose
ED	Effective dose
I_{sp}	Specific impulse (N-s/m^2)
I_e	Effective radiation intensity (kW/m^2)
k_1, k_2	Constants in the relation between Probit and dose given in Eq. 14.8
LD	Lethal dose
p	Probability function
t	Time (s)
t_e	Effective time (s)
x	Stimulus
Y	Probit defined by Eq. 14.5
ϕ	Cumulative distribution function
σ	Standard deviation
μ	Mean
Δp	Overpressure (Pa)

Further Reading

1. Crowl, D. A. and Louvar, J.F., *Chemical Safety: Fundamentals with Applications*, Prentice Hall, Eaglewood Cliffs, NJ, 2002.
2. Boddurtha, F. T., *Industrial Explosion Prevention and Protection*, McGraw Hill, New York, 1980.
3. Davies, J. K. W., The application of box models in the analysis of toxic hazards by using the Probit dose–response relationship, *Hazardous Materials*, 22, 319–329, 1989.
4. Vallero, D., *Fundamentals of Air Pollution*, 4th ed., Elsevier/Academic Press, London, 2008.

5. Ferradas, E. G., Alonso, F.D., Minaro, M. D., Aznar, A. M. , Gimeno, J. R. and Alonso, J. M., Consequence analysis to humans from bursting cylindrical vessels, *Process Safety Progress*, 26, 289–298, 2007.

Exercise

14.1 An overpressure of 40 kPa is experienced in a market place due to an explosion whose epicenter is some distance away. Determine the total percentage of people in the market place whose hearing gets impaired due to this overpressure of 40 kPa. The damage to the ear drum is by the overpressure, and the dose is specified in terms of N/m^2 of overpressure. The Probit parameters k_1 and k_2 for damage to ear drum from overpressure are $k_1 = -15.6$ and $k_2 = +1.93$. A linear fit for the Response (R) versus Probit (Y) curve gives the response $R = 38.2Y - 141$.

Appendix A
Dependence of Sound Speed on Temperature

A sound wave propagating in a stationary gas medium at pressure p and density ρ causes small perturbations in pressure, density, and particle velocity. These small perturbations are represented as dp, $d\rho$, and dV in pressure, density, and velocity of the gas, respectively, and is shown in Fig. A.1. The wave travels at the speed of sound a. We arrive at the mass balance by considering a frame of reference with the wave at rest (Fig. A.2):

$$A\rho a = A(\rho + d\rho)(a + dV) \tag{A.1}$$

For small values of $d\rho$ and dV, Eq. A.1 becomes

$$dV = -a\frac{d\rho}{\rho} \tag{A.2}$$

The momentum balance (Fig. A.2) is

$$dp = -\rho a \, dV \tag{A.3}$$

Combining Eqs. A.2 and A.3, we get

$$\frac{dp}{d\rho} = a^2 \tag{A.4}$$

For the isentropic conditions of the medium associated with the very small perturbations (adiabatic and reversible process):

$$\frac{p}{p^\gamma} = C \tag{A.5}$$

Fig. A.1 Propagation of
sound wave at velocity a in a
medium

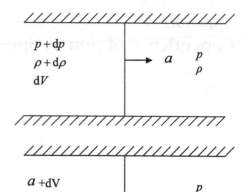

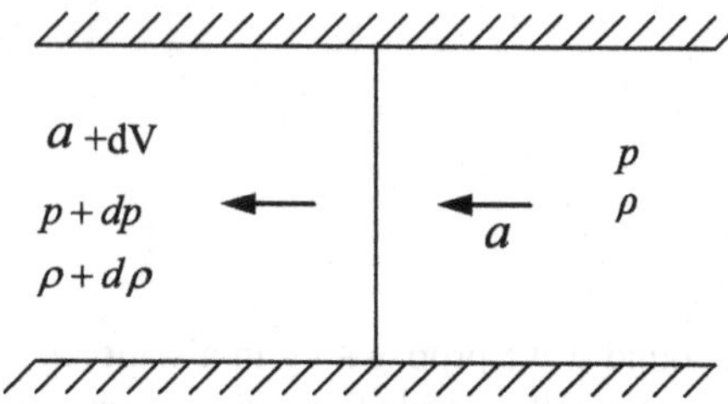

Fig. A.2 Properties in frame
of reference with sound wave
stationary

Here C is a constant and γ is the ratio of the specific heats. Differentiating the above
we get

$$\frac{dp}{d\rho} = \gamma C \rho^{\gamma-1} \tag{A.6}$$

Substituting the value of constant C from Eq. A.5 gives

$$\frac{dp}{d\rho} = \frac{\gamma p}{\rho} \tag{A.7}$$

The equation of state for an ideal gas is

$$p = \rho R T \tag{A.8}$$

where R is the specific gas constant and T is the temperature.
   Equation A.8 is substituted in Eq. A.7 to give

$$\frac{dp}{d\rho} = \gamma R T \tag{A.9}$$

Using Eqs. A.4 and A.9, we have

$$a = \sqrt{\gamma R T} \tag{A.10}$$

The sound speed a therefore increases as the square root of the temperature.

Appendix B
Mach Number M_{S1} Behind a Shock Propagating at Constant Mach Number M_S

Let a shock wave propagate at a velocity $\dot{R}_S$ in a stationary medium of density ρ_0, pressure p_0, temperature T_0, the sound velocity being a_0. Let the medium behind the shock follow it with a velocity u with density ρ_1, pressure p_1, and temperature T_1. The sound velocity in the medium behind the shock wave is a_0. In the frame of reference of the shock wave, the undisturbed medium at conditions ρ_0, p_0, and T_0 moves towards the shock at velocity $\dot{R}_S$ while the shocked gases leave the shock with a velocity $u_1 = \dot{R}_S - u$ at conditions ρ_1, p_1 and T_1.

The mass, momentum, and energy equations across the shock in the frame of reference of the shock are:

$$\rho_0 \dot{R}_S = \rho_1 u_1 \tag{B.1}$$

$$p_0 + \rho_0 \dot{R}_S^2 = p_1 + \rho_1 u_1^2 \tag{B.2}$$

$$h_0 + \frac{\dot{R}_S^2}{2} = h_1 + \frac{u_1^2}{2} \tag{B.3}$$

Here h_0 and h_1 denote the specific enthalpy of the medium ahead and behind the shock, respectively.

Assuming the medium to be a perfect gas wherein $h = C_p T$, the energy Eq. (B.3) is written as

$$C_p T_0 + \frac{\dot{R}_S^2}{2} = C_p T_1 + \frac{\dot{R}_S^2}{2}$$

Further since for a perfect gas

$$C_p = \frac{\gamma R}{\gamma - 1}$$

where γ is the specific heat ratio and R is the specific gas constant, the equation reduces to

$$\frac{\gamma R T_0}{\gamma - 1} + \frac{\dot{R}_S^2}{2} = \frac{\gamma R T_1}{\gamma - 1} + \frac{u_1^2}{2} = \text{constant}$$

provided that the process is adiabatic. If we define the conditions corresponding to the critical condition by '*', the critical velocity is u^*, and the critical sound velocity is a^*. The Mach number at critical condition M^* is 1; i.e., the sound velocity at critical condition a^* is equal to the critical velocity u^*. The energy equation becomes

$$\frac{\gamma R T_0}{\gamma - 1} + \frac{\dot{R}_S^2}{2} = \frac{\gamma R T_1}{\gamma - 1} + \frac{u_1^2}{2} = \frac{\gamma R T^*}{\gamma - 1} + \frac{a^{*2}}{2}$$

Here T^* is the temperature corresponding to the critical conditions.

Simplifying the above equation by noting that the sound speed in the undisturbed medium is given by $a_0^2 = \gamma R T_0$ and in the shocked medium by $a_1^2 = \gamma R T_1$, we have

$$\frac{a_0^2}{\gamma - 1} + \frac{\dot{R}_S^2}{2} = \frac{a_1^2}{\gamma - 1} + \frac{u_1^2}{2} = \frac{a^{*2}}{\gamma - 1} + \frac{a^{*2}}{2} = \frac{a^{*2}}{2}\left(\frac{\gamma + 1}{\gamma - 1}\right) \tag{B.4}$$

From Eq. B.4 we get

$$a_0^2 = \frac{\gamma + 1}{2} a^{*2} - \frac{\gamma - 1}{2} \dot{R}_S^2 \tag{B.5}$$

$$a_1^2 = \frac{\gamma + 1}{2} a^{*2} - \frac{\gamma - 1}{2} u_1^2 \tag{B.6}$$

The momentum Eq. (B.2) can be written in the form

$$\frac{p_0}{\rho_0 \dot{R}_S} + \dot{R}_S = \frac{p_1}{\rho_1 u_1} + u_1$$

giving

$$u_1 - \dot{R}_S = \frac{p_0}{\rho_0 \dot{R}_S} - \frac{p_1}{\rho_1 u_1} \tag{B.7}$$

Equations B.5 and B.6 can be combined by dividing them by $\gamma \dot{R}_S$ and γu_1, respectively, as

$$\frac{a_0^2}{\gamma \dot{R}_S} - \frac{a_1^2}{\gamma u_1} = \left(\frac{\gamma + 1}{2}\frac{a^{*2}}{\gamma \dot{R}_S} - \frac{\gamma - 1}{2}\frac{\dot{R}_S^2}{\gamma \dot{R}_S}\right) - \left(\frac{\gamma + 1}{2}\frac{a^{*2}}{\gamma u_1} - \frac{\gamma - 1}{2}\frac{u_1^2}{\gamma u_1}\right) \tag{B.8}$$

The left side of the above Eq. B.8 gives

$$\frac{a_0^2}{\gamma \dot{R}_S} - \frac{a_1^2}{\gamma u_1} = \frac{p_0}{\rho_0 \dot{R}_S} - \frac{p_1}{\rho_1 u_1}$$

which from the momentum Eq. B.2 reduces to

$$u_1 - \dot{R}_S$$

The right side of Eq. B.8 is simplified as

$$\left(\frac{\gamma+1}{2}\frac{a^{*2}}{\gamma R_S^*} - \frac{\gamma-1}{2}\frac{\dot{R}_S^2}{\gamma \dot{R}_S}\right) - \left(\frac{\gamma+1}{2}\frac{a^{*2}}{\gamma u_1} - \frac{\gamma-1}{2}\frac{u_1^2}{\gamma u_1}\right)$$

$$= \frac{\gamma+1}{2\gamma}\frac{a^{*2}}{\dot{R}_S u_1}(u_1 - \dot{R}_S) - \frac{\gamma-1}{2\gamma}(\dot{R}_S - u_1)$$

Equating the simplified forms of the left and right sides gives

$$u_1 - \dot{R}_S = \frac{\gamma+1}{2\gamma}\frac{a^{*2}}{\dot{R}_S u_1}(u_1 - \dot{R}_S) - \frac{\gamma-1}{2\gamma}(\dot{R}_S - u_1) \tag{B.9}$$

Hence, we get

$$\frac{\gamma-1}{2\gamma} + \frac{\gamma+1}{2\gamma}\frac{a^{*2}}{\dot{R}_S u_1} = 1$$

and on simplification

$$a^{*2} = \dot{R}_S u_1 \tag{B.10}$$

The relationship provides the connect between the velocity of the medium upstream of the shock, and the velocity downstream in the frame of reference of the shock and is known as Prandtl relation.

Defining characteristic Mach numbers of the flow upstream and downstream of the shock as:

$$M_0^* = \frac{\dot{R}_S}{a^*}; \quad M_1^* = \frac{u_1}{a^*}$$

we get from Eq. B.10 the relation

$$M_0^* M_1^* = 1 \tag{B.11}$$

The relationship between $M_S = \dot{R}_S/a_0$ where a_0 is the sound speed in the medium ahead of the shock and $M_{S1} = u_1/a_1$ for which a_1 is the sound speed in the medium downstream of the shock can be derived from the relationship between M_0^* and M_1^* using the energy equation. The sum of $h + (u^2/2)$ is the stagnation enthalpy, which is a constant at all points in the flow. Hence

$$\frac{\gamma R T_0}{\gamma-1} + \frac{\dot{R}_S^2}{2} = \frac{\gamma R T_1}{\gamma-1} + \frac{u_1^2}{2} = \frac{\gamma R T^*}{\gamma-1} + \frac{u^{*2}}{2}$$

and therefore in terms of sound speed

$$\frac{a_0^2}{\gamma-1} + \frac{\dot{R}_S^2}{2} = \frac{a_1^2}{\gamma-1} + \frac{u_1^2}{2} = \frac{a^{*2}}{\gamma-1} + \frac{a^{*2}}{2} = \frac{a^{*2}}{2}\left(\frac{\gamma+1}{\gamma-1}\right)$$

This reduces to

$$a_0^2 + \frac{\dot{R}_S^2(\gamma - 1)}{2} = \frac{\gamma + 1}{2}a^{*2}; \quad a_1^2 + \frac{u_1^2(\gamma - 1)}{2} = \frac{\gamma + 1}{2}a^{*2} \tag{B.12}$$

and

$$\frac{2}{M_S^2} + (\gamma - 1) = \frac{\gamma + 1}{M^{*2}_0}; \quad \frac{2}{M_{S1}^2} + (\gamma - 1) = \frac{\gamma + 1}{M^{*2}_1}$$

We therefore get

$$M_S^2 = \frac{2}{\frac{\gamma+1}{M^{*2}_0} - (\gamma - 1)}; \quad M_{S1}^2 = \frac{2}{\frac{\gamma+1}{M^{*2}_1} - (\gamma - 1)} \tag{B.13}$$

and

$$M^{*2}_0 = \frac{(\gamma + 1)M_S^2}{2 + (\gamma - 1)M_S^2}; \quad M^{*2}_1 = \frac{(\gamma + 1)M_{S1}^2}{2 + (\gamma - 1)M_{S1}^2} \tag{B.14}$$

Since from Eq. B.11

$$M_0^* M_1^* = 1$$

we have substituting the values in terms of M_S and M_{S1} from Eq. B.14

$$\frac{(\gamma + 1)M_S^2}{2 + (\gamma - 1)M_S^2} = \frac{2 + (\gamma - 1)M_{S1}^2}{(\gamma + 1)M_{S1}^2} = \frac{2}{(\gamma + 1)M_{S1}^2} + \frac{\gamma - 1}{\gamma + 1} \tag{B.15}$$

Rearranging the terms, we have

$$\frac{(\gamma + 1)M_S^2}{2 + (\gamma - 1)M_S^2} - \frac{\gamma - 1}{\gamma + 1} = \frac{2}{(\gamma + 1)M_{S1}^2}$$

The above gives

$$\frac{2}{M_{S1}^2} = \frac{(\gamma + 1)^2 M_S^2}{2 + (\gamma - 1)M_S^2} - (\gamma - 1) = \frac{4\gamma M_S^2 - 2(\gamma - 1)}{2 + (\gamma - 1)M_S^2} \tag{B.16}$$

Hence

$$M_{S1}^2 = \frac{1 + \frac{\gamma - 1}{2}M_S^2}{\gamma M_S^2 - \frac{\gamma - 1}{2}} = \frac{M_S^2 + \frac{2}{\gamma - 1}}{\frac{2\gamma}{\gamma - 1}M_S^2 - 1} \tag{B.17}$$

Hence for $M_S > 1$, $M_{S1} < 1$. It is to be noted that the Mach number M_{S1} is of the medium in the frame of reference of the shock which progresses in an undisturbed medium at a Mach number M_S.

Appendix C
Reflection of a Constant Velocity Shock Wave Normal to a Rigid Surface

Consider an incident shock wave, traveling at constant velocity $\dot{R}_S$ in quiescent atmosphere, to hit against a rigid surface shown in Fig. C.1. Let the pressure, density, temperature, and sound velocity in the quiescent medium be p_0, ρ_0, T_0, and a_0, respectively. Let us presume that the incident shock totally rebounds from the rigid surface.

The magnitude of the pressure, density, temperature, and sound velocity of the gases behind the incident shock are denoted by p_1, ρ_1, T_1, and a_1. These are shown in Fig. C.1 in the inertial frame of reference and in Fig. C.2 in the frame of reference of the incident shock wave. The values of the parameters p_1, ρ_1, T_1 and velocity u_1 behind a constant velocity shock wave traveling at Mach number M_S were related to the free stream quiescent values p_0, ρ_0, and T_0 in Chap. 2 by the following expressions:

$$\frac{p_1}{p_0} = \frac{2\gamma}{\gamma + 1} M_S^2 - \frac{\gamma - 1}{\gamma + 1} \tag{C.1}$$

$$\frac{\rho_1}{\rho_0} = \frac{\gamma + 1}{\gamma - 1 + 2/M_S^2} \tag{C.2}$$

$$\frac{T_1}{T_0} = \left(\frac{2\gamma}{\gamma + 1} M_S^2 - \frac{\gamma - 1}{\gamma + 1} \right) \left(\frac{\gamma - 1 + 2/M_S^2}{\gamma + 1} \right) \tag{C.3}$$

$$\frac{u_1}{\dot{R}_S} = \frac{\gamma - 1 + 2/M_S^2}{\gamma + 1} \tag{C.4}$$

M_S in the above equation equals $\dot{R}_S/a_0$. The value of particle velocity u_1 behind the shock wave was in the frame of reference of the shock stationary as shown in Fig. C.2.

The incident shock gets reflected from the rigid surface at the same velocity $\dot{R}_S$. The reflected shock propagates in a medium which is already processed by the incident shock and whose sound speed is a_1 and whose pressure, density, and temperature are p_1, ρ_1, T_1, respectively. The Mach number of this reflected shock is

© The Editor(s) (if applicable) and The Author(s), under exclusive license to Springer Nature Switzerland AG 2021
K. Ramamurthi, *Modeling Explosions and Blast Waves*,
https://doi.org/10.1007/978-3-030-74338-3

Fig. C.1 Properties ahead
and behind an incident shock
wave traveling at velocity $\dot{R}_S$
in the inertial frame of
reference

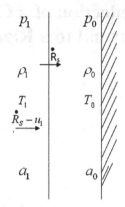

Incident Shock (Traveling)

Fig. C.2 Properties in the
frame of reference of the
incident shock as it
approaches the rigid surface

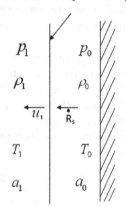

Incident Shock (Stationary)

therefore $\dot{R}_S/a_1$ and would be less than the Mach number of the incident shock wave since $a_1 > a_0$.

The reflected shock travels in the medium which is reflected by the rigid surface and moves in the direction of the reflected shock with velocity u_1 and whose pressure, temperature and density are $p_1, T_1,$ and ρ_1, respectively. This is shown in Fig. C.3 in the frame of reference of the reflected shock. The net approach velocity of the medium in the frame of reference of the reflected shock (reflected shock considered stationary) is $\dot{R}_S + u_1$ at pressure, temperature, and density of $p_1, T_1,$ and ρ_1. Denoting the Mach number of the medium approaching this reflected shock wave in the frame of reference of the reflected shock wave being stationary as M_{S1}, we have:

$$M_{S1} = \frac{\dot{R}_S + u_1}{a_1} \qquad (C.5)$$

Fig. C.3 Properties in the frame of reference of the reflected shock

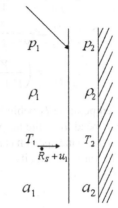

Reflected Shock (Stationary)

The pressure p_1, density ρ_1, and temperature T_1 ahead of the reflected shock get enhanced over and above these values to p_2, ρ_2, and T_2, respectively. The sound speed behind the reflected shock is now a_2. We wish to determine the value of pressure behind the reflected shock to that ahead of it, viz. (p_2/p_1).

Assuming the specific heat ratio does not change with temperature, the values of pressure p_2, density ρ_2, temperature T_2 are:

$$\frac{p_2}{p_1} = \frac{2\gamma}{\gamma+1}M_{S1}^2 - \frac{\gamma-1}{\gamma+1} \tag{C.6}$$

$$\frac{\rho_2}{\rho_1} = \frac{\gamma+1}{\gamma-1+2/M_{S1}^2} \tag{C.7}$$

$$\frac{T_2}{T_1} = \left(\frac{2\gamma}{\gamma+1}M_{S1}^2 - \frac{\gamma-1}{\gamma+1}\right)\left(\frac{\gamma-1+2/M_{S1}^2}{\gamma+1}\right) \tag{C.8}$$

The magnitude of the pressure ratio behind the reflected shock (p_2/p_1) is determined in the following for two specific cases corresponding to strong incident shock and a weak incident shock.

a. Magnitude of Pressure p_2 When the Incident Shock is Strong

Let us consider a strong incident shock wave of Mach number $M_S = 5$ traveling through air. If the temperature of the quiescent air is 300 K, the specific heat ratio is 1.4 and the specific gas constant is 286 J/(kg K), the sound speed $a_0 = \sqrt{\gamma R T_0} = 348$ m/s. The value of:

$$\frac{p_1}{p_0} = \frac{2\gamma}{\gamma+1}M_S^2 - \frac{\gamma-1}{\gamma+1} = 29 \tag{C.9}$$

$$\frac{\rho_1}{\rho_0} = \frac{\gamma+1}{\gamma-1+2/M_S^2} = 5 \tag{C.10}$$

$$\frac{T_1}{T_0} = \left(\frac{2\gamma}{\gamma+1}M_S^2 - \frac{\gamma-1}{\gamma+1}\right)\left(\frac{\gamma-1+2/M_S^2}{\gamma+1}\right) = 5.8 \tag{C.11}$$

The temperature T_1 behind the incident shock is therefore $5.8 \times 300 = 1740$ K. The sound speed a_1 in the air processed by the incident shock is $a_1 = \sqrt{\gamma R T_1} = 837.6$ m/s. The sound speed has increased from a value of 348 m/s in the quiescent gas.

The value of velocity u_1 in the frame of reference of the incident shock stationary from Eq. 2.45 is:

$$\frac{u_1}{\dot{R}_S} = \frac{\rho_0}{\rho_1} = \frac{1}{5}$$

The shock velocity $\dot{R}_S = 5 \times 348 = 1740$ m/s. This gives the value of u_1 as $u_1 = 348$ m/s. The value of $M_{S1} = (\dot{R}_S + u_1)/a_1$. Substituting the values of $\dot{R}_S$, u_1, and a_1, the Mach number of the reflected shock in the frame of reference of the medium ahead of it being quiescent is:

$$M_{S1} = \frac{\dot{R}_S + u_1}{a_1} = 2.49$$

The pressure behind the reflected shock wave is:

$$\frac{p_2}{p_1} = \frac{2\gamma}{\gamma+1}M_{S1}^2 - \frac{\gamma-1}{\gamma+1} = 7.09 \tag{C.12}$$

The pressure behind the reflected shock therefore increases by a factor of 7.09 over and above the value of the increase in pressure behind the incident shock wave.

If the incident shock wave were to have a Mach number of 10, the value of (p_2/p_1) would be 7.74.

The value of the pressure jump behind the reflected shock wave tends to a value of about 8 as the incident shock Mach number increases to large values.

b. Magnitude of Pressure p_2 when the Incident Shock is Weak

Let us consider a weak incident shock wave of $M_S = 1.2$. In this case $M_{S1} = 1.812$ and (p_2/p_1) becomes equal to 3.67. The value of (p_2/p_1) progressively decreases as the incident Mach number is further reduced. In the limit of a sound wave (p_2/p_1) equals 2.

Magnitude of Pressure Jump behind Reflected Shock for any given Mach Number of Incident Shock

The ratio of the pressure across a reflected shock is seen to increase with increase of the incident shock velocity. For a very strong shock wave the ratio tends to about 8,

Fig. C.4 Variation in
pressure ratio across a
reflected shock for different
Mach number of incident
shocks

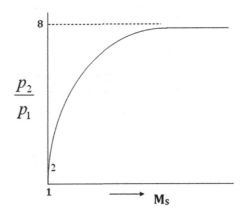

whereas for a sound wave the ratio is 2. Figure C.4 shows the trend of variation of
the pressure ratio across a reflected shock as a function of the incident shock Mach
number when the incident shock gets totally reflected at a rigid wall.

The wall pressures due to the reflection of the incident shock will therefore vary
with the strength of the incident shock wave. The ratio (p_2/p_1) across the reflected
wave is known as the shock reflection coefficient and is denoted as C_R. An expression
could be derived for it as a function of the pressure ratio (p_1/p_0) across the incident
shock from Rankine–Hugoniot relations. The pressure ratio (p_2/p_1) is

$$\frac{p_2}{p_1} = \frac{(3\gamma - 1)(p_1/p_0) + (\gamma + 1)}{(\gamma - 1)(p_1/p_0) + (\gamma + 1)} \tag{C.13}$$

In the limit of a very weak shock wherein $(p_1/p_0) = 1$, we get from Eq. C.13 the
value $(p_2/p_1) = 2$. In the limit of a very strong shock for which $(p_1/p_0) \to \infty$, we
get

$$\frac{p_2}{p_1} \approx \frac{3\gamma - 1}{\gamma - 1}$$

which for air having $\gamma = 1.4$ gives $(p_2/p_1) = 8$.

When the medium in which the blast wave propagates is not a perfect gas, the
upper limit of the reflection coefficient C_R gets to be much greater than 8 due to the
higher values of energy absorption in the medium. The high value of the reflected
blast overpressures for strong blast waves results in drastic increase of blast loading
of unyielding structures.

Appendix D
Shock Reflection and Transmission: Mechanical Impedance and Acoustic Impedance

The mechanical impedance Z of a given medium was defined in Chapter 3 as the ratio p'/u', viz. the ratio of the pressure change associated with a shock wave to the velocity change in it. If we were to consider the blast wave to be weak, which in the limit would be an acoustic wave, an expression for the impedance is readily possible. The impedance of the sound wave in a given medium is spoken of as acoustic impedance of that medium.

An expression for the acoustic impedance of a medium is derived in the following by considering a sound wave propagating in a quiescent medium having pressure p_0 and density ρ_0. The speed of propagation of the wave in the undisturbed medium is the sound velocity in the medium and is denoted by a_0. The pressure and density behind the wave are denoted by $p_0 + p'$ and $\rho_0 + \rho'$ while the velocity of the medium behind the wave is u'. Here p', ρ' and u' are small. A schematic of the properties with the wave propagating at the sound velocity a_0 is shown in Fig. D.1.

The ratio p'/u' is to be determined. The frame of reference in Fig. D.1 is the inertial frame of reference with the wave moving at velocity a_0. If we consider the frame of reference of the wave (i.e., the wave being stationary), i.e., the observer is on the wave as shown in Fig. D.2, the observer now perceives the flow coming towards him with sound velocity a_0. The fluid behind the wave has a velocity $(a_0 - u')$ at a pressure $p_0 + p'$ and density $\rho_0 + \rho'$.

Denoting the cross section of the flow as unity, the momentum balance equation, which gives the rate of change of momentum as equal to the net force, can be written as:

$$(p_0 + p') - p_0 = \dot{m}[a_0 - (a_0 - u')] \tag{D.1}$$

Here $\dot{m}$ is the mass flow rate. From the mass balance equation, we have:

$$\dot{m} = \rho_0 a_0 = (\rho_0 + \rho')(a_0 - u') \tag{D.2}$$

Substituting the value of $\dot{m}$ from the mass balance Eq. D.2 in the momentum Eq. D.1 and simplifying, we get:

$$p' = \rho_0 a_0 u' \tag{D.3}$$

© The Editor(s) (if applicable) and The Author(s), under exclusive license
to Springer Nature Switzerland AG 2021
K. Ramamurthi, *Modeling Explosions and Blast Waves*,
https://doi.org/10.1007/978-3-030-74338-3

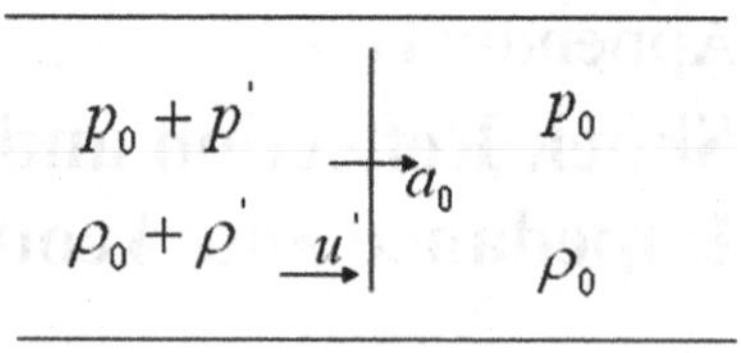

Fig. D.1 Schematic of wave propagation

$$p_0 + p' \qquad\qquad p_0$$
$$\Big|_{a_0}$$
$$\rho_0 + \rho' \underset{u'}{\longrightarrow} \qquad \rho_0$$

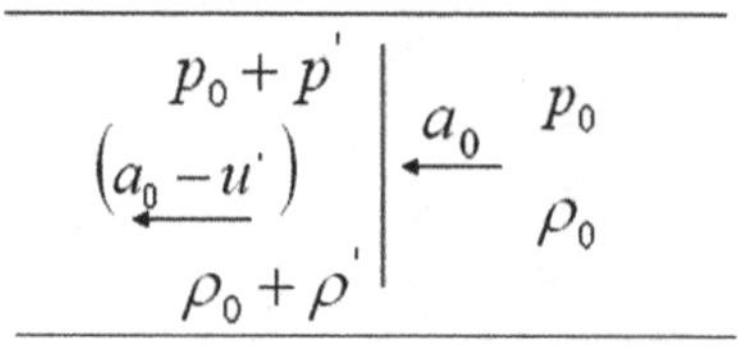

Fig. D.2 Schematic of wave propagation in the frame of reference of the sound wave

$$p_0 + p' \qquad\qquad p_0$$
$$(a_0 - u') \quad \overset{a_0}{\longleftarrow} \qquad \rho_0$$
$$\rho_0 + \rho'$$

This implies that the acoustic impedance, which is the ratio of p' to u', equals:

$$Z = \frac{p'}{u'} = \rho_a a_0 \qquad (D.4)$$

The acoustic impedance of a medium is therefore given by the product of the density of the medium and the value of the sound speed in that medium. It is a thermodynamic property being a function of the density and sound speed in the medium. It is peculiar for a given medium and is a measure of its stiffness.

The isentropic compressibility of a medium was seen in Chap. 1 to be given by $\kappa_S = 1/\rho_0 a_0^2$. The acoustic impedance is therefore $1/\kappa_S a_0$, i.e., the inverse of the product of isentropic compressibility of the medium and the sound speed of the medium in which the wave propagates.

For atmospheric air at 300 K, the density ρ_0 is p_0/RT, where R is the specific gas constant for air. With $p_0 = 100$ kPa and the specific gas constant of air $R = 286$ J/(kg K), $\rho_0 = 1.16$ kg/m^3. The sound speed in air is $a_0 = \sqrt{\gamma RT}$, where γ is the ratio of the specific heats and is 1.4. The value of a_0 is therefore 348 m/s. This gives the value of the acoustic impedance of air as $1.16 \times 348 = 403$ Pa-s/m.

The momentum and mass balance equations did not assume whether the medium was a gas, a solid or a liquid, and therefore, the definition of acoustic impedance holds good for any medium. The acoustic impedance for any medium is therefore the product of its density and the sound speed in it or alternatively the inverse of the product of its isentropic compressibility and sound speed. The values of the acoustic impedance of different medium or materials of construction are given below (Table D.1).

Since the sound speed is proportional to the internal or molecular energy of the medium, the acoustic impedance represents how much pressure is generated in a given medium by the molecular velocity changes.

Table D.1 Acoustic impedance

S. No.	Medium	Acoustic Impedance Z N-s/m^3 or Pa-s/m
1	Air	420
2	Water	1.5×10^6
3	Fat tissue	1.33×10^6
4	Muscles in body	1.7×10^6
5	Bone	6.6×10^6
6	Brick	7.4×10^6
7	Pyrex Glass	13×10^6
8	Steel	46×10^6
9	Tungsten	101×10^6

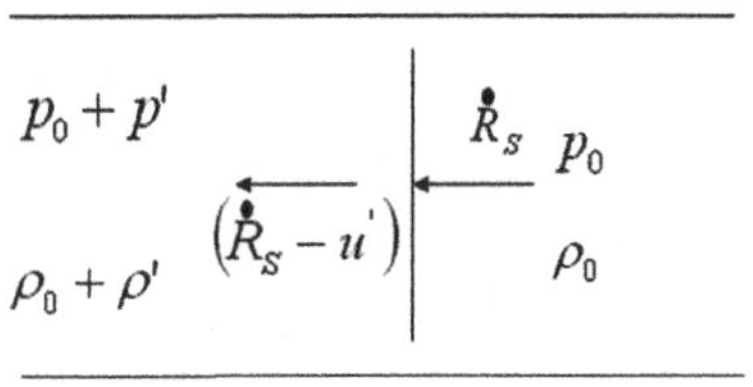

Fig. D.3 Schematic of shock wave propagation in a medium

When a strong shock wave is considered, Fig. D.2 in the frame of reference of the wave gets modified to Fig. D.3 where we now have shock speed $\dot{R}_S$ instead of sound velocity a_0. The momentum equation for any medium whether solid, liquid, or gas becomes:

$$(p_0 + p') - p_0 = \rho_0 = \rho_0 \dot{R}_S [\dot{R}_S - (\dot{R}_S - u')] \tag{D.5}$$

This gives the shock impedance as

$$Z = \frac{p'}{u'} = \rho_0 \dot{R}_S \text{ N-s/m}^2$$

From the expressions derived for the reflected and transmitted waves, it is seen that the overpressures of the reflected and transmitted shock waves are a function of the ratio of impedances of the two media and does not depend on the absolute value of the impedance of a particular media. Acoustic impedance values are therefore used instead of shock impedance to determine the transmitted and reflected shocks.

The change in acoustic impedance at the interface is known as impedance mismatch. It is the impedance mismatch that leads to reflection and transmission of blast waves at the interface of two different media.

Appendix E
Energy Release in a Chemical Reaction assuming Thermodynamic Equilibrium: Illustration for the Explosion of TNT and a Fuel-rich Gaseous Mixture

New species are formed in a chemical reaction. The energy associated with the constituents in the reactants and products (viz. their standard heats of formation) decides the energy released in the reaction. The constituents or the chemical species could also exist in different phases such as a gas, a liquid, or a solid. The species and their phases are determined assuming equilibrium between them at the pressure and temperature of the explosion. In this Appendix, the product species formed and the energy release in the chemical reaction is determined under the condition of thermodynamic equilibrium of the products.

The method of obtaining the species under equilibrium conditions is justified as process of equilibration is much faster than the chemical reaction processes in the explosion. Thermodynamic equilibrium thus forms the basis of several commercial computer codes for determining the products and the energy release. It is necessary to realize the number of assumptions involved and the limitations of the procedure before using the energy release predicted in these computer codes.

E.1 Gibbs Free Energy and Equilibrium

The thermodynamic property Gibbs Free Energy G provides a measure of the maximum amount of work which can be obtained from a system at constant pressure p and temperature T. It is used for determining the product species, which can coexist with each other in equilibrium in a chemical reaction. The Gibbs Free Energy G for a single component gas is defined as:

$$G = U + pV - TS \tag{E.1}$$

Here U is the internal energy, V is the volume, and S is the entropy. In a differential form, it becomes

$$dG = dU + pdV + Vdp - TdS - SdT \tag{E.2}$$

From the first law of thermodynamics, $dU = dQ - pdV + d\xi$. Here $d\xi$ designates work done on the system other than pressure volume work pdV and dQ denotes the heat supplied. Since $TdS = dQ = dU + pdV$, the value of dG in Eq. E.2 becomes

$$dG = Vdp - SdT + d\xi \qquad (E.3)$$

At constant temperature T and pressure p corresponding to the temperature and pressure of the combustion products, Eq. E.3 gives $dG = d\xi$ and represents the non-displacement or non pdV work of the chemical reactions. If dG is positive, work is done on the system of chemical reactions at the constant temperature T and pressure p, whereas if dG is negative work is done by the chemical reactions. When $dG = 0$, no work is possible, and hence, the chemical reactions are in equilibrium. The change in Gibbs Free Energy, whether positive or negative is therefore an indication of the direction of change in the chemical reactions at the given pressure and temperature. The Gibbs Free Energy is useful for determining equilibrium of chemical reactions.

E.2   Equilibrium of Species and Equilibrium Constants

When we have a number of species in the products of a reaction, we need to incorporate all of them to determine the change in Gibbs Free Energy. If in a generalized way, we define the number of species (M_i) in the products to be $i = 1, \ldots, N$ and their number of moles to be n_i to be $i = 1, \ldots, N$; i.e., the products are given by $\sum_i n_i M_i$, i.e., $n_1 M_1 + n_2 M_2 + \cdots$, the value of dG in the single component gas in eq. B3 when extended for the changes in the number of species dn_i becomes

$$dG = Vdp - SdT + \left(\frac{\partial G}{\partial n_1}\right)_{T,p,nj;j\neq 1} dn_1 + \cdots + \left(\frac{\partial G}{\partial n_i}\right)_{T,p,nj;j\neq i} dn_i \qquad (E.4)$$

for all values of j from 1 to n. The quantity

$$\left(\frac{\partial G}{\partial n_i}\right)_{T,p,nj;j\neq i}$$

represents the increase in Gibbs Free Energy when one mole of component i is added to an infinitely large quantity of the mixture keeping all other species same at constant values of temperature and pressure. The addition of dn_i moles does not significantly change the overall composition. It is known as the chemical potential (μ_i) and represents the driving tendency of chemical systems to equilibrium. The change in Gibbs Free energy is therefore written as

$$dG = Vdp - SdT + \sum_{i=1,\ldots N} \mu_i dn_i \qquad (E.5)$$

The Gibbs Free Energy G for a system having 'N' species following Eq. E.1 becomes

$$G = U + pV - TS + \sum_{i=1}^{N} \mu_i n_i \qquad (E.6)$$

E.3 Chemical Potential and Equilibrium Products in a Reaction

If we consider equilibrium product of a reaction wherein specie A changes to specie B in the products

$$A \Leftrightarrow B \qquad (E.7)$$

the reduction in the number of moles of specie A equals the number of moles of specie B formed. Denoting the change of moles in A as $-dn_A$ and in B as dn_B, we have:

$$dn_A = dn_B = d\chi \qquad (E.8)$$

where dn_A moles A are converted into dn_B moles B. $d\chi$ is the magnitude of the change in the number of moles. We have from Eq. E.5 the change in Gibbs Free Energy at a given value of pressure and temperature being kept the same as

$$dG = -\mu_A dn_A + \mu_B dn_B \qquad (E.9)$$

giving

$$dG = (\mu_B - \mu_A)d\chi$$

The progress of the reaction at a given temperature T and pressure p is given by

$$(dG/d\chi)_{T,p} = \mu_B - \mu_A \qquad (E.10)$$

When dG becomes zero, no further change is possible and the species A and B are in equilibrium. This is given by:

$$dG/d\chi = 0 \quad \text{or} \quad \mu_A = \mu_B \qquad (E.11)$$

Changes in the amount of species A and B proceeds until G reaches the minimum value for which $dG/d\chi = 0$, giving $\mu_A = \mu_B$. The species in the products of combustion, when in equilibrium, thus have the same chemical potential.

E.4   Extension to Multiple Species in the Products of the Reaction

For a single component or specie system, we had

$$dG = Vdp - SdT \qquad (E.12)$$

giving

$$(dG/dp)_T = V \qquad (E.13)$$

Assuming the species to be ideal gases, we have from the equation of state for an ideal gas $pV = nR_0T$, where R_0 is the universal gas constant and hence

$$(dG/dp)_T = nR_0T/p \qquad (E.14)$$

The above expression on integration between a standard condition of 1 atmosphere and p atmospheres gives $G = G^0 + nR_0T \ln p$, where G^0 is the value of G under the standard reference condition of one atmosphere. For a multi-species system, if the different species follow the perfect gas law, we can write for the ith component $G_i = G_i^0 + R_0T \ln p_i$, and

$$\mu_i = \mu_i^0 + R_0T \ln p_i \qquad (E.15)$$

where p_i is the partial pressure of component i and μ_i^0 is its chemical potential under standard conditions.

E.5   Equilibrium Constant K_p and Determination of Species

The progress of the reaction between species $A \Leftrightarrow B$ at a given temperature T and pressure p can be written by substituting Eq. E.15 in Eq. E.10 to give:

$$(dG/d\chi)_{T,p} = \mu_B^0 - \mu_A^0 + R_0T \ln(p_B/p_A) = \Delta G^0 + R_0T \ln(p_B/p_A) \qquad (E.16)$$

Here, p_B and p_A are the partial pressures of B and A, respectively. The reaction proceeds with Free Energy changes under conditions specified by the partial pressures p_B and p_A. The term $-(dG/d\chi)_{T,p}$ is called affinity of the reaction and is indicated by the ratio of partial pressures p_B/p_A. At equilibrium for which $(dG/d\chi)_{T,p} = 0$, we get

$$\Delta G^0 = -R_0T \ln(p_B/p_A) \qquad (E.17)$$

The value of p_B/p_A at equilibrium is known as equilibrium constant for the reaction and is denoted by K_p, giving

$$\Delta G^0 = -R_0 T \ln(K_p) \tag{E.18}$$

The above equation tells us that if ΔG^0 in a chemical reaction is negative, $\ln(K_p) > 0$ giving $p_B > p_A$ and hence more products B are formed.

From Eq. E.18 we also have

$$\ln(K_p) = -\Delta G^0 / R_0 T \tag{E.19}$$

The equilibrium constant is independent of the total pressure and is seen to be function of only the temperature. It can be readily determined from the standard values of Gibbs Free Energy for the different substances undergoing reactions such as the substance forming products at a given temperature using the change in the standard Gibbs Free Energy. The standard values of Gibbs Free Energies are available in standard texts on thermodynamics. Using the change in the standard Gibbs Free Energy in a given reaction, the equilibrium constant for the reaction at the given temperature can be determined from Eq. E.19. The partial pressure and hence the number of moles of the species in the products can be obtained from the equilibrium constant using Eq. E.17. The procedure is illustrated in Sects. E.6 and E.7 of this Appendix.

E.6 Example of Using Equilibrium Constants to Determine the Products and Energy Release in the Explosion of TNT

The use of the equilibrium constant for determining the number of moles of the species in the products and the energy release in a chemical reaction is illustrated by the example of the explosion of the explosive Tri Nitro Toluene [TNT $-$ C_7H_5 $(NO_2)_3$]. TNT is a fuel-rich explosive since it has 6 oxygen atoms while for complete oxidation of C and H atoms in it, it requires $16^1/_2$ atoms of oxygen. Let us presume that b_1 moles of H_2O, b_2 moles of C, b_3 moles CO, b_4 moles H_2 and b_5 moles N_2 are formed in the reaction. Since oxygen is in short supply during the chemical reaction of the explosive TNT, we cannot have oxygen in the products and further the fully oxidized products like H_2O and CO_2 can only be partially formed leaving the formation of under-oxidized or non-oxidized species like CO and H_2. The reaction could thus be represented by:

$$C_7H_5(NO_2)_3 \rightarrow b_1 H_2O + b_2 C(s) + b_3 CO + b_4 H_2 + b_5 N_2 \tag{E.20}$$

C(s) represents carbon in solid phase. The atom balance for the four elements C, H, O, and N gives:

$$C: \quad b_2 + b_3 = 7 \tag{E.21}$$

$$H: \quad 2b_1 + 2b_4 = 5 \tag{E.22}$$

$$O: \quad b_1 + b_3 = 6 \tag{E.23}$$

$$N: \quad 2b_5 = 3 \tag{E.24}$$

Equation E.24 gives the value of b_5. We cannot determine the values of the other four variables b_1, b_2, b_3, and b_4 from the remaining three Eqs. E.21, E.22, and E.23. We require one more equation which is the equation for equilibrium for reaction

$$C + H_2O \Leftrightarrow CO + H_2$$

that links C, H_2O, CO, and H_2. The equation for the equilibrium constant K_1 for the reaction is

$$K_1 = \frac{p_{CO} \times p_{H_2}}{p_{H_2O}} \tag{E.25}$$

In the above equation, p denotes the partial pressure of the particular specie. The solid carbon in Eq. E.21 does not figure in Eq. E.25 for the equilibrium of the products as it does not contribute to the partial pressure. We need to convert the partial pressures into the number of moles. The total pressure of the products of combustion is:

$$p_{H_2O} + p_{CO} + p_{H_2} + p_{N_2} = p \tag{E.26}$$

The solid carbon, it may be noted, does not contribute again to pressure in the above equation. If the sum total of the moles of gaseous species in the products is b i.e., $b_1 + b_3 + b_4 + b_5 = b$, the moles of b_1, b_3, and b_4 are:

$$b_1 = \frac{p_{H_2O}}{p} b \tag{E.27}$$

$$b_3 = \frac{p_{CO}}{p} b \tag{E.28}$$

$$b_4 = \frac{p_{H_2}}{p} b \tag{E.29}$$

Substituting in Eq. E.25, we get:

$$K_1 = \frac{b_3 \times b_4}{b_1} \times \frac{p}{b} \tag{E.30}$$

Equation E.30 is the chemical equilibrium equation linking the values of b_1, b_3, and b_4. Solution of Eqs. E.21, E.22, E.23 and B30 gives the moles of the products b_1, b_2, b_3, and b_4, and we can get the composition of the products.

E.6.1 Initial Assumption for Total Number of Moles in Products

The value of pressure p in Eq. E.30 is specified. However, the value of b in the equation, which is the total number of moles in the products, is not known. It is therefore necessary to assume its value and improve upon it as the sum of $b_1 + b_2 + b_3 + b_4 + b_5$ in subsequent iterations. For the initial value that could be assumed for b, we consider the case when completed products of combustion, i.e., CO_2 and H_2O being formed for which the reaction is:

$$C_7H_5(NO_2)_3 \rightarrow 2.5H_2O + 1.75CO_2 + 5.25C(s) + 1.5N_2 \qquad (E.31)$$

This gives the initial value of b that could be assumed as $2.5 + 1.75 + 1.5 = 5.75$.

E.6.2 Inputs for Solving Equilibrium Relation and Iteration for Temperature of Products

The value of equilibrium constant K_1 depends on temperature and is obtained from the standard values of Gibbs function at the temperature of the products. The temperature of the products, however, is not known and has to be initially assumed. It can be corrected and improved on depending on the heat release in the reaction. If we assume initially the temperature of the reaction say as 2500 K, we have the values of the standard Gibbs function as:

$$G^0_{CO} = -327.245 \text{ kJ/mole}$$
$$G^0_{H_2} = 0$$
$$G^0_{C} = 0$$
$$G^0_{H_2O} = -106.55 \text{ kJ/mole}$$

Hence ΔG^0 for the reaction $C + H_2O \Leftrightarrow CO + H_2$ as $-327.245 - (-106.55) = -220.695$ kJ/mole. The value of

$$\ln(K_1) = \frac{-\Delta G^0}{R_0 T} = \frac{220.695}{(8.314 \times 10^{-3} \times 2500)} = 10.618$$

This gives the value of $K_1 = 40863.55$ at 2500 K.

The large value of K_1 is suggestive of more CO and H_2 being formed compared to H_2O and C in the chemical reaction $C + H_2O \Leftrightarrow CO + H_2$. The values of b_1, b_2, b_3, b_4, and b_5 are now determined from atom balance and chemical equilibrium conditions provided that the temperature of the products is 2500 K, and the sum of the moles of the gaseous products is 5.75 as assumed earlier. The sum of the values of b_i's is thereafter checked with value of $b = b_1 + b_3 + b_4 + b_5$ and if different from the assumed value, the revised value is now taken and the calculation repeated.

After suitable convergence, the heat release in the reaction is calculated using the heats of formation of the individual species in the products. The temperature of the products is determined from the heat release using the number of moles and the molar specific heats of the products. Denoting the enthalpy change per mole for each of the product species between the temperature of the products T_b (K) and the initial temperature at the standard value of 298 K as H, we have:

$$[b_1 H_{H_2O} + b_2 H_C + b_3 H_{CO} + b_4 H_{H_2} + b_5 H_{N_2}]_{(Tb-298)} =$$
$$- \left[b_1 \Delta H^0_{fH_2O} + b_2 \Delta H^0_{fC(S)} + b_3 \Delta H^0_{fCO} + b_4 \Delta H^0_{fH_2} + b_5 \Delta H^0_{fN_2} - \Delta H^0_{fTNT} \right]$$

$$(E.32)$$

Here the enthalpy change between T_b and 298 K denoted as H per mole includes for water the sensible heat, latent heat of evaporation, and the superheat, and similarly for the carbon the heating of the solid carbon to the sublimation temperature, the heat of sublimation, and the heat of the vapor, respectively. The right side of the equation gives the net heat released in the reaction at the standard state with ΔH^0_f denoting the standard heat of formation. The values of ΔH^0_f for the naturally occurring elements C(S), H_2, and N_2 are zero.

The iterative procedure is sketched in the flowchart given below:

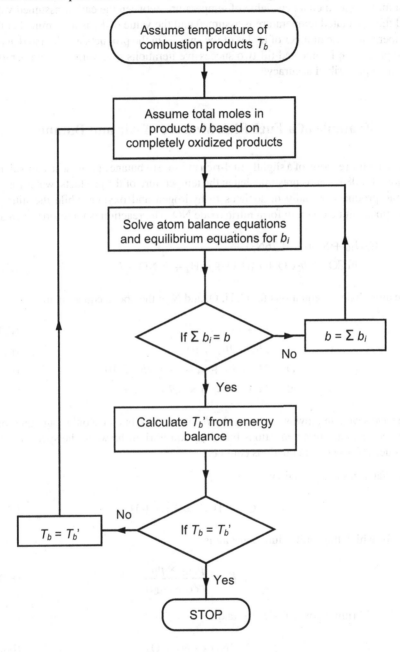

Flow Chart for the Computations

If this temperature T_b, calculated from the enthalpy change is different from the assumed value, then a new value of temperature between the earlier assumed value and the calculated temperature is assumed and the value of K_1 is determined at this temperature. The number of moles of the species in the products is calculated again. The procedure is repeated till two successive iterations give values of temperature within a prescribed accuracy.

E.7   Example of a Fuel-Rich Mixture of Air and Butane

Consider the reaction of a slightly fuel-rich mixture of butane and air at an equivalence ration of 0.083. Let us presume that at the temperature of the products, water formed in the products partially dissociates to hydrogen and oxygen while the nitrogen combines with oxygen to form nitric oxide NO. The reaction is therefore given as:

$$C_4H_{10} + 5(O_2 + 3.76N_2) \rightarrow$$
$$b_1CO_2 + b_2CO + b_3H_2O + b_4H_2 + b_5NO + b_6O_2 + b_7N_2 \qquad \text{(E.33)}$$

The atom balance equations for C, H, O, and N in the above equation are:

$$C: \quad b_1 + b_2 = 4 \qquad \text{(E.34)}$$
$$H: \quad 2b_3 + 2b_4 = 10 \qquad \text{(E.35)}$$
$$O: \quad 2b_1 + b_2 + b_3 + b_5 + 2b_6 = 10 \qquad \text{(E.36)}$$
$$N: \quad b_5 + 2b_7 = 5 \times 3.76 \qquad \text{(E.37)}$$

We have seven unknowns b_1, b_2, b_3, b_4, b_5, b_6, and b_7, but have only four equations. We need another three equations based on equilibrium between the species in the products. These three equations could be:

1. Water gas shift reaction:

$$CO + H_2O \Leftrightarrow CO_2 + H_2 \qquad \text{(E.38)}$$

for which the equilibrium constant is:

$$K_1 = \frac{p_{CO_2} \times p_{H_2}}{p_{CO} \times p_{H_2O}} \qquad \text{(E.39)}$$

2. Equilibrium between NO, N_2, and O_2:

$$2NO \Leftrightarrow N_2 + O_2 \qquad \text{(E.40)}$$

The equilibrium constant K_2 for the above reaction is:

$$K_2 = \frac{p_{N_2} \times p_{O_2}}{p_{NO}^2} \tag{E.41}$$

3. Dissociation of steam to H_2 and O_2:

$$H_2O \Leftrightarrow H_2 + {}^{1}/{}_{2}O_2 \tag{E.42}$$

For which the equilibrium constant K_3 is:

$$K_3 = \frac{p_{H_2} \times p_{O_2}^{1/2}}{p_{H_2O}} \tag{E.43}$$

Equations E.39, E.41, and E.43 can be expressed in terms of moles instead of partial pressures assuming the products to be ideal gases following the discussions in the last section. If the total pressure is p and the total number of moles in the products is b, the above three equations for equilibrium become:

$$K_1 = \frac{b_1 \times b_4}{b_2 \times b_3} \tag{E.44}$$

$$K_2 = \frac{b_7 \times b_6}{b_5^2}, \quad \text{and} \tag{E.45}$$

$$K_3 = \frac{b_4 \times b_6^{1/2}}{b_3} \times \left(\frac{p}{b}\right)^{1/2} \tag{E.46}$$

In Eq. E.46, we presume that water exists in the products as steam. The pressure at which combustion takes place is given. If the total number of moles in the products is initially assumed based on completed products to be formed, and a value of the temperature of the combustion products is assumed, we can from the standard values of the Gibbs Free Energies determine the values of equilibrium constants K_1, K_2, and K_3. We now have 7 Eqs. (E.34, E.35, E.36, E.37, E.44, E.45, and E.46) and 7 unknowns $b_1, b_2, b_3, b_4, b_5, b_6$, and b_7. We can therefore determine the values of $b_1, b_2, b_3, b_4, b_5, b_6$, and b_7 for the assumed values of b and temperature. These can be iterated and improved upon till the value of b's in two successive iterations is about the same. The heat generated in the reaction determined from the moles of the each of the specie in the products and their standard heats of formation and is given by:

$$Q = -\left[\sum b_i \Delta H_{f,i}^0 - \sum \Delta H_{f,C_4H_{10}}^0\right] \tag{E.47}$$

The temperature is determined based on the sensible, latent heat and gas phase heating of the products and compared with the assumed value of temperature. The temperature is further iterated till it converges to the required degree of accuracy. At the final iterated value of the temperature of combustion products, the moles of each of the species in the products and the heat release are determined.

E.8   Phases of the Species

The moles of the product species obtained in the above computations could be in the gas, liquid, or solid phases. Their phase can be readily determined using the vapor pressure of the specie at the particular temperature and the boiling temperature. The liquid and solid species do not contribute to the pressure. As an example, if we consider the reaction among the products

$$H_2 + {}^1\!/_2 O_2 \Leftrightarrow H_2O$$

the H_2O formed could either be in liquid phase as water or in gas phase as steam provided that the temperature exceeds the boiling temperature of H_2O at the stated pressure. The total moles of H_2O is the sum of the moles in the liquid and gas phases. The amount of H_2O in the gas phase corresponds to the partial pressure of it. The partial pressure is determined with the procedure outlined earlier, and the sum of the partial pressures equals the total pressure for an ideal gas. Thus the amounts in the different phases could be determined.

E.9   Correction for Pressure in the Dense Gases: Fugacity

The expression for chemical potential $\mu_i = \mu_i^0 + R_0 T \ln p_i$ was derived in Sect. E.3 assuming an ideal gas. The above computations are therefore strictly valid only if the products are ideal gases. However, at the high pressures and densities of an explosion, this may not be true. The higher values of pressure will cause the molecules to be nearer to each other, and the specific volume cannot be approximated as for an ideal gas by the expression $v = R_0 T / p$. The molecules tend to stick together, and the Gibbs Free Energy is less than that of an ideal gas at the given value of pressure. A corrected value of pressure known as fugacity is therefore used instead of pressure. The fugacity is defined as the corrected value of pressure such that the chemical potential is the same for a real gas when the expression of ideal gas is used by substituting for , viz.

$$\mu = \mu_0 + R_0 T \ln \frac{f}{f_0}$$

The use of fugacity provides a fundamental way of determining equilibrium composition of dense gases in chemical reactions.

E.10   Minimization of Gibbs Free Energy

Instead of using equilibrium constants, the Gibbs Free Energy could be minimized to determine the species. We had seen that the equilibrium conditions for a reaction correspond to minimum of Gibbs Free Energy. At a given temperature and pressure, the Gibbs Free Energy for a N species system is:

$$G = \sum_{j=1}^{N} n_j \mu_j \tag{E.48}$$

The values of $n_j, j = 1, \ldots, N$ are to be determined under chemical equilibrium conditions for which G is a minimum.

The minimization of G is subject to the constraint that the atoms of all elements taking part in the chemical reaction are conserved. The moles of the species in the products are thus obtained. The temperature is determined using the procedure outlined in the flowchart. Computer codes such as NASA SP273 and COSMIC use this minimization method.

Appendix F
Parameters of Chapman–Jouguet Detonation

The equation of the reaction Hugoniot (Eq. 6.2) derived in Chap. 6 is

$$\frac{\gamma}{\gamma - 1}\left(\frac{p}{\rho} - \frac{p_0}{\rho_0}\right) - \frac{1}{2}(p - p_0)\left(\frac{1}{\rho_0} + \frac{1}{\rho}\right) = Q \qquad \text{(F.1)}$$

where Q is the heat released in the reactions per unit mass. The equation to the Rayleigh line (Eq. 6.3) in Chap. 6 is

$$\frac{dp}{d(1/\rho)} = \frac{p - p_0}{1/\rho_0 - 1/\rho} \qquad \text{(F.2)}$$

The Chapman–Jouguet state corresponds to the point of tangency of the Rayleigh line to the reaction Hugoniot. Hence at the Chapman–Jouguet state, the slope of the reaction Hugoniot would be the same as that of the Rayleigh line. The slope or gradient of the reaction Hugoniot is therefore

$$\frac{d}{d(1/\rho)}\left[\frac{\gamma}{\gamma - 1}\left(\frac{p}{\rho} - \frac{p_0}{\rho_0}\right) - \frac{1}{2}(p - p_0)\left(\frac{1}{\rho_0} - \frac{1}{\rho}\right) - Q\right] = 0 \qquad \text{(F.3)}$$

Here p and ρ are the pressure and density at the Chapman–Jouguet state. The above gives

$$\frac{dp}{d(1/\rho)} = \frac{(p - p_0) - \frac{2\gamma}{\gamma - 1}p}{\frac{2\gamma}{\gamma - 1}\frac{1}{\rho} - \left(\frac{1}{\rho_0} + \frac{1}{\rho}\right)} \qquad \text{(F.4)}$$

The tangency condition at the Chapman–Jouguet state is therefore obtained by equating Eq. F.2 with F.4 as given below.

$$\frac{dp}{d(1/\rho)} = \frac{(p - p_0) - \frac{2\gamma}{\gamma - 1}p}{\frac{2\gamma}{\gamma - 1}\frac{1}{\rho} - \left(\frac{1}{\rho_0} + \frac{1}{\rho}\right)} = \frac{p - p_0}{\left(\frac{1}{\rho_0} - \frac{1}{\rho}\right)} \qquad \text{(F.5)}$$

K. Ramamurthi, *Modeling Explosions and Blast Waves*, https://doi.org/10.1007/978-3-030-74338-3

Solving Eq. F.5 gives

$$2(p - p_0)\frac{1}{\rho}\left(\frac{\gamma}{\gamma - 1} - 1\right) = \frac{2\gamma}{\gamma - 1}p\left(\frac{1}{\rho_0} - \frac{1}{\rho}\right)$$

On simplification, we get

$$\frac{p - p_0}{\frac{1}{\rho_0} - \frac{1}{\rho}} = -\gamma p \rho \tag{F.6}$$

If we denote the Chapman–Jouguet velocity by v_{CJ} and the velocity of the products of detonation in the frame of reference of the detonation as v_1, the continuity equation gives

$$\rho_0 v_{CJ} = \rho v_1$$

and we get the Rayleigh line from Eq. 6.3 in Chap. 6 as

$$\rho_0^2 v_{CJ}^2 = \rho^2 v_1^2 = \frac{p - p_0}{\frac{1}{\rho_0} - \frac{1}{\rho}} \tag{F.7}$$

Comparing the above with Eq. F.6, we have

$$v_1^2 = \frac{\gamma p}{\rho} \tag{F.8}$$

Substituting the values of the slope $dp/d(1/\rho)$ obtained by differentiating Eq. F.1 as $-\gamma p \rho$ (Eq. F.6), we get

$$\frac{\gamma}{\gamma - 1}\left(p + \frac{1}{\gamma}(-\gamma p \rho)\right) = \frac{1}{2}\left(\frac{1}{\rho_0} + \frac{1}{\rho}\right)(-\gamma p \rho) + \frac{1}{2}(p - p_0)$$

With the detonation pressure p being very much higher than the pressure p_0, we can neglect p_0 in comparison with p and obtain

$$\frac{\gamma p}{\gamma - 1}(1 - \gamma) = \frac{1}{2}\left(\frac{1}{\rho_0} + \frac{1}{\rho}\right)(-\gamma p \rho) + \frac{p}{2} \tag{F.9}$$

The above equation simplifies to

$$-\gamma p = \frac{1}{2}\gamma p - \frac{1}{2}\gamma p \frac{\rho}{\rho_0} + \frac{p}{2} \tag{F.10}$$

Solving for ρ/ρ_0 we have

$$\frac{\rho}{\rho_0} = \frac{\gamma + 1}{\gamma} \tag{F.11}$$

The density ratio for a strong shock wave was $(\gamma + 1)/(\gamma - 1)$. Hence the density ratio across a Chapman–Jouguet detonation is lower than for a shock due to the heating by the chemical reactions.

The Chapman–Jouguet velocity of a detonation v_{CJ} from Eq. F.6 is

$$v_{CJ}^2 = \frac{1}{\rho_0^2} \frac{p - p_0}{\frac{1}{\rho_0} - \frac{1}{\rho}}$$

We can simplify it to give

$$v_{CJ}^2 = \frac{p_0}{\rho_0} \left[\frac{p/p_0 - 1}{1 - \rho_0/p} \right] \tag{F.12}$$

Substituting the value of $\rho_0/\rho = \gamma/(\gamma + 1)$ from Eq. F.11, we get

$$v_{CJ}^2 \frac{\rho_0}{p_0} = (\gamma + 1) \left[\frac{p}{p_0} - 1 \right] \tag{F.13}$$

The left side of Eq. F.13 is

$$\gamma v_{CJ}^2 \frac{\rho_0}{\gamma p_0} = \frac{\gamma v_{CJ}^2}{a_0^2}$$

where a_0 is the sound speed in the undisturbed reactive gas. Hence it equals γM_{CJ}^2 where M_{CJ} is the Mach number of the Chapman–Jouguet detonation. Hence the pressure ratio across a Chapman–Jouguet detonation is

$$\frac{p}{p_0} = \frac{\gamma M_{CJ}^2}{\gamma + 1} \tag{F.14}$$

For a strong shock we had the relation

$$\frac{p}{p_0} = \frac{2\gamma M^2}{(\gamma + 1)}$$

and hence the pressure in a Chapman–Jouguet detonation is about half that in a shock wave of the same Mach number.

For the Chapman–Jouguet detonation the value of

$$\frac{p}{p_0} = \frac{\gamma M_{CJ}^2}{(\gamma + 1)}$$

and $\rho/\rho_0 = (\gamma + 1)/\gamma$ are known from Eqs. F.14 and F.11; we get the value of p/ρ as

$$\frac{p}{\rho} = \frac{p_0}{\rho_0} \frac{\gamma^2 M_{CJ}^2}{(\gamma + 1)^2} = \frac{\gamma v_{CJ}^2}{(\gamma + 1)^2} \tag{F.15}$$

Substituting in the equation for the reaction Hugoniot (Eq. F.1) noting that $p \gg p_0$, we get

$$\frac{\gamma}{\gamma - 1} \frac{\gamma v_{CJ}^2}{(\gamma + 1)^2} = \frac{p}{2} \left(\frac{1}{\rho_0} + \frac{1}{\rho} \right) + Q = \frac{1}{2} \left(\frac{p}{\rho_0} + \frac{p}{\rho} \right) + Q \qquad (F.16)$$

But the slope of the reactive Hugoniot at the Chapman–Jouguet state was

$$\frac{dp}{d(1/\rho)} = \rho_0^2 v_{CJ}^2 = -\gamma p \rho$$

and hence

$$v_{CJ}^2 = \frac{\gamma p \rho}{\rho_0^2} \qquad (F.17)$$

Since ρ/ρ_0 from Eq. F.11 is $(\gamma + 1)/\gamma$, we get from Eq. F.17

$$\frac{p}{\rho_0} = \frac{v_{CJ}^2}{\gamma + 1} \qquad (F.18)$$

Substituting the values of p/ρ from Eq. F.14 and p/ρ_0 from Eq. F.18 in the equation for reaction Hugoniot (Eq. F.16), we get

$$\frac{\gamma}{\gamma - 1} \frac{\gamma}{(\gamma + 1)^2} v_{CJ}^2 = \frac{1}{2} \left[\frac{\gamma v_{CJ}^2}{(\gamma + 1)^2} + \frac{v_{CJ}^2}{(\gamma + 1)} \right] + Q \qquad (F.19)$$

Simplifying we get

$$Q = \frac{v_{CJ}^2}{(\gamma - 1)(\gamma + 1)^2} \left[\gamma^2 - \frac{1}{2}(\gamma - 1)(\gamma + 1) - \frac{1}{2}\gamma(\gamma - 1) \right]$$

This leads to

$$Q = \frac{v_{CJ}^2}{2(\gamma^2 - 1)}$$

and therefore the Chapman–Jouguet velocity of a detonation v_{CJ} is related to the heat release per unit mass Q by the relation

$$v_{CJ}^2 = 2(\gamma^2 - 1)Q$$

Appendix G
Change of Concentration with Distance and Time: Solution of the Diffusion Equation

The diffusion of concentration in one dimension is given by the diffusion equation

$$\frac{\partial C}{\partial t} = D\frac{\partial^2 C}{\partial x^2} \tag{G.1}$$

Here the concentration is given by C moles/m^3 while x and t denote the distance in meters and time in seconds, respectively. D is the diffusion coefficient in m^2/s. If the concentration at location $x = 0$, which we consider as the origin, and time $t = 0$ is given as C_i moles/m^3, we are required to find the concentration at any distance x and at time t. We can assume that the concentration very far from the origin is always zero, i.e., $x \rightarrow \infty$, $C = 0$.

The concentration changes both with respect to distance and time. The parameter influencing the concentration at x and t would depend on the diffusion coefficient D and x and t. The dimensions of these three parameters are:

Diffusion coefficient, D: L^2/T
Distance, x: L
Time, t: T

Since two of the above three parameters (diffusion coefficient, distance, and time) have independent dimensions, only one non-dimensional parameter can be formed from these three-dimensional parameters as per the Buckingham π theorem. The non-dimensional parameter, denoted by η, is obtained from a straightforward dimensional analysis as

$$\eta = \frac{x}{\sqrt{Dt}} \tag{G.2}$$

The above parameter represents the combination of distance and time according to which the concentration would vary. It is known as the similarity variable. Writing the partial derivatives $\partial C/\partial t$ and $\partial^2 C/\partial x^2$ in the diffusion equation in terms of η would give an ordinary differential equation since the two parameters x and t are merged into a single parameter η.

K. Ramamurthi, *Modeling Explosions and Blast Waves*, https://doi.org/10.1007/978-3-030-74338-3

The partial derivatives $\partial C / \partial t$ and $\partial^2 C / \partial x^2$ in the diffusion equation can be written in terms of η as:

$$\frac{\partial C}{\partial t} = \frac{dC}{d\eta} \frac{d\eta}{t} = -\frac{x}{2t\sqrt{Dt}} \frac{dC}{d\eta} \tag{G.3}$$

$$\frac{\partial C}{\partial x} = \frac{dC}{d\eta} \frac{d\eta}{dt} = \frac{1}{\sqrt{Dt}} \frac{dC}{d\eta}$$

$$\frac{\partial^2 C}{\partial x^2} = \frac{\partial}{\partial x} \left(\frac{dC}{dx} \right) = \frac{d}{dx} \left(\frac{1}{\sqrt{Dt}} \frac{dC}{d\eta} \right)$$

$$= \frac{1}{\sqrt{Dt}} \frac{\partial}{\partial x} \frac{dC}{d\eta} = \frac{1}{\sqrt{Dt}} \frac{d}{d\eta} \frac{dC}{d\eta} \frac{d\eta}{dx} = \frac{1}{Dt} \frac{d^2 C}{d\eta^2} \tag{G.4}$$

The total differential equation describing the partial differential Eq. G.1 becomes

$$\frac{d^2 C}{d\eta^2} = -\frac{\eta}{2} \frac{dC}{d\eta} \tag{G.5}$$

The boundary conditions reduce to

$$\eta = 0: \quad C = C_i \tag{G.6}$$
$$\eta = \infty: \quad C = 0 \tag{G.7}$$

Equation (G.5) can be written in the form:

$$\frac{d\left(\frac{dC}{d\eta}\right)}{\frac{dC}{d\eta}} = -\frac{\eta}{2} d\eta \tag{G.8}$$

Integrating the above equation, we get

$$\ln\left(\frac{dC}{d\eta}\right) = -\frac{\eta^2}{4} + \text{constant} \tag{G.9}$$

This implies

$$\frac{dC}{d\eta} = C_1 e^{-\eta^2/4} \tag{G.10}$$

Here C_1 is a constant of integration. Integrating again from $\eta = 0$ to η, we get

$$C = C_1 e^{-\eta^2/4} d\eta + C_2$$

C_2, in the above equation, is another constant of integration.

The constants of integration C_1 and C_2 are determined below from the given values of C at $\eta = 0$ and $\eta = \infty$ given in Eqs. G.6 and G.7. We have at $\eta = 0$:

$$C_i = 0 + C_2 \text{ giving } C_2 = C_i \tag{G.11}$$

and at $\eta = \infty$:

$$0 = C_1 \int_0^\eta e^{\eta^2/4} d\eta + C_i = C_1 \frac{\sqrt{\pi}}{2} + C_i \text{ giving } C_1 = -\frac{2}{\sqrt{\pi}} C_i \tag{G.12}$$

The value of concentration at any η is therefore

$$C = C_i - \frac{2}{\sqrt{\pi}} C_i \int_0^\eta e^{-\eta^2/4} d\eta \tag{G.13}$$

where $\eta = x/\sqrt{Dt}$.

Index

Printed in the United States
by Baker & Taylor Publisher Services